TECHNOLOGIES & SOCIETY

RON WESTRUM

Eastern Michigan University

Technologies & Society
The Shaping of People and Things

Wadsworth Publishing Company
Belmont, California
A Division of Wadsworth, Inc.

Sociology Editor: Serina Beauparlant
Editorial Assistant: Marla Nowick
Production Editor: Donna Linden
Designer: Andrew Ogus
Print Buyer: Barbara Britton
Permissions Editor: Robert Kauser
Copy Editor: Alan Titche
Photo Researcher: Stephen Forsling
Technical Illustrator: Guy Magallenes
Compositor: G & S Typesetters, Austin, Texas
Cover Design: Andrew Ogus

Photo Credits: Cover: Lewis W. Hine, *The Sky Boy,* from *Men at Work.* Courtesy the
International Museum of Photography at George Eastman House, Rochester. Frontis-
piece: Berenice Abbott, *El, Second and Third Avenue Lines, Bowery and Division Street,
Manhattan.* Courtesy the Museum of the City of New York. Part One: Courtesy the State
Historical Society of Wisconsin. Part Two: Ralph Steiner, *Lewis Lozowick.* Courtesy
Murray Weiss. Part Three: Fritz Goro, *Life* Magazine © Time Inc. Part Four: Theodor
Horydczak, *Corrugating Metal Sheets at Central Alloy Steel Corp., Massillon, Ohio.* Cour-
tesy the Library of Congress. Part Five: Horace Nicholls, *British Munitions Factory.* Cour-
tesy Imperial War Museum, London.

Printed in the United States of America   19
1 2 3 4 5 6 7 8 9 10   95 94 93 92 91

Library of Congress Cataloging-in-Publication Data

Westrum. Ron. 1945–
    Technologies and society : the shaping of people and things /
Ron M. Westrum.
        p. cm.
    Includes bibliographical references.
    ISBN 0-534-13644-3
    1. Technology—Social aspects.   I. Title.
T14.5.W47   1991
303.48′3—dc20   90-12234
                                                                    CIP

I would like to dedicate this work to the memory of my friend and classmate John Paul Fitts, whose promising career in social science was cut short by a canoeing accident in 1980. I'm sure that Fitts would never have agreed with the approach taken here (he'd have said it "doesn't go far enough"), but unfortunately such a dialogue between us will never take place.

C O N T E N T S

PART   THREE

Originators & Managers of Technology   84

C H A P T E R   8

Innovation: Inventions & Institutions   150

C H A P T E R   9

The Sponsorship of Technology   171

CHAPTER 10

How Sponsors Evolve   194

PART   FOUR

Users of Technology   212

CHAPTER 11

The Technology/User Interface   215

CHAPTER   1 2

Adapting & Tinkering   237

CHAPTER   1 3

Technological Accidents   250

CHAPTER   1 4

The Distancing Effects of Technology   268

PART   FIVE

Monitoring Technology: Shaping the Present & the Future   290

CHAPTER   15

Social Control of Technologies   293

CHAPTER   16

Technology Assessment   323

C H A P T E R   1 7

Foresight & Social Intelligence   353

PREFACE

Technologies & Society is a systematic introduction to the technology/society interface, intended largely for undergraduate and graduate courses in "Science, Technology, and Society" (STS). The book has sufficient basic concepts for beginners, and it contains enough scholarly references for more advanced students. And though I hope my colleagues will take some interest in what I have written here, they are an incidental audience, for this book is written for students.

As a director of programs in technology and society for both undergraduate and graduate students, the need for a comprehensive treatment of this subject has long been evident to me. For three decades now there has been an active interest in the sociology of science; in contrast, the sociology of technology has been neglected. The reasons for this neglect are not entirely clear, but it is time that it stop. Already a thriving academic community focuses on the history of technology, but a sociological approach is equally needed.

Our role in shaping technology and its role in shaping us need exploration. We all need a good sense of how technology and society work together so that each of us can make better decisions about how we integrate technology into our lives. The alternatives of slavish adoption of technology on the one hand and violent rejection on the other are not healthy. A greater understanding will enable us to widen our options. While a number of valuable studies on specific aspects of the sociology of technology exist, an overall approach to the field is lacking. *Technologies & Society* endeavors to fill this gap.

The subject is vast; again and again an effort of will was needed to keep from writing, "to cover this topic properly would require an entire book." On some topics, the amount of information was overwhelming, and these topics have been treated in a more summary fashion than I would have preferred. I have tried to indicate where readers might find more complete treatments of these subjects.

I also found the opposite problem: a paucity of information on key topics. In some cases I was reasonably sure the information existed but had no idea where to look. The kinds of sociological studies on which such a book should be based are often lacking. Consider only a single instance: With the exception of workplace studies, there has been little systematic study of technology users. Yet we need to know more about what people do with technology once they get it. I hope this book will be a stimulus for the development of such knowledge.

About the Book

The book's contents are arranged in a logical sequence, but because different readers will approach it with different needs, various paths through the book may be appropriate. Part One defines what is meant by "technology," examines the interaction of technology and social systems, and presents basic definitions and issues.

Part Two reviews two major schools of social theory about technology—Marx and the Marxists, and the Ogburn generation—and four contemporary viewpoints. Each theory represents a distinct point of view and brings powerful intellectual tools to the study of technology. Chapter 2 discusses Marx's introduction of political concepts for analyzing technological change. Chapter 3 introduces the Ogburn school's use of social trend analysis and statistics to measure social change. Current theories of technology and society, which tend to stress society's interaction with the environment and with cognition to a greater degree, are covered in Chapter 4. This chapter is meant not only as a brief overview of current activities, but also as a rationale for the material that follows.

Part Three discusses the creators of technology—the originators and managers. I have discussed this topic in some depth, because the way technology is invented, designed, implemented, and managed is critical in shaping society. Chapter 5 discusses the personalities and behavior of inventors. In Chapter 6 the role of the inventor is placed in a wider social context. In Chapter 7 we consider the role of designers in shaping technology. Chapter 8 explores the way innovation is encouraged or resisted by social institutions. Chapters 9 and 10 introduce and develop the concept of technology sponsorship, a key concept in later chapters.

Part Four considers the users of technology. An underlying premise is that people do not receive technology passively but rather adapt it and tinker with it to shape it to their own ends. Chapter 11 focuses on user skills and the need to provide a better fit between users and their technology. We also consider users' emotional reactions to technologies. Chapter 12 shows how users tend to alter the technology they have, sometimes in creative ways, sometimes in destructive ones. Chapter 13 looks at a serious problem associated with using high-tech devices and systems: technological accidents. Chapter 14 examines how high technologies may be creating psychological distance between people in society.

Part Five addresses ethical and value issues involved in technological and social change. Chapter 15 examines the ways in which society controls technology; it will become obvious that our methods of control are crude and need refinement. Chapter 16 discusses technology assessment—the very important issue of trying to predict the social impacts that technology will have. Finally, in Chapter 17, we talk about society as a learning system for creating and monitoring technology, including its shortcomings in reacting to technology's impacts on our environment.

Throughout this book we will see the way that technology and social institutions influence each other. In looking at the wide range of interactions between things and people, we will see on the one hand the great complexity of technology/society interfaces. On the other hand we will see how we might make these interfaces work more effectively.

Acknowledgments

I have made every effort to make this book up-to-date and an accurate reflection of current practice in the subjects discussed. In addition to consulting the primary literature, I have interviewed a considerable number of technologists, such as inventors, designers, consultants, and technology assessors. When appropriate, I have quoted them or cited their work. I have drawn on my experience as a consultant to such business firms and research organizations as Lotus Engineering, General Motors, and the Rand Corporation. "Information System Failure," a seminar sponsored by the North Atlantic Treaty Organization (NATO) and led by John Wise and Anthony Debons, was an important influence, as was "Safety Control and Risk Management," sponsored by the World Bank and led by Jens Rasmussen and Roger Batstone. Writing a history of the Sidewinder missile with

Howard Wilcox brought me into contact with a wide circle of technologists and researchers.

As an undergraduate at Harvard I was fascinated by the structure of technology, a fascination encouraged by Harrison White. Reading Elting Morison's splendid *Men, Machines, and Modern Times* while I was at the University of Chicago was a turning point for me. Graduate study with Albert Wohlstetter and Duncan MacRae also pushed me in this direction. Bob Perrucci's request that I write a chapter on technology and social change for a textbook coaxed me still further. While teaching at Eastern Michigan University I was delighted to discover the graduate program "Liberal Studies in Technology," which I soon joined. I am grateful to Paul Kuwik, Alexandra Aldridge, and Felix Kaufmann for welcoming me into the program; later we were joined by Wayne Hanewicz. Working in the intensely stimulating atmosphere of the MLS/Technology program has been a genuine spur to thought. Many of the ideas in this book were first developed through my courses in the MLS program. Many of my students in the undergraduate Technology and Society program at Eastern have also contributed ideas and references: Arija Bergmann, Mike Strelecki, Kurt Seifert, and above all Steve Wilcox have done the most, but many others have helped, too. My friends Claudia Capos, Kathy Johnson, and Vanessa Vechell contributed moral support.

I have incurred a number of other debts in writing this book. Not the least has been the encouragement of the excellent team at Wadsworth, my publisher. From the book's inception through its completion, the friendly urgings ("Where is your manuscript?") of sociology editors Sheryl Fullerton and Serina Beauparlant encouraged me. Marla Nowick solicited reviews from nine referees: Michel Callon, Centre de Sociologie de L'Innovation; Richard Hawkins, Southern Methodist University; Thomas Hughes, University of Pennsylvania; J. Scott Long, Indiana University; Nicholas Mullins, Virginia Polytechnic Institute; James C. Peterson, Western Michigan University; Andrew Pickering, University of Illinois; Harriet Zuckerman, Columbia University; and Edward Woodhouse, Rensselaer Polytechnic Institute. These reviews were a great stimulus to the book's improvement. Some of the greatest contributions to improvement of the manuscript, however, came from Donna Linden, the production editor, and Alan Titche, the copy editor. Through Alan's tough editing, many of the infelicities of style and not a few errors in the book were removed. The Wadsworth team worked just as hard on this book as I did, and I am grateful.

I want to recognize several of my colleagues in the sociology of technology who have also stimulated my thoughts: Wiebe Bijker, Joseph Coates, David Collingridge, David Edge, Boelie Elzen, Bruno Latour, Donald MacKenzie, Charles Perrow, James Reason, Arie Rip, Allen Schnaiberg. Many of these people have been extremely generous with reprints and references. Marcello Truzzi, colleague and intellectual sparring partner, has also been a great help in alerting me to useful references.

Wiebe Bijker, Michel Callon, Daryl Chubin, Joe and Vary Coates, Thomas Parke Hughes, Kathy Johnson, Mel Kranzberg, Stuart Macdonald, Gary Marx, Paul MacCready, Donald MacKenzie, Robert Probst, Jacob Rabinow, Jeff Riemer, Arie Rip, Arnold Skromme, and Keith Westrum were kind enough to look over some of my chapter drafts and make helpful comments. My father, Edgar F. Westrum, Jr., urged me to start writing on a computer; he also assisted in my choice of machine and software. I cannot now imagine doing this book, as I did my last one, by typewriter—a fine demonstration of addictive technology. Although I am grateful for all these contributions, responsibility for the final product rests with me. No doubt some readers will take issue with one or another of my interpretations. I hope they will communicate their disagreements to me. This book is intended to be a progress report, and the communications of colleagues will aid in the progress!

Ron Westrum
Ann Arbor, Michigan

TECHNOLOGIES & SOCIETY

PART ONE

Introduction

The Shaping of People & Things

We live in a world that is *designed*. Though much of our existence is shaped by our interactions with people, it is also shaped by the structures, pathways, devices, and processes of the designed world. In the morning we put on our clothes and eat food cooled by a refrigerator or warmed by a stove. During the day we drive a car down a paved street to a supermarket, where we use a cart to gather products that are rung up on an electronic cash register. At night we turn out the electric light, go to sleep in a manufactured bed, and sleep until our alarm clock wakes us up. Technology is around us every moment; we constantly use devices and technological processes to achieve our personal ends. In our work we use technologies, even if only such simple ones as pencil and paper. In recreation we use technology, whether running shoes or a speedboat and water skis. We take over-the-counter drugs to shape our moods and prescription medicines to treat our diseases, and everywhere—from the salt on our food to the weedkiller on our lawns—we use chemicals to alter, preserve, and pro-tect. The world in which we live is full of technologies that are designed. Who does the designing, who makes the choices, and how do we get the technologies we use?

Technology and Society— Interacting Systems

Technology and society are forces that to-gether shape the world in which we live, shifting its contours and rearranging its parts, just as oceans move sand dunes. A single new technology may transform a primitive society or send ripples of change through an advanced one, and the cumulative im-pact of thousands of inventions and social changes continually transforms the land-scape of the social world we share. To thrive amidst these waves of change requires both a sense of direction and an ability to understand how change works.

Technology and society intertwine: When one changes, the other is likely to change as well. Social changes bring new needs, desires, and insecurities; technical changes force new choices, new adjustments. In either case change produces in-stability, and in the process of responding to each change, still more instabilities are created. In the ferment of social and technological change, it is often impossible for us to see the direction in which we are going, or to understand the trends in which we participate. Like swimmers caught up in a powerful undertow, we are swept along into new experiences and new innovations. We must work hard to keep swim-ming, or we may drown in the surging waves of change that wash around us.

Every so often, out of this ocean of changes, something is thrown up on the beach, and it comes to rest, no longer to be tossed among the waves. On my book-case is a glass milk bottle from the 1930s. Its embossed label tells me that it is from the "Superior Dairy Co., Ann Arbor, Michigan" and that the dairy produced milk and ice cream. This sturdy bottle obviously was made for indefinite recycling. On the bottom of the bottle is an enormous "S." Just as cattle brands help to distinguish cows from several herds, this mark was designed to help distinguish this bottle from those of other dairies. And it was designed very specifically for people who looked at the bottle upside down, doubtless in a rack containing many other bottles. The bottle is a splendid artifact: It is solid, simple, and useful, and once it was part of an elabo-rate system.

The existence of the bottle presupposes other bottles like it, some identical, some with different bottom marks. It presupposes dairies to collect the milk, pas-teurize it, and transport it to collecting points. There the milk was put into bottles like this one and distributed by truck to homes.[1] Empty bottles were returned, sorted by dairy, washed, and refilled. The bottle also presupposes users: drinkers of milk, cooks, and mothers who poured milk into glasses for thirsty children. And the bottle further suggests the existence of glassmakers, of milk trucks, of dairy apparatus, of refrigerators, and so on; endlessly elaborate relations within a vast commercial network, involving regulatory bodies, advertisers, and dairy workers, all of whom shared a belief that milk builds strong bodies, and very much more.

The bottle is now an antique. The forces of change have moved it from "active" to "inactive" status. The Superior Dairy is gone. Milk is seldom distributed now in glass bottles; instead it is packaged in wax-paper cartons or plastic containers. Thus a sturdy glass bottle, once a key part of the system of milk distribution, is now a piece of flotsam. Technological and social changes have transformed it from a vital cog into a piece of debris. The system into which it once fit has passed into history.

We can understand a good deal about technologies in general by considering this milk bottle. Technologies are not isolated things; rather, they are parts of systems of human action. They are linked through people to other things and are shaped by human beliefs, needs, and aspirations. When the system no longer needs them, technologies become stranded like the bottle; they remain interesting in themselves but are no longer vital. The parking meters in New York City will soon go the way of this bottle because the system of which they are a part is changing, too. Parking in New York City is now so expensive that parking meters become full quickly and must be emptied several times a day. They cannot be larger because they would become more frequent targets for vandals. So they will go, replaced by other forms of metered parking.[2]

This book is about the mutual shaping of people and things. It explores the interaction of people and technology in a changing society. It examines the social relations between people and milk bottles, parking meters, nuclear power plants, and many other technologies. It explores how people and technology shape each other, how the technologies we choose have enormous impacts on our lives, and how in turn we search for new technologies to satisfy our wants and needs. The process is never-ending.

In our examination of the process, we want to answer a series of basic questions. How is technology created? Why do we create the kind of technology that we do? Who makes decisions about it? What effects does our technology have on our individual lives and our social structure, and how do we monitor those effects? These questions are intimately related to questions about social institutions. The very structure of our society is related to the kind of technologies we have. Our institutions—our customs, values, organizations, habits of thought—are powerful forces that shape our technology in distinctive ways. We need to examine these institutions, to understand their strengths and weaknesses, to see whether or not they are fit for their purposes. Technology poses unique challenges to our society; we must determine whether our institutions can cope with them.

Personally, I fear that they cannot. As one thoughtful person put it, "We have third generation machines [but] first generation minds."[3] This mismatch between our technologies and our institutions is costly and dangerous. The costs appear in various ways: in danger, discomfort, waste, and conflict. A stranded milk bottle is one thing; a stranded town is something else. When diesel trains replaced steam locomotives, they left stranded in their wake not only a vast technology, but also entire towns whose existence had depended on the technology of wood- or coal-burning engines.[4]

Similar changes to water transport have also forced changes on cities near the sea. With the revolution in ocean shipping, which entailed containerization and automated cargo handling, many former ports have been forsaken and replaced by others more suited to handling the standardized containers:

> The container ship required deeper water, specialized gantry cranes, and acres of ground for their terminals—ground that was just not available in the upriver docks surrounded by cities. . . . The owner of these new ocean vessels took great exception to slow passages up winding rivers, hours wasted at anchor awaiting tides, and laborious docking in and out of enclosed docks. Far better if the ports to handle this new breed of ship could be built nearer to the open sea.[5]

And so older, inland ports are being abandoned, and new, more convenient ones are being constructed nearer the sea.

The commercial forces that drive these changes are powerful and sometimes very destructive. The sociotechnical system can surge forward with little regard for the "side effects." Some of these side effects, though invisible, can far outpace the benefits gained from using the technology. For instance, consider the American passion for hamburgers. Few people munching on what some consider to be the national food realize that they are contributing to the destruction of tropical rain forests. But in Latin America, whole forests are being cut down to provide grazing land for beef cattle.[6] The destruction of these forests means wiping out numerous species of plants and animals that survive only in this environment; many of these species have not yet been identified.[7] The forests are also important for cleansing the atmosphere of carbon dioxide and thereby avoiding the dangers of an atmospheric "greenhouse effect."[8] Of their own accord the commercial and political institutions we have created take advantage of opportunities such as this to make profits or pursue national development, often with little regard for the ultimate consequences.[9] These are very hazardous institutions for us because they threaten our welfare, and we need to change them. It is not only appropriate technology that we need, but also appropriate institutions.[10]

Unless we can develop such new institutions—unless we can narrow the cultural gap between the things we have created and the organizations we are using to manage them—we put ourselves and our environment in peril. High technologies require a new level of consciousness, a mental attitude very different from the current one. Already some organizations in our society have begun to operate in a new mode.[11] Yet such new modes are few and far between. For other institutions, such adjustments in attitude will likely come only with much pain and bitter, costly resistance. In the recent past we have seen a number of high-technology disasters: Bhopal, Chernobyl, and the NASA space shuttle *Challenger*. In each of these cases there existed a conspicuous mismatch between the complexity of the technology and the capabilities of the human beings responsible for managing it.[12] These events, bad enough in themselves, were important warnings, for long-term trends pose even more ominous dangers. And other changes, more numerous and more serious, will follow unless we make reforms. Understanding the nature of technology and its social supports is a first step toward such reforms.

The purpose of this book is to explore the dynamics of technology and its social institutions. It is important to realize that analysis is only a first step; ultimately analysis must be followed by institutional reform—the development of institutions that are more "fit" and more able to cope with high technology.

What Is Technology?

The most important reason to study technology is that it is impossible to understand the world we live in without understanding technology. We live in an artificial environment. Our houses, cars, clothes, books, factories, and food all reflect technologies. Technology has always been critical for humanity, but its effects have increased in scope.[13] Just as the philosopher Socrates was right in emphasizing the examination of *ethics* in labor-intensive Athens, we who live in a technology-intensive society need to give our relationship with machines a similar, searching examination.[14] But what, precisely, is technology? Let us propose a definition: Technology consists of those material objects, techniques, and knowledge that allow human beings to transform and control the inanimate world. This definition contains three essential elements:

1. Technology is *things*. These things can be simple devices like wrenches, elaborate systems like ocean liners, or huge, ramified networks, such as a pipeline system or a power grid. To get away from the clumsiness of the word "thing" and the false precision of "machine," we will call units of embodied technology "devices."[15] This term is still imprecise, but it is the best one available.

2. Technology is *techniques*. The recipes used for baking a cake, and the instructions for shooting a bow, operating a computer, or running a smelter, are also technology. In this sense, technology refers to the practical skills needed to do something; the something may be very modest (as with cooking) or quite elaborate (as with craft skills, medicine, or civil engineering). Such knowledge may include things like formulas, which can be written down, and muscular skills, which cannot. Techniques allow us to alter the recipe to fit varied situations or to embody virtuosity in performing a given procedure.

3. Technology is *abstract knowledge*. This part of technology consists of the more abstract concepts used by inventors and scientists to design and develop new things. It includes the highly calculative abilities that allow people to design an oil refinery or to build spacecraft that can land on the moon. Here technology

is closely related to science. And like science, it is an intellectual system that can be continually upgraded with new knowledge and theories.

Technology, then, is both concrete and abstract. It is both the pair of scissors I hold in my hand and the knowledge of how to make them from iron, carbon, and chromium. It is both the resonating electrical grid and the system of knowledge that allowed electrical engineers to build it. Mostly we will speak of technology as concrete objects, although technology as technique and knowledge will be important when we discuss skill, and design and invention, respectively.

Technology relates to a vast complex of human activities involving the material world. It is engineering, architecture, industrial design, and much more. Yet it is not to be considered synonymous with all forms of practical knowledge. It does not seem right to call social techniques "technology." Although running a computer and running an encounter group are both techniques, and although their invention, development, and training may have similar dynamics, "social technology" deserves its own treatment and thus will be discussed here only in passing.[16]

Technology Is Embedded
in a Social Matrix

The organizing concept for this book is the idea that technologies are embedded in and sponsored by institutions. Technology does not float freely in society; it is firmly anchored in particular groups and social relationships. This grounding of technology in institutions has vital effects on both the technology and the institutions, for their fates are intertwined: Changes in technology affect institutions, and vice versa. Impending changes either in institutions or in technologies may be blocked because of undesired consequences to the other. Let's consider a few of the ways in which technology is embedded in social life:

1. *The social division of labor.* In complex societies like our own, knowledge of technology is distributed among the many parts of the society. Some people are skilled with some technologies, some with others. Unlike the breadth of ingenuity displayed in *Swiss Family Robinson,* it would be impossible for anyone in our society to master all our technologies. Hence a basic condition for the continuation of our society is that there be a permanent division of labor. This division of labor, however, can be structured in very different ways with very different consequences.

2. *Exclusive possession of technologies.* In some societies, certain people are forbidden to use specified technologies, whereas others are permitted to do so. For

instance, in many primitive societies, only men or only women are allowed to use certain implements, such as stone tools. In the sumptuary laws of sixteenth-century England, commoners were not allowed to wear lace or to carry swords; only the nobility were allowed to do so. In many countries licensing of certain professionals gives them exclusive legitimate use of medical technologies. To-day most societies restrict the use of technology to certain categories of people on the basis of age, gender, training, or skill.

3. *Responsibility for operation and development of technology.* One of the major differences between capitalist and socialist countries concerns who owns and operates certain large technological systems like railroads and factories. In capi-talist countries, such systems are likely to be in the hands of private corpora-tions, whereas in socialist countries they are in the hands of the central government. That these differences in forms of sponsorship are important be-comes evident when we compare the experiences of capitalist and socialist countries. As communist countries continue to experiment with markets and similar capitalist institutions, the importance of sponsorship forms will emerge even more clearly.

4. *Association of social groups with technologies.* Certain groups may feel that given technologies are a part of their identity, as for instance with hippies and LSD, or rifle racks in pickup trucks for "good old boys." In some cases the association can be imposed rather than chosen. Mexicans and marijuana once went together in the public mind; English sailors were known as "limeys" be-cause of lime juice used on British ships to prevent scurvy.[17] Consider also "shanty" Irish versus "lace curtain" Irish. Ethnic groups often use particular technologies for social mobility: hence the "Chinese" laundry and the German brewery, for example.

5. *Dependence of social forms on technology.* Certain kinds of social interaction may be possible only through the intervention of technology.[18] For instance, distance-bridging technologies like ballistic missiles make long-distance war-fare a possibility. Branch banking in the form we know it today came about through the existence of the telephone.[19] Landing people on the moon would have been impossible without the computational capabilities of computers. In some cases technology is not essential, but rather provides a strong influence. Thus the small families of the late twentieth century have resulted in part from the increased availability of birth control technologies.[20]

These are merely a few of the ways that technology interweaves with social in-stitutions. What is apparent from these examples, though, is the mutual support pro-vided between technology and the social institutions in which it is embedded. Like the body's skeleton, technology is an integral part of society. The correct question to ask, then, is not how technology causes social change, but rather, what role technol-ogy and society play in shaping each other. For technology is, as we will see, as much a response to social change as it is a shaper of society. Technologies do not simply *occur;* they are chosen and designed to bring about certain results, although the results produced may not always be the ones intended.[21]

Soviets and Microchips:
You Can't Change Machines
Without Changing People

At this point an example may bring this issue of the interdependence of technology and society into sharper focus. How the Soviets worried over the microchip revolution in the pre-*glasnost* ("openness") era provides an interesting case in the relation of technology and institutions. It shows how intimately society and technology are tied together, and how changes in a technology may be intimately related to other changes in the social fabric.

In the late twentieth century computer literacy is a very important skill for commerce, military affairs, and cultural standing. In the U.S.S.R., however, computer use and availability has lagged enormously behind the West. Bills in retail shops are still calculated on abacuses, pocket calculators are rare, and even typewriters are not in great supply. Moving to widespread computer use in only a few years would represent an enormous advance, quite literally a "great leap forward." Because such an advance is necessary to improve the Soviets' commercial and military position, however, they are determined to achieve it. One institution they are determined to change is the schools. In March 1985 the Politburo of the Soviet Union announced that by 1990 the nation would have one million microcomputers in its 60,000 secondary schools. Powerful members of the Soviet Academy of Sciences rapidly took the lead in promoting this program.[22] This is, however, only one of many steps in the Soviets' attempts to modernize against very difficult odds.[23]

Whatever the economic merits of this kind of advance, at the beginning it faced severe political problems. In the past, the Soviet Union had attempted to exercise exceptional control over the minds of its citizens by limiting access to communication channels and through extensive censorship.[24] All means of communication were controlled by the state to the greatest extent possible, and private ownership of presses, filmmaking equipment, and duplicators was virtually nonexistent. Photocopying machines, when taken out of service, were broken up by special "hammer men" who made sure that they were not used for duplication of prohibited materials.[25] For decades the major form of the circulation of illicit manuscripts was *samizdat,* the use of multiple carbons by writers. Given the ability of the computer to reproduce, the presence of a larger number of microcomputers available to the public posed a nightmarish problem for the censorship apparatus.

A second problem was posed by obtaining the technology in the first place. The Soviet Union's computer industry is largely derivative of foreign technology. Virtually all Soviet computer equipment has largely copied that of the West, in whole or in part. The few quality electronic parts made in the U.S.S.R. are used largely by the military sector.[26] By recent estimates, there are only 50,000 personal computers in the country, compared to 25 million in the United States.[27] Thus the expectation that Soviet schools will be filled with personal computers in the near future seems unrealistic. At present most Soviet students learn about "informatics" through paper-and-pencil exercises.

What has fueled the enormous development of computing in the United States is a system for invention and development in which many different (and sometimes quite small) units are free to compete with each other.[28] The development of the personal computer was assisted by intense amateur interest, rapid exchange of information through clubs and periodicals, and exceptional mobility of ideas and people.[29] The development of "Silicon Valley" itself is a result of the freedom to form and dissolve enterprises rapidly—not a popular idea in the U.S.S.R. before *glasnost*.[30] Economic freedom allowed rationality to work so as to generate new ideas and products. Previously the Soviets had a more "calculative" model of rationality: Everything must be planned in advance by the central authorities.[31] Previously there were few routes by which a company like Apple Computer could be formed in the Soviet Union.

Initial ideas for Soviet use of personal computers followed traditional patterns of authority. Although students eventually might be able to enter what they wished on their computers, access to disk drives and to printers was to be tightly controlled; computers would be on the desk of the teacher, not in the hands of the students.[32] This was called "collective use of personal computers," a phrase that nicely highlighted the cultural contradictions of an information revolution in a tightly controlled society. Files of individuals were also likely to be made easy to review by central authorities. One could expect that the use of modems, which would allow access to telephone lines for data or other transmissions, would be tightly circumscribed indeed. Currently the quality of telephone lines in the U.S.S.R. is generally too poor to permit data transmission.

The Soviet Union thus faced a very uncomfortable dilemma. Failure to develop computer literacy represented a severe threat to economic growth and military superiority, yet widespread access to computers threatened to undermine central control of information. Before *glasnost,* computer growth was to be carefully managed in order to control the availability of information.

Similar problems arose with the rapid spread of videotapes and videotape players in Soviet-bloc countries. Unlike computers, the availability of videotape players has not been tightly controlled in the U.S.S.R. and the Warsaw Pact countries. Although videotape players can be bought legally, contraband tapes, including such favorites as *Rambo* and *Kramer vs. Kramer,* are smuggled in by party and nonparty members alike.[33] More often now, Eastern European countries are turning to the West for high-tech products. Often customs officials in these countries wink at the import of personal computers or other high-tech products that are supposed to be imported only from the U.S.S.R.[34]

Mikhail Gorbachev's campaign to permit limited informatic freedom has produced many changes. Now informatic freedom, the "free flow of ideas," including a reduction in censorship of newspapers, magazines, and other mass media, is permitted to a limited extent.[35] Other changes in the structure of society, called *perestroika* ("debureaucratization"), are equally important. They include expanded use of markets for consumer products, the ability of enterprises to make their own production decisions, and reduced central control of the economy. In recent years the first private associations—for environmental issues—and the beginnings of political parties have appeared. Social changes in the U.S.S.R., then, have reduced the conflict between the use of information technology and social values. The "second Russian Revolution" is creating changes whose impacts no one can foresee.[36]

One might argue that Gorbachev and his advisors realized that in the long run, the U.S.S.R. was going to be threatened not only by the superiority of foreign technology, but also by the need for the kind of social institutions that would allow this technology to be developed and deployed rapidly. It is tempting to speculate about the role that considerations of technology played in Gorbachev's new social policies.[37] Consideration of technology/society contradictions may well have provided an impetus for new social policies.

The Human Costs
of Technology

We cannot separate, then, a society and its technology. Even though the same technology will permit a variety of different social arrangements, technology constrains choices.[38] For this reason, many societies have chosen to forgo the benefits of certain technologies rather than accept their social consequences. For instance, during the sixteenth and seventeenth centuries in China, there was a reluctance to adopt artillery because it was a foreign invention—bad enough in itself—that might bring foreign customs into use.[39] The eventual result of this reluctance, however, was military humiliation by foreigners and the imposition of unwanted activities (such as the opium trade) from the outside. In a competitive world, we may be forced for our own survival to use the technologies others develop. The tools we use constrain us, but the tools others use may constrain us, too. Because technology is embedded in an intensely competitive world order, our economic welfare may force us to innovate and use technology rapidly. This "future shock" is very hard on us because it increases the pace of life, forces us to work harder with less time for leisure, provides less social "slack," and increases emphasis on economics versus aesthetics.[40]

Several social philosophers have pointed out some of the disadvantages of our extensive use of high technologies. Jacques Ellul sees technology as a threat because the kind of social order it brings is too calculative and regimented.[41] Langdon Winner has argued that some technologies impose a social order, regardless of the intentions of their sponsors.[42]

Ivan Illich has developed a critique of modern technology and its social institutions in a series of influential books.[43] He points out that modern tools (which for Illich mean not only physical but also social technologies) tend to require large amounts of expertise and that there is a tendency for this knowledge to be monopolized by experts. Because experts possess the necessary knowledge, we tend to relinquish to them control over our lives and our society. They, in turn, constantly develop more complicated technologies, which only they are allowed to direct and evaluate. There is no escaping this preeminence of experts unless we turn to tools that are less complicated and require less specialized expertise. Computers provide a strong example of this dependence on experts. Illich has suggested that if we deploy more "convivial" tools, then our lives will be happier and more directly under our own control.

These are important ideas, and we will examine them in some detail in the chapters that follow. We will see that the sociotechnical world is very complicated. Ultimately we can do much to make technology a more positive force for human welfare. If our problems are not to worsen, certain issues must be faced squarely. Understanding the social relations of technology is part of the process of confronting these problems.

Can Our Society Manage Its Technologies?

Throughout this book we will explore how social institutions shape and respond to technology. A basic principle of design is that a thing must fit its purpose. In a society with the kinds of technology we have, we need institutions that are appropriate to the complexity of the technology they must handle. Are the institutions we have, then, fit for their purposes? If they are not, what are we going to do about it? Obviously we have some alternatives: We can change the society to make it more adequate to cope with the technology; we can refuse to deploy technologies that our current social institutions can't handle; we can try to develop new sociotechnical systems that *jointly optimize technology and the human factor;* or we can simply let the system go on as it is now and suffer at some future time the consequences of a serious mismatch between complex technologies and the adequacy of our social institutions to handle them.

But are our social institutions really so inadequate? And can we redesign them? The answers to both of these questions is "yes," and the task is urgent. The issue of adequacy seems to involve four themes: complexity, reaction time, responsibility, and creativity.

Complexity. We need social institutions that are capable of responding to the complexity of the technologies with which they are confronted; often, they are not. They do not have the *requisite variety,* as it is called in cybernetics.[44] That is, the spectrum of responses at our disposal is not as broad as the variety of problems we face. Thus we may find ourselves confronting situations for which our attempts to cope are incredibly crude. For example, our efforts in using nuclear power involve a level of complexity that can produce lethal results because learning occurs through trial and error. Examination of the Three Mile Island reactor after the accident there provided a series of surprises that should give us pause.[45] It is evident that although the actual consequences of the accident were modest, the human beings monitoring the reactor did not understand what was going on, even after the crisis was over.

Computers also entail a level of complexity with which our society may be as yet unready to cope. Even while attempting to unravel the role of computer programs in the most recent stock market crash, we are installing new, more powerful computers whose reactions will be even more rapid and potentially disastrous.[46] Recently, a young computer hacker discovered that a single programming mistake meant the difference between a harmless prank and a dangerous computer virus; some 6,000 computers were affected by the rampaging computer virus as a result.[47] If this is what a prank will do, consider what the consequences of an intentionally destructive act might be.

Ecological problems present yet another complexity with which our system is ill-equipped to cope. Small interventions can produce large impacts because, in Barry Commoner's words, "everything is connected to everything else." As we will see later, important ecological problems may reveal themselves only by ambiguous signals, making our response difficult and slow. Not only are our social intelligence systems slow to detect the signals, but some technology sponsors may make active efforts to get us to disregard such signals. Problems may be disguised or even covered up by interested parties.[48] By the time the problems have been discovered, the damage may already have been done.

Reaction Time. Can social institutions react quickly enough to technological developments? When we contemplate the kinds of social changes in which technology is involved, the need to develop foresight and rapid reaction capabilities is obvious. We see the need for foresight in a number of ways: in technology's effect on the workplace, in the destruction of our natural habitat, in the way that technology is shaping our family lives, and in the technological hazards and accidents that seem more threatening every day. Much of our technological development has taken place without such foresight. The ability to project forward—to imagine what is likely to happen before it does indeed happen—is essential to a world ever more tightly packed with people who are using up natural resources at an increasing rate.

We might call this a "projective" attitude. We often call the thinking process "reflection," but reflection over what has already happened is not enough; we also need to develop a kind of "fast forward" to explore problems before we run into them. Knowledge of the dynamics of technology and society will allow us to predict consequences to a greater degree than is now possible. But without a conscious attempt to examine the changes we are making, we will not even learn from our own experience. The ozone hole problem, explored in Chapter 17, shows a learning system that is running too slowly. Our actions are outstripping our ability to understand and learn from them. We need to make our projective system more powerful and take its findings more seriously. "Learning from experience" is too slow to deal with the kinds of forces that now act on the earth and its atmosphere.

Responsibility. A society must see what it is doing and take responsibility for the decisions it makes. Nowhere is this more urgent than in the realm of high technology. Few parents would hand a loaded pistol to a ten-year-old child, but many of us are willing to give control of nuclear reactors to utilities that still operate problem-plagued systems. Many of us are willing to let the military experiment with horrifying biowarfare organisms and risk contaminating the global environment. We are willing to do as a society what we would shrink from doing as individuals.

In many different ways, our society separates those accountable for technological impacts from those with the power to create such impacts. Those who perpetrate dangers are rarely socially connected to the people they imperil. Victims may be separated by time, space, and anonymity from those who cause them injury.[49] This cannot continue. In a society that makes extensive use of high technology, we cannot allow people to feel distant from the consequences of their own actions, because these actions have enormous power to shape lives. And yet that is exactly what is happening. Responding to economic pressures, people carelessly deploy technologies whose impacts will be with us for centuries. Nuclear wastes, for instance, will persist in the environment for tens of centuries. A species made extinct can never be brought to life again. The existence of technological crimes involving perpetrators who never see their victims suggests that we need to reconsider our views of guilt, innocence, and responsibility. The factory owner who dumps toxic wastes into a stream, the nuclear power plant operator who falls asleep on the job, the teenage computer hacker whose antics lead to the death of a hospital patient—each forces us to think about the nature of responsibility in a high-tech society.

No doubt such people need to be curbed, but at issue here is the system itself. We create the system that allows these people to operate in this manner. There will always be irresponsible people; but we need not have a system that encourages them to be irresponsible and lets them get away with it. We need to link actions with the responsibility for their consequences. A social science of technology is a step in this direction.

Creativity. Much of the dialogue on technology and society casts technology as the active agent and society as the passive victim. Yet people show an enormous amount of everyday creativity in relation to technology. The creativity of inventors and innovators is obvious; what is often not so obvious is the creativity shown by the users of technology, not only in terms of skill, but also in their efforts to reshape the technology to fit personal needs. These skills in fitting, shaping, and customizing technology are valuable, and we can make them even more valuable by designing the technology so that it is more flexible in the first place. We should not allow so much of our technology to be resistant to change; nor is there any reason why the overall level of technological knowledge in our society should remain so low.

Through flexible design of technologies, then, and through better education about technology, we can amplify our natural abilities and inclination toward creativity. Much of the technology we have developed to date stultifies human creativity; it makes individuals into passive recipients of action from elsewhere. Television is a prime example. A society of people who are used to watching things, instead of doing something about them, will be less healthy. It will expect a relatively small cadre of people to administer the mess, while the individual occupies a well-grooved niche that he or she never leaves. This is not a democratic vision, and it likely will not long sustain a democratic society. If we want such a society, we had better start thinking about ways in which the design of technologies will permit more and broader options while maintaining our abilities to create and redesign. Technological creativity thrives on diversity.

Closely related to creativity is the kind of environment that allows it to flourish. The design of such environments involves a *generative rationality,* a strategy for encouraging constructive change. The sociologist Max Weber, who did so much to describe the dangers of excessive bureaucracy, despaired that it would come to dominate Western civilization.[50] For Weber, rationality meant the ability to calculate consequences, to provide accurate predictions about what would happen. He discussed organizations that behaved like computers and people who behaved like robots. We could call Weber's rationality a *calculative rationality.* But rationality has another face, the face shown by invention and creativity. Change does occur, and there is a rationality that gives such change an inner logic. This generative rationality provides a rationale for growth and for progress.[51] Although the sources of personal and social novelty are complicated, to call them "irrational" is wrong. There is a logic to creative activity, and we should seek it.

But generative rationality has its limits; it is dangerous to depend on it exclusively. In defending unfettered technological development, many people argue that it doesn't matter what we do, for science will find a way to fix it. This optimistic attitude may reflect psychological health, but it is very dangerous in that some of the problems we create are not so easy to fix. Sometimes this so-called calculative rationality is essential for keeping our actions within reasonable bounds. We need a balance between these rationalities, between calculation and creativity. To find the proper balance will require wisdom; the nature and source of such wisdom are worthy topics of further research.

Conclusion

We will encounter these themes throughout the remainder of this book. In looking at the developers, users, and evaluators of technology, we will see the extent to which current social institutions embody these values. In examining the variety of social institutions that manage technology, it will become obvious that our society is not very effective in coping with the kinds of changes we generate. Our management of technological hazards in particular leaves very much to be desired, but other problems are also important. We have overstructured work, created an impoverished family life, and replaced personal relations with electronic substitutes. We seem to be better at inventing new things than at creating a harmonious synthesis among already-existing technologies.

Yet the potential for growth, for creativity, for using technology to amplify the positive aspects of our lives exists among us. Throughout our exploration of technology and society are hopeful signs, evidence that people have learned how to use technology to enhance their lives and their societies. Just as we can learn from accidents and disasters, we can also learn from our successes, so that we may enrich our lives with their lessons.

Notes

1. For an amusing look at this now rare system of milk delivery, see Odis E. Bigus, "The Milkman and His Customer: A Cultivated Relationship," *Urban Life and Culture,* Vol. 1, No. 2 (July 1972), pp. 131–166.

2. James Brooke, "Time Is Running Out for the Coin-Operated Parking Meter," *New York Times,* Sept. 7, 1986, Section 4.

3. I owe this observation to Albert Goldberg, who told me the following story. While at a certain computer firm, Goldberg was a participant in an industrial council on customer needs and projected future uses for computers. At one point, when product innovations were being proposed, an enthusiast was touting the firm's presumably outstanding software. In response to this high praise of the product, someone else pointed out a brutal fact of life: "Well, we may have third-generation software, but those truckers out there have first-generation minds!" See also a version of this story in Albert Goldberg, "The Eclectic Technologist," *Educational Technology,* Apr. 1986, pp. 33–34.

4. See W. F. Cottrell, "Death by Dieselization: A Case Study in the Reaction to Technological Change," *American Sociological Review,* Vol. 16 (1945), pp. 358–365.

5. Michael Grey, "Ports of Haul," *TWA Ambassador,* Jan. 1987, p. 56.

6. Malcolm Gladwell, "Forestland in the Belly of the Beef," *Insight,* Dec. 1, 1986, pp. 45–46.

7. Roger Lewin, "Damage to Tropical Forests, or Why Were There So Many Kinds of Animals?" *Science,* Vol. 234 (Oct. 10, 1986), pp. 149–150.

8. For more about the possibility that the earth will overheat due to the "greenhouse effect," see Howard Wilcox, *Hothouse Earth* (New York: Praeger, 1974); and A. Ramanathan, "The Greenhouse Theory of Climate Change: A Test by an Inadvertent Global Experiment," *Science,* Vol. 240 (Apr. 15, 1988), pp. 293–299.

9. See K. William Kapp, *The Social Costs of Private Enterprise* (New York: Schocken Books, 1971); and Marshall Goldman, *The Spoils of Progress: Environmental Pollution in the Soviet Union* (Cambridge, Mass.: MIT Press, 1972).

10. See, for instance, Ivan Illich, *Tools for Conviviality* (New York: Harper & Row, 1973). Although I am unable to agree with all the author's conclusions, I feel that this is one of the most important books on technology; it merits reading by any thoughtful person.

11. See the essay by Eric Trist, "The Evolution of Socio-Technical Systems as a Conceptual Framework and as an Action Research Program," in A. Van de Ven and W. F. Joyce (Eds.), *Perspectives on Organization Design and Behavior* (New York: Wiley, 1981).

12. Along similar lines, consider the portrait of Theodore Taylor, in John McPhee's book *The Curve of Binding Energy* (New York: Ballantine Books, 1973). Taylor, an absent-minded physicist, is a genius at making nuclear weapons, although he hopes they will make war impossible. The contrast between what Taylor feels and what he is doing could not be more marked. In the same way, our society seems unable to comprehend the often damaging impacts its actions have on the environment.

13. Barry Commoner, *The Closing Circle: Nature, Man and Technology* (New York: Bantam Books, 1974).

14. Socrates often lampooned what he considered dangerous innovations in Athenian society. One might wonder what kind of searching questions Socrates would direct at today's institutions and leaders; no doubt he would be much less content than many

current philosophers with contemporary attitudes toward technology. An interesting biography of Socrates is in Xenophon, *The Anabasis or Expedition of Cyrus and the Memorabilia of Socrates* (London: Bell and Daldy, 1867). The conversations reported by Xenophon portray Socrates as a more psychologically coherent and interesting person than do the comparatively lifeless dialogues reported (or fabricated) by Plato.

15. Albert Borgman, in *Technology and the Character of Contemporary Life* (Chicago: University of Chicago Press, 1984), pp. 40–48, has a much more complicated definition of "device," which he uses throughout his book. Our usage here relates solely to the physical object itself, not to its aims or purposes.

16. On the topic of social inventions, see William F. Whyte, "Social Inventions for Solving Human Problems," *American Sociological Review,* Vol. 47, No. 1 (Feb. 1982), pp. 1–13.

17. Frederick Mosteller, "Innovation and Evaluation," *Science,* Vol. 211 (Feb. 27, 1981), pp. 881–886.

18. Similarly, the lack of appropriate technologies may require certain social adaptations. The Academie Française began in the seventeenth century because its members had no way of communicating across Paris before the telephone. To avoid missing each other, they set a definite place and time for regular meetings.

19. Fritz Redlich, "Bank Administration 1780–1914," *Journal of Economic History,* Vol. 12, No. 4 (1952), pp. 438–453.

20. Evelyne Sullerot, *Women, Society, and Change* (New York: McGraw-Hill, 1971).

21. See Donald MacKenzie and Judy Wajcman (Eds.), *The Social Shaping of Technology* (Milton Keynes: Open University Press, 1985).

22. Constance Holden, "Soviets Launch Computer Literacy Drive," *Science,* Vol. 231 (Jan. 10, 1986), pp. 109–110.

23. Maxine Pollack, "Playing Computer Catch-Up Against Incalculable Odds," *Insight* (*Washington Post* Magazine), Mar. 16, 1987, pp. 30–32.

24. Andrew Nagorski et al., "Moscow Faces the New Age," *Newsweek,* Aug. 18, 1986, pp. 20–22. Even as this is being written, censorship controls have been relaxed, and newspapers have been allowed both to write more freely and to print more copies, allowing better and more permanent access to news and opinions.

25. Philip Taubman, "The Kremlin Worries That Too Many Know Too Much," *New York Times,* Jan. 26, 1986, Section 4.

26. Maxine Pollack and Ross Alan Stapleton, "Why Ivan Can't Compute," *High Technology,* Feb. 1986, pp. 42–45.

27. Holden, "Computer Literacy," p. 110; Pollack, "Computer Catch-Up," p. 32.

28. Annalee Saxenian, "Silicon Valley and Route 128: Regional Prototypes or Historic Exceptions?" in M. Castells (Ed.), *High Technology, Space, and Society* (Beverly Hills, Calif.: Sage, 1985), pp. 85–106.

29. Paul Freiberger and Michael Swaine, *Fire in the Valley* (Berkeley, Calif.: McGraw-Hill, 1984).

30. Everett M. Rogers and Judith K. Larsen, *Silicon Valley Fever: Growth of High-Technology Culture* (New York: Basic Books, 1984).

31. Central planning has long caused problems for the economy of the U.S.S.R. See "The Stalled Soviet Economy: Bogged Down by Planning," *Business Week,* Oct. 19, 1981, pp. 72–83.

32. Pollack and Stapleton, "Why Ivan Can't Compute."

33. Nagorski, "Moscow." See also Gary Thatcher, "U.S. Experts See Dramatic Soviet Future," *Christian Science Monitor,* Apr. 7, 1987, pp. 1, 40.

34. Jackson Diehl, "Eastern Europeans Turning to West to End Technology Gap," *Washington Post,* Feb. 28, 1988, pp. H1, H18.

35. Consider two articles: Bill Keller, "Cornucopia of Heresies Bursts Into Soviet Press," *New York Times,* June 9, 1988, p. 6; and David K. Shipler, "Now, Dear Moscow Editor, About the Socialist Plague . . ." *New York Times,* Apr. 24, 1988, pp. 1, 6.

36. Between 1987 and 1989, events in the U.S.S.R. were moving so rapidly that this chapter had to be rewritten frequently. In a remarkable interview with *Newsweek* (May 30, 1988), Gorbachev stated that "we together with our society are seeking answers to all questions."

37. Mark Crawford, "Soviets Pin Economic Hopes on Technology," *Science,* Vol. 238 (Dec. 18, 1987), p. 1644.

38. Langdon Winner, "Do Artifacts Have Politics?" *Daedalus,* Vol. 109 (1980), pp. 121–136.

39. Carlo M. Cipolla, *Guns, Sails and Empires* (New York: Funk & Wagnalls, 1965), pp. 116–122.

40. Alvin Toffler, *Future Shock* (New York: Random House, 1970).

41. Jacques Ellul, *Technology* (New York: Knopf, 1964).

42. Winner, "Do Artifacts Have Politics?"

43. The most important of Ivan Illich's books is *Tools for Conviviality* (New York: Harper & Row, 1973). Some others are *Deschooling Society* (New York: Harper & Row, 1971) and *Medical Nemesis* (New York: Bantam Books, 1973).

44. W. Ross Ashby, *An Introduction to Cybernetics* (New York: Wiley, 1966), pp. 202–218.

45. William Booth, "Postmortem on Three Mile Island," *Science,* Vol. 238 (Dec. 4, 1987), pp. 1342–1345.

46. Kurt Eichenwald with John Markoff, "Wall Street's Souped-Up Computers," *New York Times,* Oct. 16, 1988, Business Section, pp. 1–9. See also David Sanger, "Wall Street's Tomorrow Machine," *New York Times,* Oct. 19, 1986, pp. 1, 30; and Mark Crawford, "Computers Amplify Black Monday," *Science,* Vol. 238 (Oct. 30, 1987), pp. 602–604.

47. John Markoff, "How a Need for Challenge Seduced Computer Expert," *New York Times,* Nov. 6, 1988, pp. 1, 13.

48. Lying by governmental and private authorities regarding contamination from nuclear-weapons production facilities should make us suspicious of the honesty of nearly any technology sponsor; see Matthew L. Wald, "Nuclear Arms Plants: A Bill Long Overdue," *New York Times,* Oct. 23, 1988, Section 4, p. 1. Among the disturbing details are reports of governmental officials' failures to demand compliance with safety guidelines and their willingness to reward unsafe behavior related to high productivity; see Keith Schneider, "Operators Got Millions Despite Hazards at Atom Plants," *New York Times,* Oct. 26, 1988, pp. 1, 7. This is merely one of several instances in which scientists working for the government have recklessly exposed people to the dangers of radiation. Perhaps it is not surprising that many ordinary citizens, assured that biotechnologies such as DNA synthesis are safe, are reluctant to believe public officials.

49. See Ron Westrum, "Vulnerable Technologies: Accident, Crime, and Terrorism," *Interdisciplinary Science Reviews,* Vol. 11, No. 4 (Dec. 1986), pp. 386–391.

50. Max Weber, "Bureaucracy," in Hans Gerth and C. Wright Mills (Eds.), *From Max Weber: Essays in Sociology* (New York: Oxford University Press, 1948).

51. See Ron Westrum and Tim Clark, "Paradigms and Ferrets," *Social Studies of Science,* Vol. 17, No. 1 (Feb. 1987), pp. 3–34.

Theory

Marx's Theory of Technology

Over the long history of social thought many theorists have puzzled over the role that technologies play in society. In one sense the development of thought about society and the technical influences on it have been parallel. During the Renaissance, contemporary technical achievements—the invention of gunpowder, the compass, the telescope—had encouraged many intellectuals to believe in human progress and in the superiority of modern society over ancient society.[1] The philosopher whose writings gave rise to many of the early scientific associations, Francis Bacon (1561–1626), considered technical progress essential to the progress of the human spirit. Bacon's writings encouraged the founders of the Royal Society of London to pursue scientific knowledge through experiment.[2]

It was not until the Enlightenment of the eighteenth century that technical progress began to be widely celebrated as an essential part of a developing society. The interest in technology originated as part of the development of the field of study called political economy (known to us simply as "economics"). Clearly, the introduction of new technologies was visibly changing the fabric of social life.

Important changes related to the growth of factory production (later called "the Industrial Revolution") were visibly associated with novel technologies.[3] The idea of progress, long associated with changes in human relations, came to be associated with changes in science and technology. In his *Essay Toward an Historical Portrait of Progress of the Human Mind,* the Marquis de Condorcet (1743–1794) included technology as one of the indications of social progress.[4] Auguste Comte (1798–1857), often referred to as the "father of sociology," asserted that society developed through theological and philosophical stages into a "positive" stage characterized by a scientific and technological approach to both natural and human phenomena.[5] But more than anyone else, the thinker who moved technology toward a central position in social theory was Karl Marx.

Karl Marx: An Overview

Karl Marx (1818–1883) was one of the most influential intellectuals of all time.[6] Few social theorists have so shaped the course of our world. His theories and concepts have stirred imaginations, started movements, led to revolutions, and spawned novel forms of government and economic systems. The various forms of philosophy that identify themselves as Marxism, the various movements that bear his name, and the different governmental systems that derive from his ideas testify to the enormous power and influence of Marx and his ideas. Our interest here is to understand Marx's thought and its impact on the social relations of technology. Even though it is difficult to disentangle Marx's ideas from the various groups and causes to which these ideas have become attached, that is exactly what we must do, for if we wish to grasp Marx's contribution to the understanding of technology, we must try to see clearly not only the many strands of his theory relating to technology, but also how these ideas have influenced social institutions.

Marx included thoughts on the role of technology in many different writings. Technology was not his major concern, and hence those interested in his thoughts on the subject must piece together his theory of technology from portions of larger works. Our major concern here is not to develop an exact account of Marx's thought, a task better left to historians and philosophers.[7] Rather, we wish to act as social theorists, looking for ideas that will help us understand the phenomena in which we are interested. Rather than being "for" or "against" Marx, we wish to develop a dialogue with Marxian theory, to integrate what is worthwhile in it and reject what is not.

Marx was both philosopher and social scientist. He wrote at a time when social science existed only in the form of "political economy," a field from which sociology, economics, and political science all derive. Although a contemporary of Auguste Comte (1798–1857), Marx did not consider himself a contributor to the new intellectual field of "sociology," as Comte had named it. Nor did Marx feel that he was creating political economy, but rather that he was replacing it with something better. Furthermore, like Comte, he felt that the application of his ideas was just as important as their development.

Marx held strong humanistic views on the meaning of human life and the way it should be lived, but unfortunately these views tended to recede in the face of immediate political issues. Many of his humanistic concerns were neglected in the erection of "Marxist" social structures, which are often brutally rigid in practice and pose a contrast to the flexibility Marx extolled in some of his earlier writings. Marx felt that communism would allow the full and free development of humanity. Ironically, however, many communist regimes repressed rather than encouraged the flowering of the human spirit.

Marx was a brilliant thinker and a powerful writer. Whereas others, notably Ivan Illich, have since dealt systematically with the same issues, few have covered so broad a canvas and touched on the enormous variety of issues relating social classes, work, politics, technology, and economics.[8] This does not mean that Marx's theories were correct; it does mean that he had a lot to say. Marx was able to institutionalize his ideas in powerful social movements that would later found governments and translate his ideas (with major adjustments) into practice. It is not surprising, therefore, that Marx's influence should be so far-reaching. In coming to terms with his ideas, then, we must confront the applications of his ideas as well as the ideas themselves.

Marx's enormous intellectual strengths were accompanied by severe weaknesses. He was educated as a philosopher, a subject in which he earned a doctorate. His attempts to develop and apply his philosophy put him into conflict with the political authorities of his time and eventually led him into active socialist politics. Marx read widely and was very well informed, not only about events of his own time, but on the history of Europe as well. His intellectual interests and energy were prodigious. But his deep involvement in intellectual and political controversies imparted to his writings a polemical flavor that makes dispassionate evaluation of them difficult. The quality of his writings accurately reflects a personality that was at once idealistic, incisive, and profound, yet capable of dogmatism, cynicism, and vindictiveness.[9]

Marx's human ideal was an intellectually and emotionally "whole" person living in a society that would support the development of such an integrated personality. Under capitalism, Marx felt, people were fragmented by the work and ownership processes and were reduced to less than full human beings. A communist society, by putting the means of production in the hands of everyone, would restore each person's self in its fullness. The individual would no longer be forced into a single occupational role, but could become a kind of Renaissance man or woman. To bring about a society that would allow such development—that could give its citizens meaningful work and a sense of controlling their own destinies, and could end poverty and injustice—Marx proposed a social system he called Communism.

Marx was the finest critic capitalism has had; yet he was virtually silent about the structure of the society that he expected would replace it. His brilliant critique did not include serious planning for a future that hundreds of millions of people would find all too specific. That communist society would be better, however, Marx did not doubt. Once classes were abolished, he felt, most of the world's injustices would disappear.[10] But Marx did not foresee that new forms of injustice might arise in such a "classless" society.[11] Clearly he was a better analyst of the ills of contemporary society than a visionary or social planner.

Because he had a pungent wit, Marx sometimes wrote things that sounded convincing but could not withstand serious examination. Many of his more casual statements were to have serious consequences later on.[12] Unhappily, many of Marx's intellectual heirs did not have minds as able as his, and much of his writing that deserves criticism has been accepted uncritically, both in his own time and in ours.[13] As his thought became institutionalized in powerful movements and governments, it found vigorous exponents who could silence many of its critics. Marx did not live to see how his ideas were put into practice in the Soviet Union, China, Cambodia, and elsewhere; it is difficult to believe that he would have approved of some of their applications if he had.[14]

Marx on Technology

For Marx, technology was a weapon in the struggles between classes. Thus for him technology under capitalism has a necessarily political character. This point is very important because it means that technology cannot be neutral in a class society; it can be used either by the owning class to oppress or by the revolutionary class to rise up. Yet Marx concentrated nearly all his effort on the first category—on technology used by the owning class (the bourgeoisie) to oppress the working class (the proletariat). Technology could regain its value, Marx felt, only by a total transformation of society that would put control of technology into the hands of everyone. How he proposed to achieve this transformation was never clearly spelled out, which meant that later "Marxists" would have to design the future society for themselves. As a result, collective ownership of the "means of production" has been interpreted in a variety of ways in the U.S.S.R., China, and Eastern Europe.

Marx is best known for his analysis of factory work, but he saw technology in a much broader context. Marx was personally fascinated by technology and expended considerable effort understanding it. He was impressed by the enormous development of commerce in the nineteenth century. Roads, bridges, canals, railways, and ships not only moved people around; they also moved goods. Marx saw in this great advancement of civil and marine engineering an expression of the energies of commercial capitalism. In *The Communist Manifesto,* Marx and Engels describe the impressive achievements of these public and private works:

> The bourgeoisie, during its rule of scarce one hundred years, has created more massive and more colossal productive forces than have all preceding generations together. Subjection of nature's forces to man, machinery, application of chemistry to industry and agriculture, steam navigation, railways, electric telegraphs, clearing of whole continents for cultivation, canalization of rivers, whole populations conjured out of the ground—what earlier century had even a presentiment that such productive forces slumbered in the lap of social labor?[15]

But this ferment of development did not advance civilization so much as it served the extension of the capitalist system of production and commerce.[16] The achievements were public, but the motivation behind them was to aid the growth of private capital—to increase the wealth of a small segment of society.

Nowhere was this passionate energy for development more evident than in the numerous colonial ventures of the developed countries. As a correspondent for the *New York Tribune,* Marx wrote of the enormous changes brought about in colonies like India, where pressure for development was destroying traditional cultures even as it was creating networks of roads and centers of commerce. The stressful impacts of technological and commercial changes were felt by peasants in India, by factory workers in Hong Kong, and by slave laborers in Africa. Everywhere, wrote Marx, the restless energies of capitalism are changing, eradicating, erecting, and shifting the structures of societies, all the while seeking to foster the growth of capitalism.

In the twentieth century these trends have become even more striking. Agriculture, including the so-called green revolution, has changed the pattern of food production all over the world. Increasingly, people farm for income rather than to feed themselves.[17] Multinational corporations surpass anything that Marx observed in the extent of their operations and the myriad ways in which they affect global society.[18] The arms trade takes modern weapons to every corner of the globe.[19] And the mass media carry advertising for capitalist firms everywhere. The entire world is competing economically, desperately trying to defend national interests against other countries who seek innovations in order to outsell them.[20] The world has become a giant economic pressure cooker, and the pressure is building. This situation is not healthy, and it may get worse.

Marx's solution to burgeoning technology was not to halt it, but rather to alter its character. To Marx the development of civilization was positive. It is not clear whether he would have favored what is now called "appropriate technology"—small-scale, energy-saving systems, such as solar power—over the energy-intensive and highly centralized alternatives. What needed to be changed, he contended, was ownership and control. The great changes that were taking place should not benefit a small number of capitalists, but rather humanity as a whole. Marx proposed to change the character of modernization by shifting the ownership of the means of production and development from the capitalist class to the whole society; in short, he proposed the development of socialism. That Marx developed neither a proper theory of this transition nor a blueprint for the ultimate form of the resultant new government was a serious defect. The consequence has been the appearance of a great variety of strategies for socialist economic development.

Marx on Economic Development

Marx predicted that shifting ownership from private interests to the public sector would relieve the pressure of economic competition. He believed that economic competition was a consequence of capitalism (true) and that communist countries would not compete economically (debatable). Currently we find socialist countries that are just as "imperialist" and competitive as their capitalist counterparts. But, of course, these are not yet "true" communist societies!

In many cases socialist development has meant an immense deployment of public welfare, education, and health resources. What the Chinese have called "the iron rice bowl" has many counterparts in other socialist societies. Generally, however, the first concern of socialist countries has been national development of heavy industry and weaponry. In many cases this rapid economic development has deprived millions of people of freedoms that Marx too easily scorned, and it has perpetrated cruelties equal to those imposed by the worst of the colonial regimes he described in the nineteenth century.[21] Marx was correct about many of the moral ills of capitalism.[22] He did not realize, however, that the concentration of power in a socialist government would allow it not only to bring about economic development, but also to ignore individuals' needs in favor of national development. Although he accurately delineated the cruelties and contradictions of capitalism, Marx failed to imagine those that socialism might bring.

A common response to this criticism by thoughtful Marxists is that none of the societies that call themselves communist or socialist are "really" communist.[23] A communist society would be a just one, these critics argue, and current regimes clearly are not. How such a socialist society would work, however, is seldom described. In the absence of such descriptions, it is not possible to know how viable a truly socialist society would be. Certainly, though, we should regard the variety of current socialist forms of government as a range of experiments providing lessons we can study and selectively incorporate into our own future. We should not simply choose between capitalism and socialism; rather, we should identify the features of both that most enhance freedom, democracy, and the development of the individual, and then adopt those features.

Marx on Class Interests and the Sponsorship of Technology

One of Marx's major contributions was his observations on the sponsorship of technology. We have already noted that those who act as the managers of technology shape its development and application. We will discuss the sponsorship of technology at greater length in later chapters. For our purposes here, we can define a sponsor as a group or organization that exercises a dominant role in the direction of technological change. To the question, "In whose interest has technology been developed?" Marx replied, "In the interests of capitalists." But in a socialist society, he contended, technology would be used in the interests of the people, of the society as a whole. Let us examine this question of "interests" for a moment, for this point is very important for much of the discussion that follows.

Technologies are ways of getting things done. They are, in this sense, means to accomplish goals that people set. But people, according to Marx, do not have only individual ends; they have goals *as a group*. Marx expressed this idea in the concept of "class interest." When a group of people share a common role in the economic process, then they possess a common interest as a class. Only if they became conscious of this class interest, Marx argued, could they act on it. Thus, if workers could become conscious of how oppressed they were by the capitalist system, then they could join together and overthrow it. The technology formerly developed by the capitalists could then be used by the workers *in the interests of the society as a whole*. As rhetoric, this is a splendid turn of phrase. But what does it mean in terms of social structure and governance?

In this matter we must first consider the issue of goals, in particular how diverse goals might be for different people in a society. Although Marx initially celebrated the diversity of individuals and wanted individuals to be able to perform a great variety of jobs, this interest in diversity is less apparent in his later writings. Increasingly, he seemed to take the position that differences among individuals were not important; indeed, the eradication of some differences (for example, between city and country) was to be an important feature of socialist life. Thus the goals of socialist society increasingly became *societal* goals, the identity of which would be most clearly visible to those guiding it. How the individual was to aid in setting the direction of society Marx never made clear.

Some later Marxists, however, interpreted Marx's ideas in a frankly elitist manner. In the writings of the Russian Bolsheviks, particularly those of Vladimir I. Lenin (1870–1924), this elitist direction becomes strongly evident. Here the party nucleus guides the party, and the party in turn guides the masses.[24] It is doubtful that Marx would approve of the way his ideas were interpreted by the first Marxist regime in the U.S.S.R. or by Lenin, to say nothing of Stalin.[25]

Marx also proposed the intriguing idea that a group may not be conscious of its interests, just as a child may not be conscious of what is really good for it. Who, then, would know what the "real" interests of a group were? To Marx, the answer was obviously those who understood the nature of the class struggle; the leaders of the communist movement. But we might instead contend that the issue of what is good for people is neither evident nor indisputable, and thus that "interests" may emerge only in the process of defining and acting upon them.[26] One of the goals of democracy is to educate the people about the process of choosing society's direction. If someone other than the members of the group is to decide what is good for them, who is it to be? Obviously, Western democracy has found one set of solutions in capitalism and in representative institutions.[27] Socialism has found another in state control and technocracy. Thus the question of who is to set the societal goals is a major issue.

Technology enters into the issue of goals because choices about technology affect fundamental conditions in a society. When a government chooses to develop heavy industry instead of producing consumer products, this technological choice has enormous impacts on the lives of those in the society. A choice about nuclear versus solar power is also a choice about the kind of society that will result.[28] It is interesting to note that in American society, most of our institutions—the Congress, the courts, even the Constitution—were developed with eighteenth-century technology in mind.[29] To take merely one example, Article Two of the Bill of Rights (1791) states that "a well-regulated militia, being necessary to the security of a free State, the right of the people to keep and bear arms, shall not be infringed."[30] To the founding fathers, defending the United States with a militia seemed a sound defense policy; it is no longer so. Clearly the United States, in the nuclear age, can no longer be protected effectively by a militia. Today the vision of the long-barreled rifle above the mantelpiece seems woefully inadequate for national defense, yet this was clearly the sort of scenario envisioned by the writers of the Constitution, and Article Two is still part of our Bill of Rights. The gun enthusiasts who so frequently cite it ignore the context in which it was originally written.

And just as in the days of the founding fathers, decisions about defense technology are decisions about what kind of society we are going to have. When we decide that the United States is going to be defended by submarine-launched ballistic missiles with multiple independent reentry vehicle (MIRV) warheads, this is not simply a technological decision, but also a decision about the nature of our society. So the question obviously becomes, who will make such a decision, and how? In Marx's view, a socialist society would always make such a decision—or one about electrification, sewage treatment, or the manufacture of bubble gum—in the interest of the whole population. But how would such an interest be assessed, and by whom?

Interestingly enough, as this book was being written socialist countries all over the world were adopting market mechanisms to make economic development more rapid and moving away from socialist central control. This is one of the most dramatic social changes of the twentieth century, more important than either of the world wars, and it is occurring with very little theoretical change. Now many choices about technology will occur more through market mechanisms than through central planning.[31] Although eventually this may mean fewer waiting lines in the Soviet Union or more refrigerators in China, one wonders what kind of knowledge is being lost. The lessons of communist planning are worth close study. Now they are in danger of being lost in the rush to install market-oriented economies.

Marx on Factory Work

Marx's strongest contribution to the social science of technology was his analysis of human work based on his extremely wide-ranging knowledge about the factory conditions of his time. In England, where he lived most of his adult life, Marx conducted historical research in the Reading Room of the British Museum. Much of his information on factory conditions came from reading the blue books of British factory inspectors.[32] He was also very much influenced by his friend and coauthor Friedrich Engels (1820–1895), author of *The Condition of the Working Class in England* and the son of a factory owner.

In the middle of the nineteenth century, when Marx began writing, the typical factory was a textile mill, where men, women, and children labored 12 or more hours a day for pitiable wages.[33] Technology in this context seemed to Marx, as it might well seem to us, to be an instrument of class oppression. Furthermore, the relatively high cost of labor and the pressure of competition meant that factory managers were constantly trying to replace their workers with machines that would do the work more rapidly and with less resistance. For Marx the factory was a setting in which one set of men, the capitalists, drove another set of people, the workers, to create surplus value out of raw materials.

Such industrial exploitation, however, was merely one example in a series of such social juxtapositions over the course of human history. Marx sought to explain not only the dynamics of the factory, but also how factories came into being. Ordinarily, industrial sociology takes the existence of factories and labor markets for granted and explores their internal dynamics; it does not try to explain how they come about or vanish.[34] But Marx contended that we should take a longer view and see the institutions not as something fixed, but rather as part of a dialectical process of social change, a process dominated by struggles between groups. In his own time, Marx identified an ongoing "labor process," the effects of which were extremely detrimental not only for workers, but for the society as a whole. Three parallel trends were operating:

1. *Deskilling.* The worker was losing skills and expertise as the factory's division of labor advanced; the worker was losing an active role in production and was becoming merely a cog in a machine.

2. *Powerlessness.* The worker had a diminishing grasp of the overall process in which he or she was engaged. The worker's consciousness narrowed with increasingly narrower job definitions. As a result, the worker became less human and experienced an increasing sense of helplessness.

3. *Degradation.* The worker was getting poorer, for incomes were forced down as managers made workers more replaceable by the unskilled and by machines. Workplace conditions made workers into drudges. Low wages, starvation, and misery seemed inescapable.

The net result of these changes, Marx concluded, was to deprive workers not only of a bond with what was produced, but also to reduce their roles as people and citizens. Just as the worker lost control over what was produced and over the processes of production, the worker's ability to act as a full-fledged member of society declined. With a smaller sense of personal power and lower income, the worker lost control over society as well.

In the next three sections we will consider deskilling, powerlessness, and degradation in light not only of Marx's theory, but also in terms of current research, for the issues Marx raised are very much with us today.

Deskilling. Deskilling occurs in the labor process whenever the system removes initiative and skills from workers and places them in the hands of management. One of the most important observations Marx made was that factory work was alienating because it was fragmentary, rigidly imposed, and degrading. The worker makes only a small part of the final product and as a result is unable to identify either with the overall process of production or with the product that comes off the line at the end. Furthermore, because the job does not require skill, the worker lacks any joy in its execution. Work in a capitalist factory, Marx noted, takes something that belongs to the worker (his or her labor) and transforms it into a product over which the worker can exercise no control. Furthermore, the machines introduced into the factory were an alien force brought there to push the worker to a faster pace, and as such they represented to the worker how hopeless resistance was to the forces of capitalism. This vision of work has been extremely influential, and it is hardly an exaggeration to say that hundreds of studies have been undertaken based on its premises. Let us examine it in more detail.

In Marx's time, as in our own, skilled workers tended to use their bargaining power to extract higher wages and other concessions from management. Skilled workers, not only in the factory but in the marketplace in general, can charge more for their work than can unskilled workers. At the same time, there is a constant desire on the part of those who require the products and services workers offer to find ways of lowering the wages they pay for labor. Henry Bessemer, the English engineer and inventor, describes in his autobiography the motive for his development of an automatic machine to make brass powder. The craftsmen who had a monopoly on the manufacture of this powder charged very high prices for it, prices that Bessemer felt were unfair. By developing a "self-acting" machine, Bessemer eliminated the need for hiring these craftsmen and broke their monopoly in the process.[35] The use of skilled mechanics and craftsmen in factories was fraught not only with financial problems for the factory owner, but with managerial problems as well. Skilled workers worked when they wanted to, demanded high wages, and kept secrets of production from their employers.[36] Managers did not so much employ these workers as contract with them on a temporary basis. This situation and its uncertainties, including strikes, led managers to seek greater control.

One solution to the problem was to hire women and children, who would work for lower wages; thus skilled male craftsmen, willful and proud, were often replaced by more tractable workers. Another solution was to alter the machinery itself so that people with fewer skills could operate it. Such workers could resist managerial pressures less easily and cost employers less money. Machinery, then, became one means by which the skills of the labor force were degraded.

This circumvention of skilled labor began early in the industrial revolution and continues still.[37] Adam Smith's classic description of the diversity of operations in the manufacture of pins shows that such practices were already well entrenched in the eighteenth century.[38] In the nineteenth century the division of labor in textile factories became quite elaborate. Mass production facilities tended to be designed for workers with few skills, and often they employed women and children. By the end of the nineteenth century, metal-working industries began to adopt equally subdivided work practices. In the United States, these trends were carried further by the invention of the assembly line, developed by William Knudsen and carried to its ultimate extreme by Henry Ford.[39] Another change was embodied in "scientific management" and its later forms, such as "efficiency management" and "industrial engineering."[40] The many immigrants who entered the United States during the nineteenth century were yet another factor, as were two world wars (requiring rapid hiring of unskilled laborers) during the twentieth century.

Fragmentary production is not universally disliked, but there is no question that many workers hate the fact that they produce only a small part of the finished product. Such work provides workers no glory and requires little skill. Even if the factory were owned by the workers, many people would be bored and alienated by putting the single nut on the single bolt, time after time, day after day. Marx was quite correct that workers who do this kind of labor are often unskilled, which enables employers to hire women and nonunion workers.[41] However, other workers enjoy this kind of work just because it is highly structured and "mindless."[42]

Personal reaction to highly subdivided work is shaped not only by the job itself, but also by management attitudes, organizational culture, and prospects for advancement. Even assembly-line workers can be shifted from one job to another; in such circumstances they slowly acquire skills in a piecemeal fashion, becoming "utility men."[43] Laborers who work in such plants, however, tend to emphasize external benefits and usually have little intrinsic love of their work as such. This attitude has been revealed by surveys that ask workers whether they would follow the same career if they could start over; very few factory workers would be factory workers if they had it to do over again.[44]

Deskilling did not develop simply from the desire of greedy capitalists to gain control over workers; it may also be a kind of economic and technological imperative. In a study of Newark, New Jersey, Susan E. Hirsch has documented the changeover from craft to industrial work there between 1800 and 1860.[45] Newark was a craftsman's town, populated and run by workers proud of their trades and skills. Workers traditionally used their own hand tools, gradually acquired complex skills through apprenticeship, and worked largely at a pace set by custom. Frequent breaks and work rituals softened the pressures for production. The artisan or "mechanic" looked forward to becoming a master, with the accompanying possibility of self-employment. The worker was a craftsman who took pride both in the work and in the associated life-style.

But technological innovations, often by the workers themselves, altered the forms of work.[46] Skilled artisans used their skills to design production machines and their personal savings to build larger companies. Increasingly, a division of labor replaced the traditional skills of the all-around mechanic. Machines began to do the jobs that workers did, and with greater speed and efficiency. Human power was replaced by steam power. Apprenticeships became less necessary (and therefore less common) as workers' skills came to be embodied in the machines. Craft work survived only for the carriage trade. Increasingly the mechanic was replaced by the unskilled worker.

In many respects, what happened in Newark was only a microcosm of developments all over the United States.[47] As factories drew laborers from beyond their immediate vicinity (thanks in part to the expansion of transportation), they grew larger, as did the firms of which they were a part. Increasingly, mergers took place between firms, and others went out of business. The growth in communications allowed managers to control parts of the organization at increasing distances. Within a business, new statistical controls and management skills facilitated larger plants and the integration of far-flung operations. The new controls also seemed to work better if the labor force could be made predictable. Pressure for skills in "managing people" grew apace with skills in managing technology.

Deskilling is still very much in evidence in the twentieth century, spurred by interest in "scientific management" and its successors. Academic interest in this topic has been very strong for the last 15 years, following the publication of Harry Braverman's book on labor and monopoly capital in 1974.[48] Through an examination of management writings, Braverman showed how in the United States interest in deskilling had developed around the ideas of Frederick Taylor, the father of "scientific management." Braverman's *Labor and Monopoly Capital* sparked a deluge of research, theorizing, and debate.[49] Virtually the entire current Marxist sociology of work has emerged as a response to this book. Braverman's book was itself a response to the observations of earlier mainstream sociologists of work, such as Robert Blauner, who felt that automation might lead to a more skilled labor force because it requires a higher degree of intellectual skill.[50] On the contrary, Braverman argued, capitalists have degraded and continue to degrade the skill level of their employees' jobs.

It is evident from the outpouring of sympathetic studies that supported his thesis that Braverman had picked a vital issue.[51] Deskilling, particularly when associated with computers in offices, is obviously a serious problem for many workers, and study after study has demonstrated that Braverman's point was well taken.[52] Even though, as Donald MacKenzie has since pointed out, deskilling is only one way in which capitalists might dispossess workers of their labor, work is obviously central for Americans, and deskilling strikes at the heart of an important value: the quality of working life.[53] Jobs are being deskilled, and the trend should concern us for several reasons. Not only is deskilling damaging to the human spirit, but it also fails to equip society with workers who understand the technologies they operate.[54] This in turn may have consequences for the workers' roles as citizens, consumers, and creators.

A number of remarks need to be made about Braverman's book. On one hand, Braverman makes some valuable observations. There is no question that much of his criticism of managerial philosophies is correct. His extension of Marx's thesis to office work is particularly important. On the other hand, the book is very frustrating to read because it slides over difficult issues. Braverman often develops a line of reasoning with apparently strong factual support; only upon reflection is it evident that he has left a great deal out of the analysis.

For instance, he takes great pains to show how much worse the factory worker's lot has become since the nineteenth century. Yet no one familiar with factory conditions at the turn of the century is likely to agree with this assertion,[55] for in spite of deskilling, there have been important gains in other areas, thanks in part to unions and enlightened managements. The shortening of work hours, the increase in factory cleanliness, and the increased rights of workers to basic human amenities have very much improved, as have wages. Braverman says none of this, perhaps because he does not think it important. It is true that a great deal of the work people in our society do is boring, mindless, and a waste of talent. And much of it is dangerous.[56] With this we cannot disagree; Braverman is absolutely on target. But to workers, these extrinsic qualities of the job that Braverman ignores are not trifles.[57]

A more important problem is Braverman's underlying theory: that capitalism is to blame for the degradation of work. Because much of this degradation has occurred under capitalist conditions, the basis for this major assertion is apparent. But why, under "nominally socialist" conditions, do we often see the same thing?[58] Braverman's reply to this criticism is shallow and unconvincing: Socialism imitated capitalism.[59] But why, after 60 years, is socialism *still* imitating capitalism? Why have democratic work arrangements appeared in capitalist countries as often as socialist ones? What is the motive for this drive for the simplification of work and the deskilling of the worker? One hypothesis is that the basic motive is a particular kind of rationality, which I call "calculative rationality" and which is typical not so much of capitalism, but of bureaucracies in general. If this is correct, then any bureaucratic society is likely to show this same degradation of work. We will explore calculative rationality and its alternatives in a later chapter.

The failure of Braverman and many other Marxist writers to use the experience of the socialist countries as a control group for their assertions about capitalism is serious. If capitalism is not the problem, then eliminating capitalism will not improve the quality of work. We need to know more about the psychology of people's need and desire for the kind of managerial control made possible by "scientific management" and similar monitoring. If calculative rationality is indeed the cause, then we will find that such methods will proliferate in situations in which the accounting mentality (so common in American organizations) thrives.[60] But in any case, if Braverman and other Marxists are wrong about the cause-effect relation between capitalism and the degradation of work, then the remedy is not creation of a "socialist" society, but rather one in which a different kind of rationality becomes the norm.

Powerlessness. Another vital issue in Marx's "labor process" is that of power. Workers want to exert influence over their jobs and even over the firms for which they work. Yet often they have very little say about what they and their companies do. There are several bases for worker influence: skills and bargaining power are clearly two of them. But also important are knowledge of the operation and forms of participation in decision-making.

Many workers have very little idea of the dynamics of the overall operation of which they are a part. They do not know where they fit into "the big picture" and thus cannot understand the importance of what they do. Such knowledge would enhance their ability to produce a quality product.

It has always amazed me that a firm's new workers are not given a two-week course in how the whole operation works. This is such a simple and obvious measure that helps the worker to feel part of a team and understand where his or her job fits into the company's operations.[61] One of the earliest studies of supervision identified the one most successful supervisor as a person who did just this kind of orientation for his part of the operation.[62] Yet many firms reinforce the worker's insignificance by keeping their employees uninformed about what its ultimate aim is; some even ignore suggestions for improvements that come from the workers.[63] There is no reason that democratic work arrangements would necessarily prove less profitable; indeed, there are instances in which they have increased profits for firms. But apparently many plant managers would rather have ignorant workers than high production. The belief that workers' knowledge of the whole operation would not increase productivity is a self-fulfilling prophecy; by keeping them ignorant, management assures that workers' contributions will be limited.

Workers are equally "kept in their place" by the design of workplace technologies. Marx and many who have followed him have noted that managers will often go to surprising lengths to break the power of workers. A prime example was Cyrus McCormick's introduction of new pneumatic molding machines to his Chicago reaper plant in the 1880s.[64] These machines actually produced moldings that were inferior to the machines they replaced, and in fact they were abandoned after three years of use. But they had one important feature for McCormick: They required many fewer skills to operate than had their predecessors. Because McCormick could hire relatively unskilled workers to operate them, he was able to break the National Union of Iron Molders, which threatened to exert economic power over employers like him. Factory technology, then, is not only used to build products, but also to empower or disempower those who use it.

In a study of the introduction of numerical (machine) control, David Noble has shown that some American managers have purposely picked systems that would keep control of automation in the office rather than on the shop floor.[65] This disregard of workers' skills in the design of the system is not a technical necessity, as experience with numerical control in American industry has shown. Rather, it reflects a conscious desire to remove the direction of operations from workers' control. The proliferation of "expert" computer systems no doubt results in part from managers' desires to take the workers' knowledge out of their heads and store it in computer programs. Not only does this strategy make the worker less powerful in negotiations with management; it also produces more stable, calculable output. Yet such changes do not reflect simple technical needs. Although incorporating expertise in the hardware is valuable, it may also be valuable to leave some of it in the human beings. A study by Håkon Andersen of numerical control in Norway has shown that it is entirely possible to incorporate human skill into numerical control operations and that it may be advantageous to do so.[66] The decision about where to locate the expertise has important implications for the empowerment of workers.

Workers' powerlessness is associated with their remoteness from the locus of managerial decision. As management becomes more remote, so does a worker's sense of influence on decisions. One factor that has reduced workers' influence is the size of the organization itself. Marx did not believe that large-scale activity in itself was bad. In fact, in *The Communist Manifesto* he and Engels called for the formation of "industrial armies." More recent empirical studies, however, show that worker morale tends to decrease with increases in workplace size.[67] It is very likely that this increasing alienation is due to an increased sense of powerlessness. A small worker in a giant plant is unlikely to feel very powerful in shaping decisions, regardless of other aspects of the corporate culture.

Another factor that has reduced workers' influence is the geographical locus of control. As companies went from single-plant to multiplant operations, absentee control of operations became common. This problem is made worse by takeovers of smaller businesses by larger ones, which then install managers who are simply paid employees—without real power to make independent decisions. As small businesses are taken over by larger ones, some of which may be huge conglomerates, control shifts from plant manager to company headquarters, often in a city hundreds of miles away.

The negative impacts of such absentee control on workers' sense of influence are palpable.[68] First, the company president or CEO no longer belongs to the same community, is neither a member of local organizations nor subject to community norms. Second, workers are unable to go directly to the head of the organization; all interaction must be through channels. Finally, the local operation is now only one of a string of operations, subject to termination at a moment's notice for the greater good of the company.[69] All these factors increase workers' sense of helplessness by removing control of local operations from local people.

The impact of organizational size should be considered in evaluating Marx's argument that ownership of the means of production should be transferred to the people as a whole. Remoteness of the centers of power can occur under Communism, just as it does under capitalism. Ownership of the means of production may mean nothing if the worker does not have a large enough voice in how his or her own factory is managed. Over 80 years ago the political scientist Robert Michels observed that even under democratic conditions, power tends to shift into relatively few hands. Michels was so impressed by this same tendency in socialist parties that he propounded the "iron law of oligarchy" to express the observation.[70] The key issue, then, is not who owns the factories (capitalists vs. the masses), but who runs them. How are those who are in control made accountable to those who work?

One significant form of workers' power has been unions. Marx said very little about unions, although he recognized their existence. Clearly unions have done a great deal to shape the workplace: They have provided workers with significantly increased bargaining power resulting in higher wages, enhanced benefits, and greater job security and safety. And yet unions have largely left to management the prerogative of making decisions about job roles. Unions have largely ignored the nature of jobs, and in fact in some cases have acted against workplace participation because it might involve collaboration with management.

When John Leitch proposed a system of worker participation in the 1910s, unions opposed and destroyed it.[71] Leitch's system of cooperation was violently opposed by the unionists as a "company union" scheme that threatened to put the union in a less aggressive posture toward management, and if workers and managers got along, what would one need a union for? Thus unions, like managers, may fear worker empowerment in ways not connected with the union. A salve for unions' fears may be the possibility that the union could promote worker cooperation with management, as happened through the "Scanlon Plan," sponsored by the Steel Workers Organizing Committee.[72] The "Scanlon Plan" involved use of a suggestion system and a dividend for cost-savings by workers. It was unusual in that it was promoted by the union. Usually unions have retarded movement toward democratic work arrangements because of this fear that cooperation will lead to workers' integration with management.

In some cases the presence of unions may have favored deskilling. Marxist theory usually contends that capitalism leads to automation, which then leads to deskilling, to alienation, and ultimately to class struggle. But this is not necessarily the order of causation. In Hawaii, for instance, the development of unions in the sugar industry raised management's labor costs, leading to more automated materials-handling procedures.[73] Thus the class struggle itself may have led to automation, rather than the reverse. Interestingly enough, the jobs that remained in the sugar industry tended to require higher skills than those eliminated by automation. Finally, one of the most interesting results of the union's radical demands was an employers' response that moved the union toward more conventional attitudes toward wages and benefits.

Similarly, unions may have reinforced the division of labor in factories by developing work rules that discouraged cross-training and multiskilling. In many large industrial plants there are certain tasks that only "skilled trades" workers are allowed to do, and then only those of the appropriate "trade" can do them. As unions became large administrative systems concerned with protecting the prerogatives of jobs, they came to reinforce divided roles and a continuum of competences. Rather than tinker with the structure of the factory, unions chose to make sure that virtually no structure could be arbitrarily reshuffled, and that workers would get decent wages and benefits.[74] On one hand, this has protected many workers' jobs; on the other hand, it has meant a great inflexibility that has prevented workers from acting as participants in decision-making. The very size of unions themselves, along with low rates of participation, has meant that rather than empowering the average worker, they have often instead provided another powerful outside force with which the worker must reckon.

Efforts over the years to provide workers increased participation in decision-making have enjoyed momentary success, only to sink again into obscurity.[75] In the United States, for example, "industrial democracy" has included workable participation plans since the turn of the century.[76] Among these are John Leitch's system of "industrial democracy," the "Scanlon Plan," "democratic leadership," Rensis Likert's "participative management," and more recently the quality-of-worklife movement.[77] All these attempts to increase worker participation met with considerable resistance from management, the "Scanlon Plan" was no exception. Although there was considerable publicity given to the plan, the number of firms that adopted it remained small. Participative management usually is advocated on the grounds that it will make the enterprise more profitable. Even though there is some evidence that this is true, particularly from experience with "Scanlon Plan" plants, managers in the United States are still very hesitant to try such experiments.[78] Even the "Japanese challenge" has done little to promote genuine change in American management styles, in spite of glowing examples and advice from social scientists.[79] Thus many of Marx's comments on the nature of work in capitalist societies still apply.

The struggle for dignity and participation in the workplace still continues in the United States. Although on the one hand there are experiments featuring increased "team-building" and worker participation, there are also many instances of worker deskilling and big-brother-like monitoring, thanks to computers and similar high technology.

Degradation. Marx did not anticipate that various social factors would improve the worker's lot. Although he was correct about deskilling, he certainly did not anticipate many of the changes that proved so crucial in improving conditions for factory and service workers: unionization, the Progressive movement, enlightened management, mass-produced goods, and universal education.[80] Shorter work hours, better factory conditions, stringent laws about child and sweatshop labor, inexpensive foodstuffs, automobiles, and televisions have made the twentieth-century worker's life better. But the greatest factor has been the movement of the labor force out of factories and into white-collar jobs. Indeed, most "workers" today are not factory workers. Society, no doubt much to the dismay of many Marxists, changed without a revolution, and the workers' lot improved in a material sense. But in other ways work is just as soulless as it was in Marx's time, and for some of the same reasons.

For instance, the worker today still feels alienated from the things that he or she produces. Subdivided jobs, lack of identification with the firm, lack of technical knowledge, and lack of participation in decisions all contribute to alienation. When the job is done, where is the proof to show that the worker contributed? Here is how one of Studs Terkel's interviewees, a steelworker, describes the negative self-image his job gives him:

> Pyramids, Empire State Building—these things don't just happen. There's hard work behind it. I would like to see a building, say, the Empire State, I would like to see on one side of it a foot-wide strip from top to bottom with the name of every bricklayer, the name of every electrician, with all the names. So when a guy walked by, he could take his son and say, "See, that's me over there on the forty-fifth floor. I put the steel beam in."[81]

In looking over the very useful collection of interviews that make up Terkel's book— a better portrait of work than many of the more scientific surveys—we find that many workers are proud of what they do. But very few *factory* workers express this pride. It is not surprising, then, that in the absence of such intrinsic compensations, workers tend to emphasize other, extrinsic, features of a job. Many workers identify a "good job" as one that pays well or has good "benefits," even though it is not intrinsically satisfying. There is no question that many managers have missed an important opportunity to win the loyalty of workers by failing to make them feel "part of the team." And the result of this lack of identification with the work process is an alienation from the things produced.

Self-respect is also damaged by management patterns at work. The culture of many factories is heavily punitive, and rank is used as an important incentive. Those of higher rank "kick around" those of lower rank, at once building their own egos and creating intense resentment on the part of those who work for them. The words "beat up," a very common expression in the vocabulary of American office workers, express the feelings of many workers who feel demoralized by their managers. The concept, although expressed in different words, is equally common on the shop floor.

In Marx's view, capitalism leads inevitably to the degradation of the worker. The "labor process" under capitalism would necessarily cause the lot of workers to decline; their skills would diminish along with their incomes and self-respect. The only solution for this situation, Marx contended, was a revolution that would overturn capitalist society and put in its place a just communist society.[82] Yet experience with actual communist practice has been mixed. In a number of countries communist revolutions have taken place, but the kind of society that Marx would have welcomed exists in few of them. The labor process is often just as degrading in "socialist" countries as it is in capitalist ones; furthermore, it lacks the civil liberties or the material prosperity enjoyed by many workers in capitalist countries.[83] But there are some important exceptions.[84] For instance, for some time, Yugoslavia has tried to bring about democracy in factories. The results of this experience are interesting.

Yugoslavia under communist rule elected very early (in 1950) to move toward self-management of factories and political life generally.[85] This commitment to democracy has been reflected not only in industrial practices, but also in national laws that recognize the right of the workers to a share in factory and community governance. Joint governance of factories by workers and management is now taken for granted there, even though managers remain much more influential. Although there is still a considerable gap between communist ideals and reality, the system has come closer than many critics have expected.[86] There have been attempts both to make the organization serve the goals of the community and to respond to the needs of the workers in it. Managerial prerogatives and salaries have been substantially reduced in comparison to those of the workers. Instituting self-management while building up an economy and introducing extensive technological changes at the same time, however, has proved to be very difficult. Nonetheless, Yugoslavia has managed to move very much toward a society that embodies at least some of the principles that Marx deemed important.

In addition to structural changes, intellectual and emotional changes are necessary for workers' participation at work. It is interesting to note the conditions required for management by the workers, as outlined by one Yugoslav sociologist, Bogdan Kavcic:

1. Each worker must be interested in a broad spectrum of problems of whatever organization (factory, community, etc.) he or she wishes to influence;

2. The worker must possess adequate knowledge and abilities for responsible management; and

3. Each worker must have the necessary organizational facilities for participation in decision-making.[87]

Further, although the management process is still under refinement,

> the role of workers in their environment has undergone a complete change. Theoretically and legally no limits are set to workers' participation in the management of social affairs. Nevertheless, there are limits of a practical nature—the scope of workers' attention, the time at their disposal, etc. are certainly limited quantities. Besides, workers do not only enjoy self-management rights, they are also responsible for management of their organization and of the whole society. Experience has shown that self-management as a right asserts itself much faster than self-management as a responsibility.[88]

It is important to recognize that self-management is a skill that can grow only with time and practice. The Mondragon experiment in Spain, involving a large network of interlocking cooperatives, and one of the most impressive examples of humanistic self-management, has not had a smooth development, even though overall it has been a success.[89] There have been disagreements and struggles, and even a major strike. Worker participation, like democracy generally, takes time, faith, and ingenuity. Marx did not sufficiently emphasize how important learning and the use of experience were to the success of such experiments. Yet the willingness to experiment is very much needed if worker participation is to work. It is no exaggeration to view systems like the Yugoslav one or Mondragon as a part of a vast experiment in worker participation that is taking place all over the world. The results of this experiment are worth studying.

Marx's Legacy

Marx's major legacy regarding technology was his emphasis on the role it played in relations among people. In his view, technology is a tool in human struggles; when the character of technology was changed by shifting it from private to public hands, it would assist in freeing humanity. Marx was not a technological determinist—he did not argue that the technology shaped the society—but he did contend that certain kinds of technology had been used to shape relations in the factory. A new society would not do away with technology or factories, but it would change the relationship between the individual and the factory. People would still do industrial work, but the direction of that work would be in their hands. The products of their labor would belong to them; they, not impersonal economic forces, would control their own fortunes. And classes would vanish.

How all this would happen Marx did not say, and it is not clear whether he ever gave much thought to the structure of communist society. He did give a lot of attention to the principles that would guide this society, but he did not address how they were to be put into practice. He did not examine what sorts of problems large technologies of any sort would pose for democratic control; nor did he anticipate that electric power grids or gas pipelines might necessitate huge bureaucracies that are insensitive to individuals and their needs. He neither foresaw that valid differences of opinion might arise nor considered how they might be resolved. These are not small matters, and they have provoked intense conflict for many of the societies faced with implementing Marx's ideas.

Marx's real contribution was to emphasize that technologies do not stand alone. They are always surrounded by social institutions, they are brought into being to achieve social ends, and they shape the contours of social life because they are part of human activities that give structure to relationships. A technology brings with it not only people to maintain it, but designers, users, and above all, sponsors. Technology is directed—it is designed, is intended for some purpose—and Marx emphasized how important it was to understand its purpose. He also understood the dynamics of the workplace. But his mistake was to lay the blame for workplace alienation entirely on the shoulders of capitalism. He failed to consider that other factors, such as bureaucracy or workplace size, might be important. He certainly did not understand the problems that socialism would face.

Conclusion

Karl Marx's philosophy of technology raised some fundamental questions: Who designs it for what purpose, and who controls it? As an analyst of the social impacts of capitalism and its technologies, Marx was a true pioneer. Yet the subsequent emphasis on scrutinizing "technology under capitalism" has led to a serious lack of good theory and research on technology under socialism.[90] Why, under socialist regimes, is it still difficult to democratize technology, especially in larger factories and national technological systems? Is the pressure to control other human beings merely a side effect of capitalism, or does it have other causes? Why is calculative rationality present under socialism, and how can it be dealt with? And finally, under what conditions, technological or otherwise, can people live as Marx felt they ought to? What are the real sources of alienation from work, from the products of labor, and from others? Marx's virtue was in raising these questions; his answers to them, however, are no longer sufficient for us.

Notes

1. Paolo Rossi, *Philosophy, Technology, and the Arts in the Early Modern Era* (New York: Harper & Row, 1970).

2. Paolo Rossi, *Francis Bacon: From Magic to Science* (Chicago: University of Chicago Press, 1968).

3. Paul Mantoux, *The Industrial Revolution in the Eighteenth Century* (New York: Harper & Row, 1965). Some scholars present evidence that suggests an earlier industrial revolution; for example, see John Nef, *Industry and Government in France and England 1540–1640* (Ithaca, N.Y.: Cornell University Press, 1964). For criticisms of the Nef thesis, see Sidney Pollard, *The Genesis of Modern Management* (Baltimore: Penguin Books, 1965), p. 320.

4. Marquis de Condorcet, *Esquisse d'un Tableau Historique des Progres de l'Esprit Humain,* Monique et Francois Hincker (Eds.) (Paris: Editions Sociales, 1963).

5. Frank Manuel, *The Prophets of Paris* (New York: Harper & Row, 1962), pp. 249–296.

6. Much of this chapter is based on my own reading of Marx. I have found three essays to be particularly helpful in understanding Marx on technology: Nathan Rosenberg, "Karl Marx on the Economic Role of Science," in his *Perspectives on Technology* (Cambridge: Cambridge University Press, 1977), pp. 126–138; Nathan Rosenberg, "Marx as a Student of Technology," in his *Inside the Black Box: Technology and Economics* (Cambridge: Cambridge University Press, 1982), pp. 34–54; and Donald MacKenzie, "Marx and the Machine," *Technology and Culture,* Vol. 25, No. 3 (July 1984), pp. 473–502. I also read David McLellan, *Karl Marx: His Life and Thought* (New York: Harper & Row, 1977).

7. For some idea of how complicated such a task can be, see MacKenzie, "Marx and the Machine."

8. Ivan Illich, *Tools for Conviviality* (New York: Harper & Row, 1973).

9. McLellan, *Karl Marx.*

10. Karl Marx and Friedrich Engels, *The Communist Manifesto* (Harmondsworth, England: Penguin Books, 1979), p. 29.

11. New forms of injustice have indeed been experienced by those living in societies without the "classes" about which Marx worried. See Milovan Djilas, *The New Class* (New York: Praeger, 1957).

12. I think it is fair to hold philosophers responsible when the vagueness of their ideas does not provide sufficient protection for those who are forced to live according to them. If Marx had insisted on freedom of speech (and similar rights) instead of ridiculing it, he would have helped protect the rights of many who were to live out his ideas. In this respect, the writers of the American *Federalist Papers,* James Madison, Alexander Hamilton, and John Jay, were head and shoulders above Marx. Note the comments in Federalist Paper #51:

> If angels were to govern men, neither external nor internal controls on government would be necessary. In framing a government which is to be administered by men over men, the great difficulty lies in this: you must first enable the government to control the governed; and in the next place oblige it to control itself. A dependence on the people is, no doubt, the primary control on the government; but experience has taught mankind the necessity of auxiliary precautions. (*The Federalist or the New Constitution* [London: J. M. Dent, 1926], p. 264)

Marx's omission of these "auxiliary precautions" and similar considerations has been disastrous for many who have had to live under Marxist regimes.

13. This was a problem shared by Sigmund Freud, whose ideas were also brilliant and (like all good ideas) needed revision as time passed. His successors, however, have never seemed intellectually up to making such revisions.

14. As the defender of the peasants against aristocratic oppression in Germany, Marx would never have sanctioned the starvation of the 13 million "Kulaks" who died during Stalin's forced collectivization campaign in the early 1930s; see Craig R. Whitney, "Back in the U.S.S.R., the New Ideas Are Visionary, but Not Very Visible," *New York Times,* Jan. 29, 1989, Section 4, p. 2. The world never became as outraged over the destruction of the Kulaks as it did over the Nazi Holocaust, perhaps because this incident was soft-pedalled by reporters with conflicting loyalties. Eugene Lyons, an astute American reporter sympathetic to Stalin, explains how he managed not to report the full extent of the horrors, about which he knew, in his book *Assignment in Utopia* (New York: Harcourt, Brace, 1937), pp. 279–292).

15. Marx and Engels, *The Communist Manifesto,* p. 12.

16. A recent effort to chronicle the early history of capitalism is Fernand Braudel's superb "Civilization and Capitalism" series, published in the United States by Harper & Row: *The Structures of Everyday Life* (1981), *The Wheels of Commerce* (1982), and *The Perspective of the World* (1984).

17. Michael Perelman, *Farming for Profit in a Hungry World* (Montclair, N.J.: Allenheld, Osmund, 1979).

18. Louis Turner, *Invisible Empires: Multinational Companies and the Modern World* (New York: Harcourt Brace Jovanovich, 1970); Graham Bannock, *The Juggernauts: The Age of the Big Companies* (Harmondsworth, England: Penguin Books, 1973); and Ahmed Idris-Soven, Elizabeth Idris-Soven, and Mary K. Vaughn (Eds.), *The World as a Company Town* (The Hague: Mouton, 1978).

19. For instance, see "The Infernal Cycle of Armament," a special issue of *International Social Science Journal,* Vol. 28, No. 2 (1976). But to understand the real scope of the weapons trade, the reader is advised to look through one of the giant volumes of *Jane's Weapons Systems* and see the vast number and variety of weapons produced in, and used by, different countries.

20. Michael Shanks, *The Innovators: The Economics of Technology* (Harmondsworth, England: Penguin Books, 1967); James Botkin, Dan Dimancescu, and Ray Stata, *The Innovators: Rediscovering America's Creative Energy* (Philadelphia: University of Pennsylvania Press, 1986); and Christopher Freeman, *Technology Policy and Economic Performance: Lessons from Japan* (London: Pinter, 1987).

21. Examples of these cruelties include the liquidation of the Kulaks in Russia under Stalin, the political excesses of the Cultural Revolution in China, and the massacres and forced starvation under the Pol Pot regime in Kampuchea.

22. It was noted in J. Larry Brown, "Hunger in the U.S.," *Scientific American,* Vol. 256, No. 2 (1987), that there are 20 million people in the United States who are chronically underfed, 12 million of them children. That this situation should exist in a country where "surplus" food is regularly destroyed is scandalous. Recently, a panel of the National Academy of Sciences assigned to study the plight of the homeless in the United States urged reforms of welfare and child assistance. Ten members of the panel were so disturbed, however, that they wanted to go beyond the usual guidelines of the Academy and call for even broader action. When the members of so conservative an institution call

for such changes, the situation must be grave indeed; see Connie Leslie, "Heartbroken," *Newsweek,* Oct. 3, 1988, p. 10.

23. Thus in Donald MacKenzie and Jucy Wajcman, *The Social Shaping of Technology* (Milton Keynes: Open University Press, 1985), p. 17, the authors describe the Soviet Union and China as "nominally socialist" societies. Often these countries are termed "state socialist" or are said to be in the "dictatorship of the proletariat" phase. I feel very uncomfortable with this sort of language. Because over half the world's population lives in these nominally socialist societies, a term for them that is analogous to "capitalist" and "socialist" would be useful. It is possible that for economists like Milton Friedman, the United States is only a "nominally capitalist" society!

24. Nathan Leites, *A Study of Bolshevism* (Glencoe, Ill.: Free Press, 1953).

25. See Hedrick Smith, *The Russians* (New York: Quadrangle, 1976).

26. There are a number of problems with the idea that a group has "interests" when it has not consciously defined what these interests are. The most obvious problem is the size of the group: What is the unit of analysis? A person's class, family, or political party could all be considered valid units of analysis. A second problem is whether to deem material, psychological, or social interests to be most important. Note that when adults discipline children "in their best interests," these often turn out coincidently to be what is in the adults' best interests, too. Finally, there is the problematic relationship between ends and means. Groups have strategies for achieving consciously held goals; but when they have not consciously formulated a goal, should strategies for achieving it be set from the outside? Clearly Marx thought so.

27. Clearly one of the greatest challenges for democracy is the management of giant corporations. This issue may represent a greater threat to the United States than many of the entities we typically consider our "enemies."

28. In spite of heavy criticism of his ideas, I tend to agree with Amory Lovins on this point with respect to "soft energy paths." See Amory B. Lovins, *Soft Energy Paths* (New York: Harper & Row, 1979); Amory B. Lovins et al., *The Energy Controversy: Soft Path Questions and Answers* (San Francisco: Friends of the Earth, 1979); and "Sociology of the Environment," special issue of *Sociological Inquiry,* Vol. 53, No. 2–3 (Spring 1983).

29. On this matter with respect to the courts, see Nicholas Wade, "When Judges Must Know More Than Law," *New York Times,* Dec. 27, 1987, Section 4.

30. Quoted from Richard D. Heffner, *A Documentary History of the United States* (New York: Mentor, 1963), p. 34.

31. I cite here only one of the thousands of articles on this subject: Melanie Freeman, "Perestroika: The Gorbachev Revolution," *Christian Science Monitor,* Dec. 2, 1987, pp. 1, 16.

32. Lewis A. Coser, *Masters of Sociological Thought* (New York: Harcourt Brace Jovanovich, 1971), p. 64.

33. The epithet "dark satanic mills" was apparently well deserved. See E. J. Hobsbawm, *Labouring Men* (Garden City, N.Y.: Doubleday, 1967), especially pp. 75–148; also see E. P. Thompson, *The Making of the British Working Class* (Harmondsworth, England: Penguin Books, 1981).

34. One exception to this view can be found in Reinhard Bendix, *Work and Authority in Industry: Ideologies of Management in the Course of Industrialization* (New York: Harper & Row, 1963).

35. Sir Henry Bessemer, *Autobiography* (London: Engineering, 1905), pp. 34–35.

36. Herbert G. Gutman, *Work, Culture and Society in Industrializing America* (New York: Random House, 1976).

37. Perhaps no one has written better on this matter than Georges Friedmann, in his *Industrial Society: The Emergence of the Human Problems of Automation* (Glencoe, Ill.: Free Press, 1950) and in *Le Travail en Miettes* (Paris: Gallimard, 1964); the latter book has been translated under the title *Piecework.*

38. Adam Smith, *An Inquiry into the Origins and Nature of the Wealth of Nations* (New York: Modern Library, 1937), p. 5.

39. Norman Beasley, *Knudsen: A Biography* (New York: McGraw-Hill, 1947), pp. 19–56; and David A. Hounshell, *From the American System to Mass Production 1800–1932* (Baltimore: Johns Hopkins University Press, 1984), pp. 217–411.

40. No one has written a proper history of these disciplines. Harry Braverman, in his *Labor and Monopoly Capital: The Degradation of Work in the Twentieth Century* (New York: Monthly Review Press, 1974), touches some of the important points, but by no means all of them. More precise, but also more limited, is Daniel Nelson's *Frederick W. Taylor and the Rise of Scientific Management* (Madison: University of Wisconsin Press, 1980).

41. S. McKee Rosen and Laura Rose, "Machines and the Worker: A Case Study of the Cigar Industry," in their *Technology and Society: The Influence of Machines in the United States* (New York: Macmillan, 1941). As an early text in the sociology of technology, this book makes interesting reading.

42. Arthur N. Turner and Paul R. Lawrence, *Industrial Jobs and the Worker* (Boston: Harvard Graduate School of Business Administration, 1965).

43. See the very interesting portrait of such a "utility man" in Studs Terkel's *Working* (New York: Avon, 1975), pp. 232–239.

44. Special Task Force Report to the Secretary of Health, Education and Welfare, *Work in America* (Cambridge, Mass.: MIT Press, n.d.), p. 16.

45. Susan E. Hirsch, *The Roots of the American Working Class: Industrialization of Crafts in Newark 1800–1860* (Philadelphia: University of Pennsylvania Press, 1978).

46. Susan E. Hirsch, "From Artisan to Manufacturer: Industrialization and the Small Producer in Newark, 1830–1860," in S. W. Bruchey (Ed.), *Small Business in American Life* (New York: Columbia University Press, 1980).

47. Alfred Chandler, *The Visible Hand: The Managerial Revolution in American Business* (Cambridge, Mass.: Harvard University Press, 1977).

48. Braverman, *Labor and Monopoly Capital.*

49. For a summary of all the interest in deskilling, see Paul Thompson, *The Nature of Work: An Introduction to Debates About the Labour Process* (London: Macmillan, 1983).

50. Robert Blauner, *Alienation and Freedom: The Factory Worker and His Industry* (Chicago: University of Chicago Press, 1964). One of the less agreeable aspects of Braverman's book is his many snide remarks about Blauner and Georges Friedmann, both capable industrial sociologists.

51. For instance, see Andrew Zimbalist (Ed.), *Case Studies in the Labor Process* (New York: Monthly Review Press, 1979).

52. Joan M. Greenbaum, *In the Name of Efficiency: Management Theory and Shopfloor Practice in Data-Processing Work* (Philadelphia: Temple University Press, 1979).

53. MacKenzie, "Marx and the Machine," p. 493.

54. After all this deskilling, perhaps it is not surprising that American factory managers cannot now find the skilled machinists needed to operate the new, complicated machines.

55. See, for instance, Daniel Nelson, *Managers and Workers: The Origins of the New Factory System in the United States 1880–1920* (Madison: University of Wisconsin Press, 1975).

56. Paul Brodeur, *Expendable Americans* (New York: Viking Press, 1974).

57. See, for instance, Daniel T. Rodgers, *The Work Ethic in Industrial America 1850–1920* (Chicago: University of Chicago Press, 1978).

58. Thompson, *The Nature of Work,* pp. 217–227.

59. Braverman, *Labor and Monopoly Capital,* pp. 12–14.

60. Unfortunately, American business schools tend to turn out "numbers-oriented" managers, and I suspect that with the increase of interest in market efficiency in socialist countries, they too will begin to turn out "number crunchers."

61. Richard Balzer, "The Importance of Feeling Important," in his *Clockwork: Life In and Outside an American Factory* (Garden City, N.Y.: Doubleday, 1976), pp. 134–144.

62. See Elton Mayo, *Social Problems of an Industrial Civilization* (Boston: Harvard Graduate School of Business Administration, 1945), pp. 105–110.

63. Clinton S. Golden and Harold J. Ruttenberg, *The Dynamics of Industrial Democracy* (New York: Harper & Brothers, 1942) discuss management resistance to ideas initiated by workers under the "Scanlon Plan."

64. This example is taken from Langdon Winner, "Do Artifacts Have Politics?" *Daedalus,* Vol. 109 (1980), pp. 121–136.

65. David Noble, *Forces of Production: A Social History of Industrial Automation* (New York: Knopf, 1984).

66. Håkon With Andersen, "Technological Trajectories, Cultural Values, and the Labor Process: The Development of NC Machinery in the Norwegian Shipbuilding Industry," *Social Studies of Science,* Vol. 18, No. 3 (Aug. 1988), pp. 465–482.

67. Blauner, *Alienation and Freedom,* p. 49; and Geoffrey K. Ingham, *Size of Industrial Organization and Worker Behavior* (Cambridge: Cambridge University Press, 1970). Many Marxists would argue, of course, that alienation measured by opinion polls is not "real" alienation. It is certainly true that people may not be aware of their own alienation, but this still leaves us with the issue of how "real" alienation is to be measured.

68. One of the most thorough studies of the impacts of the decline of local control was part of a community study of Newburyport, Massachusetts: the classic "Yankee City" study. See W. Lloyd Warner and J. O. Low, "Yankee City Loses Control of its Shoe Factories," in *Social System of the Modern Factory; The Strike: A Social Analysis,* Vol. 4, in the Yankee City Series (New Haven, Conn.: Yale University Press, 1947), pp. 108–133. The fictional rendering of this event in John P. Marquand's novel, *Sincerely, Willis Wayde,* provides considerable insight into the psychological impact of the transfer, not just on the factory, but on the community itself.

69. John A. Young and Jan M. Newton, *Capitalism and Human Obsolescence: Corporate Control Vs. Individual Survival in Rural America* (New York: Universe Books, 1980).

70. Robert Michels, *Political Parties: A Sociological Study of the Oligarchical Tendencies of Modern Democracy* (New York: Collier Books, 1962).

71. Hence the unions' distrust for, and the eventual failure of, John Leitch's "industrial democracy" idea; see John Leitch, *Man to Man: The Story of Industrial Democracy* (Chi-

cago: McClurg, 1919). It might be noted that virtually every current "quality of work life" feature was anticipated by Leitch. It is interesting, too, that Leitch emphasized that corporate culture, not structure alone, was the key to effective worker participation. Leitch's life and management career would make an interesting study. I am indebted to my student Robert Romig-Fox, whose unpublished paper, "John Leitch and Industrial Democracy: His Contribution to Industrial Sociology," gave me much of my knowledge of this interesting man.

72. Clinton S. Golden and Harold J. Ruttenberg, *The Dynamics of Industrial Democracy* (New York: Harper & Brothers, 1942).

73. James A. Geschwender and Rhonda F. Levine, "Rationalization of Sugar Production in Hawaii, 1946–1960: A Dimension of the Class Struggle," *Social Problems,* Vol. 30, No. 3 (Feb. 1983), pp. 352–368.

74. Richard A. Lester, *As Unions Mature: An Analysis of the Evolution of American Unionism* (Princeton, N.J.: Princeton University Press, 1957).

75. See Frank Lindenfeld and Joyce Rothschild-Whit (Eds.), *Workplace Democracy and Social Change* (Boston: Porter Sargent, 1982); and Robert E. Cole, "Diffusion of Participatory Work Structures in Japan, Sweden, and the United States," in Paul S. Goodman et al. (Eds.), *Change in Organizations: New Perspectives on Theory, Research and Practice* (San Francisco: Jossey-Bass, 1982), pp. 166–225.

76. To get some idea of the duration of the struggle, see W. Jett Lauck, *Political and Industrial Democracy 1776–1926* (New York: Funk & Wagnalls, 1926); see also Leitch, *Man to Man;* and John R. Commons, "Why the Leitch Plan Makes Good," *The Independent,* Vol. 103 (July 3, 1920), pp. 7, 32.

77. See Milton Derber, *The American Ideal of Industrial Democracy* (Urbana: University of Illinois Press, 1970); Leitch, *Man to Man;* F. Lesieur, *The Scanlon Plan,* (New York: Wiley, 1958); and Rensis Likert, *New Patterns of Management* (New York: McGraw-Hill, 1961).

78. Lowell Turner, "Three Plants, Three Futures," *Technology Review,* Vol. 92, No. 1 (Jan. 1989), pp. 38–45.

79. James Botkin, Dan Dimancescu, and Ray Stata, *The Innovators: Rediscovering America's Creative Energy* (Philadelphia: University of Pennsylvania Press, 1986).

80. See Nelson, *Managers and Workers;* and J. A. Banks, *Marxist Sociology in Action: A Sociological Critique of the Marxist Approach to Industrial Relations* (Harrisburg, Pa.: Stackpole Books, 1970).

81. Terkel, *Working,* p. 2.

82. The enthusiasm of many intellectuals for revolutions often reflects a certain naïveté about the nature of revolutions. Entering upon an armed struggle is always an uncertain business, and those who emerge victorious from the revolution are not always the best people to have in charge. The conditions of some intellectuals' existences may tempt them to embrace these more extreme measures. See Nathan Leites, *A Study of Bolshevism* (Glencoe, Ill.: Free Press, 1953); Zevedei Barbu, *Democracy and Dictatorship: Their Psychology and Patterns of Life* (New York: Grove Press, 1956), pp. 208–217; and Lewis S. Feuer, "Lenin's Fantasy: The Interpretation of a Russian Revolutionary Drama," *Encounter,* Dec. 1970, pp. 23–35.

83. For instance, see Mark R. Beisinger, *Scientific Management, Socialist Discipline, and Soviet Power* (Cambridge, Mass.: Harvard University Press, 1968).

84. See, for instance, Thomas L. Blair, *The Land to Those Who Work It: Algeria's Experiment in Workers' Management* (New York: Doubleday, 1970).

85. Gerry Hunnius, "Workers' Self-Management in Yugoslavia," in G. Hunnius, G. D. Garson, and J. Case (Eds.), *Workers' Control: A Reader on Labor and Social Change* (New York: Random House, 1973), pp. 268–321.

86. Hunnius, "Workers' Self-Management in Yugoslavia," pp. 268–324.

87. Bogdan Kavcic, "Self-Management and Technology," in Peter Grooting (Ed.), *Technology and Work: East-West Comparisons* (London: Croom Helm, 1986), p. 114.

88. Kavcic, "Self-Management and Technology," p. 115.

89. William F. Whyte and Kathleen L. Whyte, *Mondragon: The Making of the Collective* (Ithaca, N.Y.: Cornell University Press, 1988).

90. But see Frederic J. Fleron (Ed.), *Technology and Communist Culture: The Socio-Cultural Impact of Technology Under Socialism* (New York: Praeger, 1977).

The Ogburn Generation

After Marx there was a long hiatus in social thought about technology. Although there were important social theorists who wrote on the subject, systematic attention to technology and its social relations was absent.[1] Since the 1920s, however, a small group of influential sociologists have pursued this subject. This chapter will discuss the rise, flowering, and demise of this group and its preeminent figure, William F. Ogburn.

Ogburn and his generation of coworkers thought deeply about technology and wrote a number of important books on it.[2] Shortly after Ogburn's death, however, the sociology of technology declined rapidly. Indeed, by the 1980s the writings of the "Ogburn generation" might be missed by a sociologist reviewing the sociology of technology.[3] The passage into obscurity of entire schools of thought can occur because the frameworks of their theories are sustained by individual scholars who work within them, publish studies, and read one another's work; when the continuity of such a group is interrupted, as it was in this case, the group's contributions may pass from sight and may fail to be integrated into subsequent studies.

The Ogburn generation flourished for over three decades, roughly from 1922 to 1959. Still, the number of people who worked within the group was small, and they failed to recruit new members; eventually they died or moved on to other fields of inquiry. Their writings remain and some are worthy of interest, but they are seldom read.

Ogburn was a seminal sociologist whose contributions covered many fields of sociology. His emphasis on statistics influenced a whole generation of new researchers. In the 1960s, however, after Ogburn had died, his coworkers either stopped publishing books on the sociology of technology or went into other fields of study. In order to understand his influence, we will look at three of the central figures of this movement who made important contributions to the study of technology's social relations.

William Fielding Ogburn

When William F. Ogburn (1886–1959) published *Social Change with Respect to Culture and Original Nature* in 1922, he was working in a well-established realm of social theory. Other social theorists, such as Auguste Comte, Emile Durkheim, Karl Marx, and Max Weber, had also attempted to explain the mechanisms of social change, and their works have enriched our present understanding. Ogburn's contribution, however, was to stress the development of *culture* as something distinct and important and potentially measurable by quantitative methods. For Ogburn, culture was a large, impersonal structure of human knowledge and practices that people create; once created, it tended to take on a life and evolution of its own. Just as Marx posited that the economic system evolved in a manner almost independent of human will, to Ogburn the technological system was largely independent of the thoughts and actions of individuals. Ogburn contended that this evolution could be measured and, to a certain extent, predicted.

Few sociologists today possess the broad perspective that Ogburn brought to his research.[4] His mind ranged easily over many topics of inquiry, and common to each was his desire to measure social change. Ogburn measured change by reference to trends—changes in social variables over time. He advocated using statistics to reveal overall trends, and he did much to interest other sociologists in using them. Ogburn preferred whenever possible to use systematic data, which often were readily available from public sources. His interests in social policy and legislation dovetailed well with his interests in social trends and statistics, and the results were some of the first attempts at technological forecasting and assessment. *Recent Social Trends in the United States,* published in 1933, was a massive attempt to measure some of the major ongoing changes of his time.[5] His book *The Social Effects of Aviation* was a pioneering study in technology assessment.[6] Ogburn in turn interested and involved S. C. Gilfillan (see below) in these efforts, and Gilfillan was later to write often and well regarding technology prediction and assessment.

Ogburn's examinations of large-scale cultural changes often took place without careful inquiry into the underlying processes,[7] which might include what went on in people's minds. The historical case-study, which recently has become so important to sociology of technology, was simply not Ogburn's style, for he distrusted people's introspections and their spoken motives for doing things. He was much more interested in what they did, which could more easily be measured and did not suffer from subjective interpretations. It is obvious today that this kind of behavioristic emphasis leads to ignoring a good deal of useful data, but it was not so obvious in Ogburn's time.

One of Ogburn's major contributions was to call attention to the social dynamics of invention. Before Ogburn, invention was considered to take place sporadically, typically by heroic inventors. Ogburn contended that on the contrary invention was constant and that social forces were as important in the inventive process as individuals. One of the ways he supported his point was by showing that many inventions, such as the steam engine, were independently invented in different places at roughly the same time.[8] This evidence argued against the idea that inventions were the creation of individual "great men" and that if great man X had not invented the device, then it would have gone uninvented for a long time. Of course, some inventions have this character; Babbage's Analytical Engine, described in a later chapter, could certainly be considered one of them. But most inventions represent an "obvious next step" when viewed from a long-term perspective.[9] And furthermore, many of them represent relatively small steps in the sense of both originality and improvement. Ogburn was very sensitive to this continual evolution and viewed invention not as a series of giant leaps but rather a series of gradual improvements.[10]

Ogburn called attention to what are usually called "the social impacts of technology." Although Marx had done this earlier in regard to industrial processes, Ogburn's treatment of the subject was more consciously theoretical. Consider the following partial list of Ogburn's principles of the social effects of technology:

1. An invention often has many effects that spread out like a fan.

2. A social change often represents the combined contributions of many inventions.

3. An invention's causes and social effects are intertwined in a process.

4. An invention produces a series of effects that follow each other somewhat like the links on a chain.

5. Groups of similar inventions have an appreciable social influence, whereas the influence of any particular individual invention may be negligible.

6. The accumulation of the influences of the smaller inventions is a significant part of the process.

7. The majority of improvements are merely slight improvements on some existing device.

8. There are both social and mechanical factors in social change; social factors in

social changes are often derived in part from mechanical inventions, and vice versa.

9. It takes time for the social influences of inventions to become fully felt.[11]

So that we can better understand the kind of detailed analysis that Ogburn preferred, let's consider the application of principle 1 to the impact of radio broadcasting on increasing interest in sports:

> When analyzed in further detail, [this impact] shows fifteen further social effects, which are as follows: The broadcasting of boxing matches and football games tends (1) to emphasize the big matches to the neglect of the smaller and local ones, (2) increasing even more the reputation of the star athletes. In the case of football (3) the big coaches become glorified and (4) their salaries become augmented. (5) The attendance at colleges specializing in football whose football games are broadcast is increased. (6) Football practice in the springtime is encouraged and (7) the recruiting of prospective star players for college enrollment is fostered. (8) The smaller the colleges or the ones with higher scholastic requirements tend to be differentiated as a class by contrast. (9) Boxing matches with big gates have accentuated trends in boxing promotion, notably the competition for large sums of money to the neglect of smaller matches. (10) Broadcasting of sports has led to a greater advertising of the climate of Florida and California, and (11) no doubt has aided a little the promotion of these two regions. (12) Broadcasting of sports has led to the development of a special skill in announcing the movement of athletes not at times easy to see, a skill rather highly appreciated. (13) Athletic and social clubs with loud speakers have become popularized somewhat on the afternoons and evenings of the matches. (14) The broadcasting of baseball games is said to have bolstered the attendance, particularly by recapturing the interest of former attendants. (15) Another effect it is said has been the reduction in some cases of the number of sporting extras of newspapers.[12]

Please note that this was written about *radio,* not television, in 1933! One wonders what Ogburn would have written about telecasts of the Super Bowl.

Ogburn's major theory of cultural evolution entails four processes: accumulation, invention, diffusion, and adjustment.[13] The evolution of culture produces an accumulation of new chunks of knowledge, which are combined into inventions by inventors. The new technical devices are then diffused (distributed) to society, and this diffusion is followed by a social adjustment to them. Continual repetition of this four-part process is the mechanism for social evolution; the process is similar to natural selection of biological species. Ogburn, however, was less interested than Darwin in the processes of selection; he focused more on the process of creation.

Ogburn on the Rate of Inventions. To get some idea of the limitations of his approach and also to see where it might lead us today, let's consider one of the key elements of Ogburn's theory of cultural change: the rate of invention.

Ogburn assumed that the number of new inventions reflected the knowledge a society possessed. He wrote:

> This accumulation [of knowledge] tends to be exponential because an invention is a combination of existing elements, and these elements are accumulative. As the amount of interest paid an investor is a function of the size of the capital he has invested, so the number of inventions is a function of the size of the cultural base, that is, the number of existing elements in the culture. . . . The reason lies in the definition of an invention as a combination of existing elements; and as the existing elements increase, the number of combinations increases faster than by a fixed ratio.[14]

We could see how this concept might be expressed mathematically in the following way. Let's assume that an invention is made up of a pair of elements. Then with two elements, we could have only one combination; with three elements, we could have three combinations of pairs; with four elements, we could have six different combinations of pairs, and so on. The general mathematical formula for the number of pair combinations (C) would be

$$C = \frac{N(N-1)}{2}$$

where N is the number of elements. Considering only pairs of elements, the implications of the addition of a new element can be enormous: With each new element, we could produce a number of new inventions equal to $N - 1$. So the thousandth new element would allow us to create 999 new pairs! Of course, this line of reasoning assumes that only pairs would be relevant, when in reality some inventions would use combinations of three, four, or more elements.[15] Thus it becomes obvious that if we consider only *the combination of elements possible,* inventions would indeed increase exponentially in the way Ogburn suggested, and we would be overwhelmed by the number of new inventions.

For a variety of reasons, though, the number of inventions does not, as Ogburn thought, depend in any simple way on the number of combinations of elements possible. There are several factors that limit the number of new inventions.[16] One limiting factor is the large number of combinations that are of little value. But a more fundamental limitation is the structure of the innovating society itself. A number of features of society serve to channel and limit invention.

For instance, the number of inventors limits the number of possible inventions. Let's say that an inventor can produce ten new inventions during his or her lifetime.[17] Then the maximum number of new inventions cannot exceed ten times the number of inventors working at any one time. And several inventors might invent the same thing, which would result in fewer than the theoretically possible number of inventions. Still, the number of inventors would determine the maximum number of inventions.

Consider another factor: Inventors are not necessarily independent; many of them will be organized into companies and larger corporations. We know that the number of inventions a corporation produces is not simply proportional to the number of its inventors; rather, it is some much smaller number because many inventions fall by the wayside in the process of focusing the group effort.[18] A new limit-

Figure 3-1 How Invention Expands the State of the Art

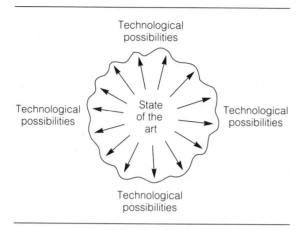

ing factor, then, becomes the number of firms in a given market. Furthermore, we know that some firms will try to limit the inventiveness of outside inventors by fair means or foul.[19] These features of social structure throttle down a society's inventiveness to a lower level.

Finally, we must consider the mental structure of the society. Imagination is usually limited. Most people's inventiveness does not consider the enormous variety of possible combinations of elements, but only those that represent a minor variation on what already exists—what is commonly called the *state of the art* (see Figure 3-1). What is considered possible is bounded by what is currently available, and what is not thought of cannot be implemented. A large variety of inventions thus never come into being because they are too far outside the conventional thought patterns of the society. As we will see, though, inventors do increase the realm of what is considered possible; they *do* widen the "envelope of possibilities."

What we can invent, then, is shaped by what has already been invented; what we can find out is shaped by what we know. But it is relevant to consider not only the extent of our intellectual resources, but also their organization. In many respects, it is not the *amount* of knowledge that is essential, but how accessible it is. Walking through a museum is very different from absorbing all the lessons it has to offer.[20]

Thus while Ogburn's assertion in its crude form is obviously incorrect, it does contain an important element of truth. Just as what we can build depends on the materials we have been given, invention depends on the intellectual resource base at our disposal. Inventors, as we will see, usually have considerable knowledge of previous technology, and this knowledge is often essential in developing their inventions. A technical education, libraries, and direct contact with technologies themselves are often critical. We also must consider, however, the often very limited institutions that society uses to exploit this knowledge.

Ogburn on Cultural Lag. Another of Ogburn's important innovations was the concept of "cultural lag." Ogburn noted that changes in one part of society often lead to changes in another part, but only after some delay. This delay is due to the relation between the needs and requirements of an innovation and its surrounding social milieu. For instance, when inventions first begin diffusion into society, it often takes a while for the technological infrastructure supporting them to develop. Just after the spread of the automobile in the United States, for example, roads that had been designed for horse-drawn carriages were no longer appropriate for motor-driven vehicles.[21] Similarly, the first owners of video cassette recorders did not have many videotapes to play or many places to get their machines repaired when they broke down; tape stores and repair operations developed a few years later. Today videotape rental is a thriving institution. Now, instead of planning to go out to "the movies," a family picks up two or three "videos" at the grocery store to watch on the home VCR over the weekend. The boom of videotape stores has even helped to reduce interest in pornographic bookstores. Instead of going to the local equivalent of the "combat zone" to get X-rated tapes, many pornophiles go instead to the video store.[22]

Ogburn usually discussed this problem in terms of the evolution of "material culture" versus that of "adaptive culture." Although he recognized that society's demands might generate innovations, in most of his examples he portrayed technology as the active element, and social practices as the passive element, in the process. A change of technology was stimulated by the accumulation of ideas and was thus "inevitable." When this inevitable technological change came, however, social practices were out of harmony with the new devices, and so (in theory) social practices had to adapt. Thus the "software"—adaptive culture—always had to adapt to the "hardware"—material culture. The resulting cultural lag, Ogburn contended, was behind many social problems and dislocations.

But are cultural lags objective or subjective? In other words, does a lag truly exist, or is it merely "in the eye of the beholder"?[23] Whereas Ogburn was firmly convinced that cultural lags were an objective fact, others might disagree. For many years consumers have routinely bought automobiles that can go 120 miles per hour. Is it a cultural lag that we have not built roads to accommodate such speeds? The development of the oral birth control pill in 1955 provided an opportunity for women to control fertility by its use. Yet Catholics are forbidden by church doctrine to use such means for controlling pregnancies. Is it a cultural lag that the church has not rushed to embrace the new technology? Ogburn assumed that in society there is a kind of moral consensus about values. Today, few sociologists would feel comfortable about asserting the existence of such a broad moral consensus. We recognize today that technological choices involve many norms and values that do not elicit universal agreement. Ogburn, however, tended to view society as more unified.

Ogburn thought that material "progress" was inevitable and therefore did not invoke moral choices. For instance, to Ogburn, once the television was invented its use was "natural." The device had been invented; it was a good thing, and it should be used. But for us there are some ambiguities. Indeed, television is a wonderful medium, but should it be used as it is? When we discover that the average American child watches six hours of television per day, we might ask, "Is this a necessary adaptation?" What if television had not been developed?[24] Would that mean more reading, a higher literacy rate, and so forth?

In Great Britain, for example, the British Broadcasting Company, a government organization, makes programming decisions for two of the channels, and a private organization, ITV, makes them for a third channel. In the United States, with the exception of the little-watched public television channels, such decisions are influenced by manufacturers who wish to advertise their products. Thus what people watch in this country is often shaped by the makers of breakfast cereal, children's toys, and laundry soap. The marked aesthetic superiority and cultural value of British programs over American ones reflect elite choices. The often shallow programs (such as soap operas and game shows) that Americans like to watch reflect popular choices. In one case we have an elite process, in the other a democratic process.[25]

Choices about television, nuclear missiles, or birth control devices are thus social decisions, not "inevitable" outcomes. It is neither "natural" to have television nor not to have it; it is a social choice that reflects values and inclinations. A "cultural lag" from this point of view is necessarily subjective and reflects the values and preferences of the observer. To Ogburn, however, this kind of relativity was foreign; it apparently did not occur to him that things might go another way.[26] If he had compared how different societies have used, say, the telephone or telegraph, he might have come to somewhat different conclusions. Our consideration of Soviets and computers in Chapter 1 has made us aware that adoption of new technologies is not automatic, even when such technologies might be of great economic value. Thus we see the value of comparative studies on the uses of technologies by different societies.

Before we leave Ogburn, it would be instructive to mention two of his major attempts to assess the impacts of technology.[27] The first of these was *The Social Effects of Aviation* (1946), which was written with S. C. Gilfillan and is a large and very thoughtful technology assessment.[28] *Social Effects* was an attempt to examine how a single industry might affect all the other sectors of society. A different approach was used in *Technology and the Changing Family* (1955, with Meyer Nimkoff), in which a social institution was the focus, and external forces affecting it, including technology, were studied.[29] Both of these works made very astute predictions regarding future impacts of technology on social institutions, and anyone interested in evaluating technology assessment may find much of interest in them. It is interesting to note, for instance, that in the book on the family neither the impact of television nor the immediate impacts of the oral birth control pill were foreseen. Such weaknesses are inevitable in technology assessment; the amazing thing is that Ogburn did such assessments at all.

Sean Colum Gilfillan

Sean Colum Gilfillan (1889–1987) was a remarkable sociologist whose scholarly achievements have been unfairly neglected. The reasons for this neglect may well have been due to his personal characteristics. Although he wrote some very stimulating books and papers, his career as a scholar was not a happy one.

Gilfillan was almost an archetype of the marginal man. Never entirely assimilated into academia, he nonetheless remained constantly on its edges. He held many odd jobs, including several "assistant-to" positions for professors, as well as memberships on government committees. On those rare occasions when he did get an academic post, usually at a small college, he was always replaced after a short time, probably because of personality clashes. It is characteristic of his eccentricities that he proposed to reform the English language by simplifying spelling; for instance, he wrote "thot" for "thought." Nor did it help that he was intensely interested in the role of genetic differences as causes of the varying performances of ethnic groups, at a time that such explanations were unpopular. Like Thorstein Veblen, another maverick sociologist, he was unwilling to compromise, and so he fought battles on many fronts. A great patron of unpopular views and grand lost causes (such as the artificial international language called Esperanto), Gilfillan was a rugged individualist. His memoirs were appropriately entitled "An Ugly Duckling's Swan Song."[30]

Gilfillan's writings reflected a fair share of rebellion toward the intellectual establishment. Yet he was a respected colleague of (and often coauthor with) Ogburn, who also was not overly respectful of institutional boundaries. His writings include several books on invention, starting with two publications from his doctoral dissertation. These became *The Sociology of Invention* and *Inventing the Ship,* both published in 1935 and both subsequently reprinted.[31] In 1970, he published *Supplement to the Sociology of Invention.*[32] In 1964 he wrote a major report on the patent system for the Joint Committee on Economics of the U.S. Congress.[33] He was also a notable futurist, and his contributions to technological forecasting have since been recognized and included in a text by one of the major authorities on the subject.[34]

Gilfillan's treatment of invention, developed as a set of principles, was very sophisticated. He viewed inventions as developing slowly, sometimes over centuries, as gradual accretions of knowledge led to new possibilities; in his words, "invention is a new combination from the 'prior art.'" Often the most important changes in technological capability are not revolutionary inventions, but rather small, incremental improvements.[35] His book *Inventing the Ship* covered hundreds of years of ship design without focusing strongly on any particular inventor.[36] Like Ogburn, Gilfillan was convinced that invention was a logical development that needed to be seen in the broad view, rather than in the details of any particular inventor's work. Thus a logical corollary to his thesis was that individual inventors were not critical for the overall trend; further, because trends are observable, it is possible to predict, within limits, when something will be invented. This is the basic principle of technological forecasting. Clearly there is much truth to this viewpoint, but other factors enter in. Gilfillan was convinced that the rotor ship, which was powered by vertical magnus-effect cylinders, would flourish in the years after 1935.[37] For whatever reasons, this did not turn out to be the case.[38] Prophecy is a tricky matter.

But Gilfillan, again like Ogburn, had hit upon an important point. If studies of invention can help us predict trends, then we ought to see where the trends are leading and plan accordingly. He saw that technological forecasting was possible and that it was important. He was later to participate in the first conference on technological forecasting at the Lake Placid Club in New York State, in May 1967.[39] But his influence on this field was modest. Again, it is likely that his personality had something to do with this; he was a pioneer, but not one who led others along his path. When he died in 1987, he left behind many achievements, but they tended to stand alone. Although he did participate in the Society for the History of Technology and left it $500 at his death, his work on the sociology of invention has been largely neglected.[40]

Bernhard J. Stern

Although Bernhard Stern (1894–1956) was an important member of what we've termed the Ogburn generation, it would not be appropriate to call him a member of the Ogburn school, for apart from an early exposure to Ogburn at Columbia University, he did not work with Ogburn very much. His research style, the journals and books in which he published his work, and the overall implications of his work, however, are very much in the Ogburn style. Like Ogburn, he had a strong belief in the importance of material forces in the development of and reaction to technology. Unlike Ogburn he was a Marxist.[41] Whereas Ogburn tended toward a view that today would be called "consensus" sociology, Stern was definitely a conflict theorist. He saw society as made up of contending groups whose struggles caused social changes.

Stern's focus was on the institutions that manage technology. Because initially he had wanted (but had been unable, due to health) to enter medicine, the topic of many of his studies was the medical community. He was particularly interested in the role that medical institutions play in enhancing and retarding invention.[42] But he was also interested in the role that powerful interests play in retarding change, particularly in a capitalist society.[43] He contended that novel technology was often suppressed because large economic interests were set against its introduction.[44] His work documented the suppression of patents (for instance, by A.T.&T.), a topic on which he was in considerable disagreement with Gilfillan, who considered patent suppression a myth.[45] For Stern, however, the behavior of organized economic interests was a more important force than "culture." He was very much interested in the behavior of sponsors, a topic we will discuss in a later chapter.

Thus Stern's work formed an important counterpoint to the ideas of Ogburn and Gilfillan, who viewed culture as a primary determinant of the fate of technology. In their views, if a technology could be invented, it would be; progress might come a few years earlier or later, but it was inevitable. In Stern's view, progress might be inevitable in the long run, but human beings lived in the short run, and in the meantime powerful institutions would steer society one way or another. Certainly in such matters as governmental policy and nuclear power, or the impact of automobile companies on public transportation, Stern seems to be more accurate. Progress is not automatic, at least not with respect to technology, and in any case advances in technology are partly a function of the resources invested in it. It is not enough to conceive an invention; it must be developed. Thus invention requires money, and money is controlled in our society by large, powerful institutions. "Trends" are important, but they may be influenced one way or another by forces other than the "mighty force of progress."

So Stern's premises and views on progress (and Marx's as well) differed very much from those of Ogburn and Gilfillan, who without excessive injustice could be called "technological determinists." As we have noted, the Ogburn group considered technology to be a largely neutral cultural force that could be used for good or evil. The importance of the sociological study of technology, then, was to allow finer societal adjustments to the technology—to bring society and technology into harmony. Although Ogburn did not explicitly say so, it is obvious from his analyses that he expected the society, not the technology, to do most of the adjusting. Ogburn suggested that society would adjust to the hydrogen bomb in the short run by dispersing population, to make it less vulnerable to attack. That the bomb constituted a danger to civilization itself and might be incompatible with it apparently did not occur to him. And in Gilfillan's view, the evolution of technology was inevitable, and therefore one might well imagine that social adjustments to it were inevitable, too.

Stern's viewpoint, however, was much less sanguine about cultural evolution. He believed not in the "invisible hand" of the market or in the blind forces of culture, but in the direct, powerful grip of large economic interests; some kind of progress might be inevitable, but progress was steered by industrial and professional establishments. We will consider this view again when we examine "network theory" in the next chapter.

The Passing
of the Ogburn Group

Ironically, one of the very cultural currents that Ogburn had done so much to study—the upheaval in social values of the 1960s—ultimately curtailed the influence of the Ogburn school. New social values replaced the ambitious conformity of the 1950s with individualistic hedonism on one hand and social action on the other.[46] The intellectual leaders of the younger generation advised the youth to "tune in, turn on, and drop out," and "We shall overcome" was the watchword of those who opposed social oppression and bigotry. The government began to be viewed as an agent of tyranny, as an instrument of war and social oppression. Big business seemed not merely boring, but evil. All this reawakening of feeling and sensitivity led to a disdain for technology; what was "natural" was good, what was artificial was bad. The slogan "Our Friend the Atom" was seen to mask the dangers of nuclear power; "Better Living Through Chemistry" (Dupont) became a synonym for drug use, or alternatively, for the production of napalm; "Progress Is

Our Most Important Product" (General Electric) became a joke to young people who believed that progress largely produced nuclear bombs and environmental pollution.

In this kind of atmosphere the dispassionate sociological study of technology did not flourish. Technology became something that needed to be assessed, to be gotten under control, to be stopped; it seemed antithetical to almost everything that was good. People became aware that much technology was not directed toward a common good, but instead existed to benefit private interests.

For instance, students of sociology were keenly aware of corporate inventions that aided the war in Vietnam. There was no longer a cultural lag, but rather a cultural gap between generations of sociologists. When the sociology of science resumed its development after many years of slumber, its European thinkers took an extremely critical stance, whereas most American researchers took the activities of science for granted, and merely studied its communication and career patterns.[47] The sociology of technology in the United States, meanwhile, virtually vanished.[48] Young people who were concerned about technology and the environment often engaged in activism rather than in academic study.

By the 1960s the members of the Ogburn group had grown old. In 1957 several members of the group wrote *Technology and Social Change,*[49] a book intended to become a standard textbook opening a new field of inquiry. Instead, it became their swan song. No similar text was to appear for three decades.[50] The group had failed to recruit new members, and as the old ones died out they were not replaced.[51] The group's ideas died out because its members vanished,[52] and in time their data became obsolete.[53]

The philosophy of the Ogburn school seemed to offer little to the new generation of sociologists.[54] While it is true that Ogburn's other contributions, particularly in statistics, were acknowledged, by the 1970s he was largely forgotten, except for his concept of "cultural lag."[55] (It is important to note that in certain *specific* areas of the sociology of technology, research not only continued but was extremely fruitful. Two such subject areas were technology in the workplace and the diffusion of innovations, both of which developed steadily.[56])

Another irony is that in the late 1950s the historical study of technology began to flourish in the United States, largely through the efforts of the Society for the History of Technology and its journal, *Technology and Culture.*[57] The Ogburnians were very enthusiastic about the Society at first, and Ogburn was its first president. Gilfillan and Francis Allen each contributed an article in the early years, but that was all.[58] Then Ogburn died in 1959. Sociologists barely participated in the new society thereafter. Out of some 431 authors of articles in *Technology and Culture* between 1959 and 1983, only 11—barely 3 percent—listed their affiliation as a department of sociology.[59] The huge majority of contributions came and still comes from historians. The historians rapidly organized, researched, and proselytized. In a relatively short time, they had built up a very impressive body of work.[60]

Why should the history of technology have flourished while its sister the sociology of technology withered on the vine? We can only speculate. First, the history of technology was strongly stimulated by older British and European scholars, who were much less impressed with the protests of youth than were their American counterparts. Second, historians as a group were less directly involved, compared with sociologists, in the cultural upheavals of the 1960s. Whatever the explanation, the history of technology developed rapidly and now has become a strong intellectual field, from which, ironically, sociologists of technology draw many of their ideas.

Conclusion

Ogburn and Gilfillan virtually created the sociology of technology. Their studies, their theories, and their suggestions for technology assessment are important. Their work had some drawbacks; their technological determinism was overdone and was justly criticized by others. Stern (and earlier, Marx) was more sophisticated about institutional analysis. "Cultural lag" may exist subjectively in the eye of the beholder, but it is a valuable (if value-laden) way of looking at the problems posed by technologies that are incompatible with cultural practices. In time, the Ogburn group's views might have received proper critiques, and reform might have resulted. Instead their ideas died out, except in bits and pieces. Technology assessments were revived in the late 1960s, in more powerful form. No longer did citizens or scholars consider technological change "inevitable." In the 1970s, new and vital theoretical approaches were to revive interest in the sociology of technology. It is to these new theoretical approaches that we turn next.

Notes

1. For a treatment of some of these thinkers, see Jay Weinstein, *Sociology/Technology: Foundations of Post-Academic Social Science* (New Brunswick, N.J.: Transaction Books, 1982).

2. Much of this chapter is based on my paper, "What Happened to the Old Sociology of Technology?", presented at the 1983 meetings of the Society for Social Studies of Science in Blacksburg, Virginia. Francis Allen was kind enough to give me a critique of this paper; this interchange contributed to the form the discussion takes here.

3. Jerry Gaston, "The Sociology of Science and Technology," in Paul T. Durbin (Ed.), *Introduction to the Culture of Science, Technology, and Medicine* (New York: Free Press, 1979).

4. For others' evaluation of Ogburn, see Francis R. Allen, *Socio-cultural Dynamics* (New York: Macmillan, 1971), pp. 180–193; Toby E. Huff, "Theoretical Innovation in Science: The Case of William F. Ogburn," *American Journal of Sociology*, Vol. 79, No. 2 (Sept. 1973), pp. 261–276; Steven L. Del Sesto, "Technology and Social Change: William Fielding

Ogburn Revisited," *Technological Forecasting and Social Change,* Vol. 24 (1983), pp. 183–196; and Jay Weinstein, "Early Academic Approaches in the United States: The Sociology of Technology of Veblen and Ogburn," in his *Sociology/Technology: The Foundations of Postacademic Social Science* (New Brunswick, N.J.: Transaction Books, 1982). See also the Introduction in Otis Dudley Duncan (Ed.), *William F. Ogburn on Culture and Social Change* (Chicago: University of Chicago Press, 1964). Francis Allen and Allan Schnaiberg have also contributed to my knowledge of this subject.

5. President's Research Committee on Social Trends, *Recent Social Trends in the United States* (New York: McGraw-Hill, 1933), 2 volumes. The introduction to this work indicates that the committee was formed at the request of President Herbert Hoover, who also was very much interested in the history of technology. I suspect that Ogburn's book *Recent Social Changes* (Chicago: University of Chicago Press, 1929) may also have been a stimulus to the committee's formation. This smaller, edited volume was reprinted from an earlier collection of articles appearing in the *American Journal of Sociology.*

6. William F. Ogburn with Jean L. Adams and S. C. Gilfillan, *The Social Effects of Aviation* (Boston: Houghton Mifflin, 1946).

7. In this respect, F. Stuart Chapin's book *Cultural Change* (Dubuque, Ia.: Brown, 1928), was markedly superior to Ogburn's *Social Change* (1922) for it elaborated on many of Ogburn's ideas. Chapin, then a professor of sociology at the University of Minnesota, was a good and prolific writer; his reputation was probably greater than Ogburn's with the general public. He was, however, neither as able a theorist nor as original as Ogburn.

8. William F. Ogburn, *Social Change with Respect to Culture and Original Nature* (New York: Dell, 1966). (Originally published in 1922 and slightly enlarged in 1950.)

9. Actually, one of the requirements for a U.S. patent is the demonstration of the *nonobviousness* of the device or process patented; see Richard L. Gausewitz, *Patent Pending: Today's Inventors and Their Inventions* (Old Greenwich, Conn.: Devin-Adair, 1983), pp. 1–22. We must separate here the perspectives of the inventor and of someone looking at a finished invention. What may have seemed like an incredibly difficult invention process to the inventor often appears natural in retrospect, a distortion produced both by the abstract nature of study as opposed to practice and by the simple fact that *once one knows how to do it,* many a difficult operation appears simple.

10. The most recent apostle of incremental improvement in technology is George Basalla; see his *The Evolution of Technology* (New York: Cambridge University Press, 1988).

11. Adapted from the article by William F. Ogburn with the assistance of S. C. Gilfillan, "The Influence of Invention and Discovery," in *Recent Social Trends in the United States, Report of the President's Research Committee on Social Trends* (New York: McGraw-Hill, 1933), pp. 122–166.

12. Ogburn, "Influence of Invention," p. 157.

13. Ogburn, *Social Change,* pp. 369–393. This section was added to the second edition of the book in 1950. Francis Allen, who called my attention to this addition, also noted that Ogburn had told him that invention merited more of the attention that was lavished instead on Ogburn's concept of "cultural lag."

14. Otis Dudley Duncan (Ed.), *William F. Ogburn on Culture and Social Change* (Chicago: University of Chicago Press, 1964), pp. 25–26.

15. William Shockley has written an interesting theoretical paper on this theme; see his "On the Statistics of Individual Variations of Productivity in Research Laboratories," *Proceedings of the Institute of Radio Engineers,* Mar. 1957, pp. 279–291.

16. Jacob Schmookler, *Invention and Economic Growth* (Cambridge, Mass.: Harvard University Press, 1966), pp. 59–63.

17. Ten is a very high number, for reasons that will become obvious later.

18. A recent *New York Times* article describes pharmaceutical scientist Solomon H. Snyder as having "seen new drugs stall at the big companies because of bureaucratic resistance." So Snyder found investors and started his own firm. See William Glaberson, "In Hot Pursuit of a Wonder Drug—and Wealth," *New York Times,* Jan. 31, 1988, Business Section, pp. 1, 16. Because of this kind of resistance, "dropouts" from large companies often spin off and form their own firms to pursue lines of technology they view as promising.

19. Edwin Sutherland, *White Collar Crime* (New York: Holt, Rinehart & Winston, 1961), pp. 95–110; and Bernhard Stern, *Historical Sociology* (New York: Citadel Press, 1959), pp. 47–101.

20. It is evident that many people, in agreement with Ogburn, tend to view the accumulation of knowledge as the key factor in creativity. However, there is a difference between knowledge accumulation and knowledge improvement; accumulation by itself is not enough. The fictional detective Sherlock Holmes told his partner Watson that he tried to forget useless knowledge:

    "You see," [Holmes] explained, "I consider that a man's brain originally is like a little empty attic, and that you have to stock it with such furniture as you choose. A fool takes in all the lumber of every sort that he comes across, so that the knowledge which might be useful to him gets crowded out, or at best is jumbled up with a lot of other things, so that he has difficulty laying his hands on it. Now the skilful workman is very careful indeed as to what he takes into his brain-attic. He will have nothing but the tools which may help him in doing his work, but of these he has a large assortment, and all in the most perfect order. It is a mistake to think that that little room has elastic walls and can distend to any extent. Depend upon it there comes a time when for every addition of knowledge you forget something that you knew before. It is of the highest importance, therefore, not to have useless facts elbowing out the useful ones." (A. Conan Doyle, "A Study in Scarlet," in *The Complete Sherlock Holmes* [New York: Dover], p. 10)

    The Danish poet-engineer Piet Hein, echoing Holmes, said in an interview that "my brain is laid out for a workroom; there is no room for a storehouse" (Anne Chamberlin, "King of Supershape," *Esquire,* Jan. 1967, p. 169). It is noteworthy that Francis Galton, a Victorian scientific genius, had relatively few books in his library, preferring for the most part to investigate for himself.

21. William F. Ogburn, "Cultural Lag as Theory," *Sociology and Social Research,* Vol. 41 (Jan.-Feb. 1957).

22. Harry F. Waters, Mark Starr, Richard Sandza, and Tony Clifton, "The Squeeze on Sleaze," *Newsweek,* Feb. 1, 1985, pp. 44–45.

23. James W. Woodard, "Critical Notes on the Cultural Lag Concept," *Social Forces,* Vol. 12 (1934), pp. 388–398.

24. One can only ponder the impact of a realistic technology assessment had one been done by the federal government in the 1930s, for in taking note of these ill effects, it might have been influential in increasing federal regulation.

25. This dilemma is explored by Gary A. Steiner in *The People Look at Television: A Study of Audience Attitudes* (New York: Knopf, 1963), pp. 226–249.

26. It also seems strange to think about a technology assessment that would allow decisions about adaptation only, but no decisions about hardware.

27. These efforts are in addition to his contributions to the study of social trends, including ten years' publication of *Recent Social Changes* and his editing of the massive two-volume *Recent Social Trends in the United States* (New York: McGraw-Hill, 1933).

28. Anyone who thinks technology assessment originated in the 1960s should examine this work. Francis Allen, a former collaborator with Ogburn, had occasion in 1983 to check the accuracy of the predictions made in this 1946 book. He found that nearly all of the predictions had come out reasonably well. The major exception was the assertion that helicopters would be more widely used (see the discussion in Chapter 16 of this book).

29. William F. Ogburn and Meyer F. Nimkoff, *Technology and the Changing Family* (Boston: Houghton Mifflin, 1955).

30. S. C. Gilfillan, "An Ugly Duckling's Swan Song: The Autobiography of S. Colum Gilfillan," *Sociological Abstracts,* Vol. 18 (1970), pp. i–xl. This paper contains a bibliography of Gilfillan's work.

31. S. C. Gilfillan, *The Sociology of Invention* (Chicago: Follett, 1935) and *Inventing the Ship* (Chicago: Follett, 1935).

32. S. C. Gilfillan, *Supplement to the Sociology of Invention* (San Francisco: San Francisco Press, 1970). According to Charles Susskind of San Francisco Press, Gilfillan wished to issue a second, larger edition of *The Sociology of Invention,* in which the new material would be included, but the copyright holder was unwilling to cooperate in this project. Hence *Supplement* was published separately.

33. S. C. Gilfillan, *Invention and the Patent System* (Washington, D.C.: Joint Economic Committee of Congress, Dec. 1964).

34. S. C. Gilfillan, "A Sociologist Looks at Technical Prediction," in James R. Bright (Ed.), *Technological Forecasting for Industry and Government: Methods and Applications* (Englewood Cliffs, N.J.: Prentice-Hall, 1968); see the comment on p. 1 of that book. Ten years later, Bright wrote that "Gilfillan in 1937 gave us the nucleus of the concept of systematic technological forecasting. He recognized the need to build a data base, the use of extrapolation, the principle of monitoring, and the principle of normative (goal-oriented) forecasting." See Bright's "Technology Forecasting Literature: Emergence and Impact on Technological Innovation," in P. Kelly and M. Kranzberg (Eds.), *Technological Innovation: A Critical Review of Current Knowledge* (San Francisco: San Francisco Press, 1978), p. 312.

35. Gilfillan, *The Sociology of Invention,* p. 5.

36. An exception was his discussion of Anton Flettner, inventor of the rotor ship. But even here, Gilfillan took pains to show that the rotor ship would have been invented by someone else, had Flettner not done it himself. See Gilfillan, *Inventing the Ship,* pp. 221–222.

37. Gilfillan, *Inventing the Ship,* pp. 211–232.

38. It is likely that the lack of appropriate sponsorship was involved in the decline of the rotor ship, although I am not sure how. Thomas Lang, an aerospace engineer and inventor of the SWATH (small waterplane area, twin hull) ship, tells me that replacement of the rotating cylinders used by Flettner by fixed cylinders with air jets set at an angle around the cylinder produces the same flow and the same propulsion more efficiently. He thinks that Jacques Cousteau has used cylinders of this kind.

39. Gilfillan, "Ugly Duckling," p. xxxix. The outcome of this conference was the volume on technological forecasting edited by Bright, mentioned in note 34. Interestingly enough, this conference occurred just about the time that technology *assessment* (as distinguished from forecasting) was beginning to be agitated in the U.S. Congress; something must have been in the air.

40. I was told about the bequest to SHOT by Mel Kranzberg. Two other important sources of information on inventors and patents connected with Gilfillan are noteworthy. Both the Patent Office Society, to which Gilfillan belonged, and the Patent, Copyright, and Trademark Society contributed in earlier years a significant amount of research on inventors. Before 1970 the journals of these societies contained an enormous amount of information that, to the best of my knowledge, has never been synthesized, though much of it is mentioned in Gilfillan's *Supplement*. Joseph Rossman, a member of the Patent Office Society, and Barkev Sanders were both active in researching inventors and patents. Both were friends of Gilfillan, and there was considerable intellectual exchange among them on this subject. Sanders was a research officer for the Patent, Copyright, and Trademark Society who wrote enough articles to fill several books. The Patent, Copyright, and Trademark Society for Research and Education also held annual conventions. But after the hiatus in research during the 1960s (see text), all of this work seems to have vanished into the stacks of libraries.

41. For Stern, the attraction of Marxism was more its humanism than its dogmatism. To appreciate the difficulties that adherence to a political faith can cause when its apparent leaders behave badly, one need only read Stern's essay "Genetics Teaching and Lysenko," in *Historical Sociology: Selected Papers of Bernhard J. Stern* (New York: Citadel Press, 1959). In this paper, written in 1949, the pressure to agree with the Stalinist party line is clearly in conflict with Stern's basic intellectual honesty.

42. See, for instance, B. J. Stern, *Social Factors in Medical Progress* (New York: Columbia University Press, 1927) and his *Society and Medical Progress* (Princeton, N.J.: Princeton University Press, 1941), pp. 150–214.

43. B. J. Stern, "Resistances to the Adoption of Technological Inventions," in National Resources Committee, *Technological Trends and National Policy,* (Washington, D.C.: U.S. Government Printing Office, June 1937), pp. 39–66, is still the basic source on this subject.

44. See Stern's articles "The Frustration of Technology" and "Restraints upon the Utilization of Inventions," in *Historical Sociology,* pp. 47–101.

45. Gilfillan, *Supplement,* p. 29.

46. Theodore Roszak, *The Making of a Counter Culture* (Garden City, N.J.: Doubleday, 1969).

47. Typical of the European approach to the sociology of science is Hilary Rose and Steven Rose, *Science and Society* (Harmondsworth, England: Penguin, 1977); see also Alain Jaubert and Jean Marc Levy-Leblond (Eds.), *Autocritique de la Science* (Paris: Seuil, 1975). Gerald Gaston wrote an apologia for the American point of view in his "Sociology of Science and Technology," in Paul T. Durbin (Ed.), *A Guide to the Culture of Science, Technology and Medicine* (New York: Free Press, 1979).

48. Allen Schnaiberg, a prominent environmental sociologist, reminded me in a letter written on July 22, 1985, that much of Ogburn's concern with technology was embodied by the Human Ecology school, whose most prominent exponent was Otis Dudley Duncan, an Ogburn student.

49. Francis R. Allen, Hornell Hart, Delbert C. Miller, William Fielding Ogburn, and Meyer F. Nimkoff, *Technology and Social Change* (New York: Appleton-Century-Crofts, 1957).

50. This statement requires qualification. Quite a number of books with titles similar to "Technology and Social Change" have appeared in the meantime, but they were usually collections of contributions from a variety of disciplines or were more philosophically inclined works. In contrast, the Allen et al. volume was a true sociological textbook with a unified approach. A literate and well-researched introduction to this subject has recently

appeared; see Rudi Volti, *Society and Technological Change* (New York: St. Martin's Press, 1988).

51. Some of them also went into other specialties, as did Delbert Miller, who became an industrial sociologist. Only Francis Allen, the youngest member of the group, maintained his focus on technology and social change, but even his work has been concerned more with social change than with technology.

52. Donald Campbell, "A Tribal System for Carrying Social Science Knowledge," *Knowledge: Creation, Diffusion, and Utilization,* Vol. 1, No. 2 (Dec. 1979), pp. 181–202. For an elaboration of this approach, see Nicholas C. Mullins, *Theory and Theory Groups in Contemporary American Sociology* (New York: Harper & Row, 1973).

53. I wrote a sociology textbook chapter on "Technology and Social Change" in 1976; see Chapter 19 in Robert Perrucci et al. (Eds.), *Sociology: Basic Structures and Processes* (Dubuque, Ia.: Brown, 1977). At the time, most of the materials I used had been written by historians. Fortunately, subsequent scholarship has been so active that much of what I wrote then is now obsolete.

54. For instance, consider the contents of the volume edited by Jack Douglas, *The Technological Threat* (Englewood Cliffs, N.J.: Prentice-Hall, 1971). Douglas's book and others by him perfectly reflected the new negative attitude toward technology, which was more value-oriented and less analytical.

55. A. J. Jaffe, "William Fielding Ogburn," in David Sills (Ed.), *International Encyclopedia of the Social Sciences,* Vol. 11 (New York: Macmillan and Free Press, 1968), pp. 277–281.

56. There is unhappily no overall survey of technology in the workplace. But see Georges Friedmann, *Industrial Society: The Emergence of the Social Problems of Automation* (Glencoe, Ill.: Free Press, 1950). For innovation research, see Everett Rogers, *The Diffusion of Innovations,* 3rd ed. (New York: Free Press, 1983). Rogers is an acknowledged expert in this field.

57. The rationale for the new society and its journal are spelled out in the Introduction to an anthology of articles from it; see Melvin Kranzberg and William H. Davenport (Eds.), *Technology and Culture: An Anthology* (New York: New American Library, 1972).

58. Francis Allen submitted a second article, but apparently it was rejected. He never submitted another article (personal communication from Francis Allen). Gilfillan also knew and was friendly with Melvin Kranzberg, the editor of *Technology and Culture.*

59. These figures are based on data researched and collected by my graduate assistant, Shalane Sheley.

60. John Staudenmaier, *Technology's Storytellers* (Cambridge, Mass.: MIT Press, 1985).

Recent Theoretical Approaches

The demise of the Ogburn school did not eliminate the sociology of technology. Other theorists have pondered the role that technology was playing, or might play, in society.[1] And as we have noted, in certain specific content areas, such as industrial sociology, studies were continued and deepened. During the 1960s and 1970s, however, a broad new theoretical approach slowly developed, and since roughly 1975 the body of theory has grown in several interesting directions. In this chapter we will explore some of these new theoretical approaches to the relationship between technology and society. As we will see, there has been a shift from the impersonal "social forces" or "trends" of Marx and Ogburn toward choices and decisions made by individuals, groups, and institutions. This shift toward considering the intent and strategies of social change agents has refocused attention on how such strategies are developed and implemented.

Energy/Environmental Sociology

The study of humanity's relationship to its physical environment has a long history in sociology and in human geography. Patterns of settlement and geographical relationships have been of interest to American sociologists since the turn of the century.[2] These topics became particularly important to the Chicago School of sociologists, centered around the University of Chicago and marked by Robert Park's book *The City* and Amos Hawley's more general ecological approach.[3] In these studies, however, technology played almost no role; awareness of the technological impact of human activities on the physical environment was almost nil.

In 1955 Fred Cottrell, a capable sociologist of technology, published *Energy and Society,* a general survey of the relationship between people and their sources of power.[4] It is tempting to think that this general survey was stimulated by Cottrell's study, conducted ten years earlier, of railroad towns isolated by the advent of diesel power, which in turn had grown out of Cottrell's studies of railroad workers.[5] Interestingly enough, Cottrell did not connect his own work with the Ogburn school.[6] *Energy and Society* did not come out of any particular sociological tradition and it did not begin a new school of thought; rather, it was an isolated effort.

Cottrell wrote before the advent of the environmental movement. Though very much interested in the role that energy played in human affairs, Cottrell did not worry about technology's impact on the environment and similar global issues. Thus, he was not driven by the same passions that were later to play such an important role in the rise of environmental sociology. The social climate of the 1960s strongly emphasized the importance of the environment, and even as technology was rejected as a focus of interest, the environment and factors affecting its preservation became a focus for study. When the environmental movement surfaced, stimulated by Rachel Carson's *Silent Spring* (1962), concerns about "the population bomb," and the Vietnam War, a very different attitude toward energy evolved.

Sociology originally played a relatively small role in the environmental movement. Sociologists tended to view the movement as simply another social movement that often had essentially elitist orientations.[7] Ironically, much of the sociological thinking on the subject was stimulated by the economist E. F. Schumacher and by the physicist Amory Lovins.[8] By the 1970s, however, a new generation of sociologists began to develop sociological approaches to the study of the issues that animated the movement. An outpouring of articles by sociologists on environmental issues began in about 1975.[9] By 1976 the American Sociological Association had an environmental sociology section.

In 1980 two major works on the environment written by sociologists were published. The first, *Overshoot: The Ecological Basis of Revolutionary Change* by

William R. Catton, was an outgrowth of the author's earlier study of public interest in the National Parks.[10] Catton was one of the first to write programmatic articles on environmental sociology. Even more important was Allan Schnaiberg's outstanding book, *The Environment: From Surplus to Scarcity*.[11] In this work Schnaiberg examines the relationships among the economy, science, technology, and the environment. In a particularly significant discussion that is reminiscent of Marx, Schnaiberg explains the relationship between scientific thought and economic interests. He shows that intellectual effort goes disproportionately into the development of new means of production, with the ultimate impact of all this increased production very much neglected.

Much of the writing on energy and the environment by sociologists is a critical examination of the arguments advanced by such environmental proponents as Barry Commoner, E. F. Schumacher, and Amory Lovins. Not surprisingly, sociologists find the contentions made by these nonsociologists to be deficient in various respects. Sociologists find that ideological disagreements, contradictions, lack of intellectual coherence or supporting data, and elitism characterize the thoughts of many of the movement's practitioners.[12] This kind of critical examination of the environmental movement is important.[13] Equally important are the far more numerous empirical studies on such subjects as attitudes regarding nuclear power, consumer responses to the energy crisis, and knowledge of technological hazards.[14] But the real issues often seem to be missed by sociologists. Few environmental sociologists, for instance, seem to spend time studying the structure of large-scale industry, the impacts of the military (in encouraging nuclear power, for instance), or global competition.[15] Some of these latter subjects are indeed studied by sociologists of science, but not by environmental sociologists; the division of labor is notable. Social movements seem to present a more comfortable target for study than do large, powerful institutions.

Cognitive Theories of Society/Technology

The next three points of view we will examine have a great deal in common, and in fact they were united in a very fine book, *The Social Construction of Technological Systems,* published in 1987.[16] All three stress connections between technology and intellectual activities, and thus they are all *cognitive theories* of society/technology. All three stress the importance of human entrepreneurship in developing and implementing technology. Here, for purposes of exposition, however, we will consider each viewpoint separately, for each lays emphasis on somewhat different aspects of the overall process.

Network Theory. Network theory is the creation of three scholars, Michel Callon, Bruno Latour, and John Law, who worked at the Centre de Sociologie de l'Innovation in Paris. It is the oldest, best developed, and most subtle of the three cognitive theories. It is also the most ambitious; it makes powerful claims that some people may find difficult to accept. Its central argument is that sociotechnical systems are developed out of heterogeneous elements, including theories, devices, and people, and that technological struggles can be seen as efforts by proponents of different systems to force their own constructions on others.

Work on these ideas began when Michel Callon attempted to make sense of the behavior of French scientists working on such technological problems as the electrical automobile. The book by the philosopher Michel Serres, *La Traduction* ("Translation"), led to Callon's development of the idea of translation in science. According to Callon, translation is

> all the mechanisms and strategies through which an actor—whoever he may be—identifies other actors or elements and places them in relation to one another. Each actor builds a universe around him which is a complex and changing network of varied elements that he tries to link together and make dependent on himself.[17]

The two key concepts here are the linking of heterogeneous elements and the attempt to control. In developing an approach to a problem, a technological or scientific entrepreneur must induce others to accept the classification (translation) that he or she makes. By doing so, not only will research proceed along favorable lines for the entrepreneur, but others will come to depend on the entrepreneur for intellectual legitimacy. A favorite exemplar for the network theory school is Niccolò Machiavelli (1469–1527), the renaissance political scientist.

Along with the basic idea of translation, network theory includes the following set of related concepts:

Actor-network: A set of entities that have been successfully translated; a set of concepts that have been successfully "sold" to others.

Black box: The solid concepts in an actor-network that cannot be deconstructed because they are accepted; those things that are no longer under discussion.

Enrollment: The process by which people, concepts, and things are made part of the actor-network; getting others to accept one's point of view and situating important concepts within one's framework.

Funnel of interests: The process by which one gets others to accept one's own research agenda by convincing them that the problems they wish to solve can be solved only by using one's own methods.

Problematization: The process of convincing others that the only issues to be resolved are those that are a problem for one's own approach; enlisting the help of others in solving problems important to one's own research/development agenda.[18]

According to network theory, then, the key to success in science and technology is to set up one's own conceptual scheme and sell it to others. By doing so one fixes the issues, the problems, and the line of approach in a manner favorable to one's own agenda. The tactics by which such maneuvering can take place have been exhaustively analyzed in a brilliant book by Bruno Latour.[19]

The implications for control of technological system development are obvious. Technological sponsors try to get others to accept their own technological program as the appropriate approach. For years, for instance, the automobile lobby has fixed attention on improvements to cars and roads; by making such improvements the focus of national transportation policy, public transportation has been neglected. Similarly, powerful sponsors of the nuclear industry and electrical utilities have made sure that only meager funds went to solar power development while huge amounts of federal funding went into the improvement of nuclear plants, and American physicians have reinforced dependence on expensive expert care to the neglect of public health measures and home care.[20]

Although network theorists do not consciously relate their work to that of Ivan Illich, his work has many similar themes. According to Illich, technologies are put together in ways that favor their powerful sponsors; many of the juxtapositions of concepts and systems are arbitrary. Further, he contends that "professionalism" often hides powerful vested interests. His work on schools, medicine, and the professions in general has emphasized the "power" issues involved, the control of legitimacy, and the harm frequently done to basic human purposes by "expert" activities.[21] Network theorists note the existence of a constant, inevitable struggle among contending approaches in society. Illich, by contrast, does not view the situation as one of contending professional groups; rather, he tends to view science/technology empire-building as a monolithic effort on the part of the sponsors of science and technology to develop monopolistic control. Furthermore, he sees such empire-building not as a natural tendency on the part of experts, but rather as the symptom of an overly bureaucratic society.

The Social Construction of Technology (SCOT)

Theory. The central idea of the "social construction of technology" (SCOT) school is that *the nature of a technology is determined by a network of interested parties*. This idea was first articulated by Wiebe Bijker and Trevor Pinch in a key article written about the evolution of bicycles.[22] The network of participants decides what is important about a given technological device or system, how it is to be evaluated, and how it needs to be improved. Changes in the device or system then follow from research and experimentation influenced by this network. The important feature here is that every device is surrounded by an "interested" network; we might say that a device exists only in so far as the network sees it as existing. Its uses, its nature, and its evolution occur only if the network considers such changes to be necessary. Similarly, when the network is content with the device its design stabilizes, even though further improvements in the device might be possible from a purely technical standpoint.

Wiebe Bijker also refers to "technological frames."[23] A technological frame is the intellectual system in which a device or a system is set; it is its place in the overall scheme of activity and thought. For instance, astronomers and admirals would find different uses for telescopes and would suggest different research activities to improve them. Such differences would stem from the respective activities in which astronomers and admirals engage—looking at stars versus looking at ships. The key distinction here from the classical (Ogburnian) point of view is that technological evolution is not automatic but guided; it is steered by the interests and demands placed on it by the groups that frame the technology.

These differences in intended uses are particularly striking in relation to the evolution of technological devices. Evaluations of technology are not objective, but instead are always made from the standpoint of a particular group. What a "good" bicycle is, Bijker and Pinch argue, depends on whether the group evaluating it consists of bicycle racers or commuters riding to work. Thus one set of influentials (sponsors) might request that tires be made softer to provide a more comfortable ride, whereas another would be more interested in superior traction—at whatever cost to comfort—for racing.[24] How the technology evolves, then, is a result of the pushes and pulls from these various groups of interested parties.

Whereas network theory concentrates on technological entrepreneurs and their activities, SCOT concentrates on networks of sponsors. Network theory emphasizes conflict among contending entrepreneurs; SCOT instead emphasizes negotiations among the elements of the network of interested parties. It is important to note that what is of greatest interest differs between the two schools. A "network" for SCOT researchers is made up of human beings, and this network of sponsors surrounding the device or technology is of primary interest. For "network theorists," however, the network of interest is made up of different, heterogeneous elements and includes the technology. Nonetheless, the similarities between (and possible integrations of) the two schools clearly exceed their differences.

Systems Theory. Thomas Parke Hughes is considered to be the major proponent of what has been called "systems" theory. Unquestionably Hughes's fine study, *Networks of Power,* has given great strength to this point of view.[25] Although Hughes's prominence is a testament to the unusual texture and quality of his work, many (but not all) of his basic insights have been shared by others, including Marx, Noble, and Langdon Winner.[26]

Networks of Power makes a series of interconnected arguments.[27] The basic concept is that of the *technological system*—an integrated network of people, technological knowledge, and devices. Some examples of systems of this kind would be railroads, electrical networks, the defense industry, the petroleum industry, nuclear power, and natural gas distribution. Not all technologies lend themselves to the kind of "system-building" that Hughes describes, for many technologies are divisible into parts and do not need to function as networks. But those that do function as networks have major impacts on society, for the following reasons:

1. Many of these networks have a natural economy of scale; that is, the larger the system, the lower the unit cost of production. Hence there is strong pressure to eliminate or merge with competitors and establish a monopoly.

2. The system is naturally dependent on the political process, and political (especially regulatory) decisions are necessarily important to it. Thus it naturally tries to influence the political process to secure favorable treatment.

3. The nature of the system strongly encourages planning. Because development takes time, its time frame is long, and often very sophisticated methods are used to predict demand, increase efficiency, and so forth. Thus "planning horizons" of ten years or more are common.

4. Technological industries employ large numbers of technically trained people. Thus they will have a significant impact on educational institutions, which they can be expected to support and from which they will draw many future employees. Furthermore, their employment of some of the most able technical people will influence employees' opinions about the value of the industries' operations and thereby shape public opinion.

Two interesting concepts derive from this overall perspective: *reverse salients* and *momentum*. Reverse salients can exist on both technical and institutional levels.[28] A technical reverse salient exists when an industry finds itself unable to continue progress because one of its technical areas has lagged behind. For instance, the audiovisual telephone's progress was halted in the 1970s because the transmission medium (metal cables) was unable to transmit the huge chunks of information that would be required. With the installation of optical fibers, however, future progress toward audiovisual transmission is likely. Institutional (social) reverse salients exist when the existing institutions hinder adoption of a new technology. Certain kinds of innovation may be very difficult in certain institutions, as we will see in the chapter on the evolution of sponsors. Both of these kinds of reverse salients, suggests Hughes, result when innovation lags, and the more ambitious inventors and entrepreneurs are likely to focus their attention on them, knowing that large profits can be made if the problems can be solved.

In some respects, reverse salients are reminiscent of Ogburn's concept of "cultural lag." However, Hughes's discussion refers entirely to lags or bottlenecks within sociotechnical systems, whereas Ogburn refers to broader adjustments between "material" and "adaptive" culture. Hughes's viewpoint is significantly different from Ogburn's because his notion of system makes the people and the technology inseparable; the technology does not operate without financiers, inventors, administrators, and others.

The concept of momentum confirms this inseparability of people and machines.[29] As the technological system endeavors to perpetuate itself, various other parts of the social system are affected and drawn into its operation. Technical experts trained in the system's technology tend to think in terms of traditional concepts and methods. The technological choices that the system has already made, furthermore, represent sunk costs—a heavy investment in doing things a certain way. For examples, consider the lasting impact of a railroad gauge or of a form of television receiver. Often severe limits on innovation are imposed by the need for compatibility of new with existing technology.[30] Thus the system's "technological style," which has been defined by these choices, tends to be stable over relatively long periods of time. Novel technologies that threaten the system's existing order are neglected or actively suppressed.

In some respects, Hughes's theory describes a system that manages people, rather than the reverse.[31] Although a technological system is invented by a definable person or set of people, it eventually develops an energy of its own. It actively collects the people it needs to exist: financiers to provide capital, administrators to manage, inventors to provide elaborations, engineers to solve routine problems, workers to carry on daily operations. The existence of the system provides a set of career opportunities for managers, workers, and technical people. It shapes choices regarding other modes of energy, transport, communication, and potential destructiveness. In time it is slowly replaced by vast new systems, but while it is dominant it is a force that must be reckoned with. However, in Hughes's view a technology does not lead to a *single* technological style; rather, different styles for utilization of the same technology will arise depending on differences in local conditions, existing technological traditions, and so on.

Hughes's vision of a technological system is reminiscent of Marx's conception of the economy as an autonomous force that is out of human control. Langdon Winner has analyzed this concept of "autonomous technology" in a much more extensive theoretical form than has Hughes.[32] In his *America by Design,* David Noble has provided a similar argument to Hughes's.[33] Considering the growth of the large networks of devices, skills, and people that Hughes, Winner, and Noble discuss, it is difficult not to agree, for technology is awesome and ramified, it is supported by powerful sponsors who often have their own way, and it often seems out of control. Sometimes it *is* out of control.

The great value of Hughes's work, however, is its retention of some sense of individual thought and action. Marx and Noble would argue that this is an illusion, that the system really carries people like Thomas Edison and Werner von Siemens along in its wake. "If not them, then others," so the argument goes. Ogburn and Gilfillan, who were more technologically determinist than any of the above, would have agreed. But human decisions make a difference, not only on the individual level but also on the group level. Hughes retains some sense of these choices, many of which, it is true, are shaped by institutions but remain choices nonetheless. It is important to remember that systems, networks, and sponsors are all built by humans—and sometimes they are also destroyed by them.

The Visible Hand: Toward an Active Theory of Technology

If we were to summarize the differences between the society/technology theories of Marx and Ogburn and those of the recent theorists, the most obvious contrast involves the agents involved in action. Marx and Ogburn viewed technical and social changes as large currents in which individuals were swept along; the active agents were various states of society, which naturally led to other states, even though the processes of change might be speeded up or slowed down. In the views of recent theorists, however, *forces* and *trends* are replaced by *strategies* and *choices*. The focus is now on the way that decision-making about technologies takes place; with different decisions, we get different directions for society.

Consider, for example, multiple independent invention—one of the sure signs, according to Ogburn, that social forces and not geniuses are responsible for technical advances. To Ogburn, studying inventors was beside the point; such an endeavor sees only trees instead of the forest. Even though it might be interesting to uncover the details of the inventive process, for Ogburn the significant issue was the social stock of knowledge that would inevitably lead to the new combinations of ideas that were the basis of new inventions. The inventors, while colorful, were merely the means by which this new crop of inventions came into being.

But studies of the history of technology have brought more attention to the details of innovation and the processes of deployment. The developers of technology now have faces; furthermore, they have personalities, motivations, and points of view that shape the systems they invent. The choices they make are not dictated by external circumstances, but rather represent "educated guesses" about the right way to do things—educated guesses that differ from one social context to another and may take different forms in the hands of different social groups.

Even more important, the new theorists have discovered the importance of *configuration*. Every technology has a protean quality at the beginning, but it can assume a variety of different shapes depending on what specific components go into the device and the way the components are linked. A technology may permit a variety of solutions, but only some will come into being. Multiple independent inventions usually mean that similar principles were used in designing the independent inventions; seldom, though, are the devices identical. This lack of identity becomes more important the larger and more complicated the device is, because *different configurations may have different social effects*.

For instance, an "automobile" may mean a large, expensive device available only to the affluent. In that case its social effects will be different than if "automobile" means a mass-produced, inexpensive, simple car like Henry Ford's Model T. A "pop bottle" may be a plastic throwaway or a recyclable glass container. In the former case, we have an addition to the city dump; in the latter we need a recycling center at every grocery store. Similarly, a "city" may be highly concentrated (like Manhattan) or very spread out (like Los Angeles); we will find different interaction patterns in the two layouts, even though they are both "cities." Clearly, the specific manner in which the technology is configured is important.

With small, simple devices, configuration may be less of a problem because a variety of alternatives may easily be invented and distributed. The larger the device and the more it forms part of a system, however, the greater the social investment ("sunk costs") in the specific configuration, the more difficult change becomes, and the more likely other social institutions are to maintain the technology in its current configuration.

Thus there is a major difference between old and new technological theorists because the older theorists did not have this sensitivity to configuration. Ogburn and Gilfillan could maintain that the invention of the steam engine was inevitable because to them any prime mover that used steam was a "steam engine." We might infer from Ogburn and Gilfillan that when a new technology comes along, social change comes about through "adaptive culture." Adaptive culture, however, may decide to adapt the technology rather than the social structure. When Hero of Alexandria (circa. A.D. 100) invented a steam engine, for instance, it did not revolutionize the Roman world because its use was restricted to purely ceremonial occasions. Similarly, many decades passed before the U.S. Army decided that soldiers should take advantage of automatic rifle technology.[34]

Furthermore, the new theorists have learned from history that technical decisions are not inevitable. In which areas does society decide to concentrate its inventive effort? In the United States it is clear, for instance, that our major priority has been the development of more elaborate weapons of war. Our largest allocations of money for research and development, and often our best scientists, have dedicated themselves to this pursuit. Yet this focus is not inevitable; it represents a choice. The direction that society takes depends on it and similar choices, not on blind social forces. Until the Meiji Restoration of 1868, the Japanese had restrained technical advances in their society; such restraint did not represent blind social forces, but rather a conscious decision. The new theorists have stressed the role of such decisions in changing both the direction and the pace of technological innovation.

Consider the central role of automobiles in the transportation system of the United States. Our emphasis on the automobile was not inevitable; in fact, it required considerable action and politicking by the automobile, rubber, and petroleum firms involved. If the railroad industry had been stronger and its leaders had made different kinds of decisions, automobiles may not have been as successful. It is clear that the shift toward automobiles was technological *policy* shaped by private concerns, and not simply a technological trend. The outcome, now viewed by some as undesirable, was not by any means inevitable, but rather represented a social choice.

In France a decision to use nuclear power to provide 60 percent of the nation's energy needs was again a conscious choice, and not one dictated by necessity or blind social trends. Although there are rational reasons for this choice—independence from foreign sources of petroleum, for example—it was not the only choice that could have been made. And once this choice had been made, many others followed it in turn: pooh-poohing of public fears about radioactivity, control by the state instead of private industry, the kind of reactors used, and so forth. Once started down the nuclear path, the decision is very difficult to alter. Large sunk costs in nuclear plants and the training of engineers and technicians, and entrenched public attitudes, all tended to maintain the system in the direction it was moving. This is an example of powerful technological momentum.

Societies may place different values on different parts of the development cycle. Some societies may emphasize design to the neglect of manufacture; in others, luxury goods may be given preference over mass-produced ones.[35] Arnold Pacey has discussed Western societies' lack of attention to maintaining the technologies they already have in place.[36] Clearly American society has given much more attention to getting technologies into operation than to getting them to work together harmoniously. Our society trains excellent analysts, but we are not very good at training synthesists.

Decisions about technologies, then, are made by people who have values and goals they wish to further, a vision of the technical future, and a certain pool of resources. Market forces and the momentum of scientific research are important, but they can be affected by policy decisions. The relatively slow pace of solar power implementation has not been dictated solely by technical factors, but also by conscious policy decisions shaped by the sponsors of current technologies: the electrical utilities, federal regulatory bodies, and the various materials industries (coal, petroleum, uranium, and so on).

Decisions about technologies in turn are shaped by the specific technical arrangements already in place. The United States has 110-volt electrical outlets; Great Britain has 220-volt outlets. Once a decision regarding voltage has been made, it is strongly entrenched by enormous investments in appliances, wires, testing traditions, and consumer preferences. This is the hand of man at work; it is not a blind technological trend. Ditto for the QWERTY arrangement of keys on the typewriter, which has survived several major changes of technology. Again, the QWERTY keyboard was a conscious decision that, once made, was very difficult to change. How are we to think about giving up automobiles now that we have invested trillions of dollars in highways, parking garages, shopping centers, and driver education courses? The very physical layout of the society we live in has been shaped by the need to accommodate automobiles, and to change the transportation mode means shifting enormous amounts of resources. The thought is staggering; hence our reluctance to do it.

Because such decisions are made by people, we need to study the people who make them. A social science of technology must pay attention to the inventors, designers, entrepreneurs, sponsors, and regulators of technology. We cannot regard such people's activities as mere details. Rather, these active agents shape our sociotechnical world as they make decisions on its nature and direction. Configurations are not dictated by the technology, but rather by the technologists. Choices about configuration provide differing payoffs to specific social groups, and the nature of these benefits influences the choices that are made.[37]

Admittedly, people's abilities to shape policy are set by the state of society and by the state of the art of science/technology. Ogburn was correct about this. We can cite occasional examples of genius attempting to push technology too far too fast, such as the attempts of the English inventor Charles Babbage to invent a computer in the 1830s. Such examples and their outcomes only reaffirm the basic observation that the state of knowledge limits inventing; it does not, however, determine inventing. Although inventors' activity may sometimes seem to be socially conditioned, a great deal of their tinkering is focused by conscious choice as much as by larger social forces; no market forces caused the development of the atomic bomb or the space program. Although we recognize the importance of these broader social forces, then, most of our attention is focused on the shapers—on individuals and groups who attempt to develop, deploy, and maintain technological systems. In later chapters we will pay more attention to the development of foresight, to the extension of vision that accompanies technology. If in the past much technology has been reactive, much of it also has anticipated the future.

Conclusion

This brief survey of contemporary approaches to social theories of technology has explored only the broad outlines. The real purpose here has been to provide some orientation for the chapters that follow. In the Introduction to *The Social Construction of Technological Systems,* the editors speak of "opening the black box of technology" and examining the contents. We have seen how energy/environmental theorists have discussed the dynamics of social institutions associated both with corporate sponsors of technology and with their opponents, the environmental movement. This discussion represents a preliminary opening of the black box. The three theoretical schools associated with the "social construction of technology" have opened it even further by contributing more sensitive models of the ways a technology's proponents advance their projects and contend with social and technological opponents. This book is a continuation of this examination of the social machinery that shapes and is shaped by technology; in subsequent chapters we will examine not only the developers of technology, but also those who deploy it, those who use it, and those who try to rein it in through analysis, regulation, and social action.

The guiding thought that shapes the approach here is simple: What is not understood cannot be controlled. If we do not understand the way technology is shaped by society, we cannot change the shaping. Very similar arguments have been made by Langdon Winner in *Autonomous Technology* and by Victor Ferkiss in *Technological Man.*[38] Before we can expand the area of intelligent human choice, we must understand how technology and society interact. It may be true, as my colleague Jay Weinstein has argued, that academic sociology is very badly suited to this task.[39] I am tempted to agree. One cannot argue, as I do here, that we have institutions that are inadequate for coping with technology without viewing sociology departments (as they are currently constituted) as inadequate for researching the kinds of issues this book raises. In the end an interdisciplinary approach to technology and society will likely be necessary to cope with these issues.

Nonetheless, sociology, whether academic or not, can help. Our major effort here will be to understand the activities of technology's creators and sponsors. Because many groups participate in shaping technologies, we need to use sociology and the variety of other social sciences to cope with this multiplicity. The key, however, is that the social sciences be adequate to the kinds of things they must study. Only in the light of a more comprehensive picture of human interactions with technology are we likely to get the kind of grasp we need.

Notes

1. Among the other theorists to consider, should a more complete treatment of this topic be written, are John Nef, Lewis Mumford, Karl Mannheim, Thorstein Veblen, Harold Innis, Marshall McLuhan, the Frankfurt School, and C. Wright Mills. Veblen, Mannheim, the Frankfurt School, and Mills all receive extensive treatment in Jay Weinstein, *Sociology/Technology: Foundations of Postacademic Social Science* (New Brunswick, N.J.: Transaction Books, 1982).

2. See typical examples of this spatial emphasis in the first American sociology text, Albion Small and George Vincent, *An Introduction to the Study of Society* (New York: American Book Company, 1894), and in the community studies done in Boston by the South End House: Robert A. Woods (Ed.), *The Urban Wilderness: A Settlement Study* (Boston: Houghton Mifflin, 1898); and Robert A. Woods (Ed.), *Americans in Process: A Settlement Study* (Boston: Houghton Mifflin, 1902).

3. Robert Ezra Park et al., *The City* (Heritage of Sociology Series) (Chicago: University of Chicago Press, 1984).

4. Fred Cottrell, *Energy and Society: The Relationship Between Energy, Social Change, and Economic Development* (New York: McGraw-Hill, 1955).

5. Fred Cottrell, "Death by Dieselization," *American Journal of Sociology,* Vol. 16 (1945), pp. 358–565. Cottrell had previously written a very interesting ethnography, *The Railroader* (Stanford, Calif.: Stanford University Press, 1940).

6. Cottrell was aware of Ogburn, of course, but he did not consider Ogburn's work to be directly related to his own. See Cottrell, *Energy and Society,* p. 5.

7. Faced with broad interdisciplinary questions, many specialists resort to their usual patterns of thought, instead of probing more deeply. Sociologists spent much more of their time analyzing the proponents of environmental action than considering the basic questions at issue. Economists, if anything, were even worse. In about 1969 I listened while a panel of University of Chicago professors, including Milton Friedman, George Stigler,

Harold Demsetz, and other intellectual heavyweights, argued that ordinary economics was sufficient for dealing with environmental issues. A hapless community college professor, who represented the contrary view, was cut to ribbons by his far more able opponents, who were nonetheless unable to envision the greenhouse effect, ozone depletion, and similar matters.

8. E. F. Schumacher, *Small Is Beautiful: Economics As If People Mattered* (New York: Harper & Row, 1973); and Amory Lovins, *Soft Energy Paths: Toward a Durable Peace* (New York: Harper & Row, 1979).

9. This observation results from surveying the dates of references in the special issue on "The Sociology of the Environment" of *Sociological Inquiry,* Vol. 53, No. 2/3 (Spring 1983). No doubt a careful, more systematic study of sources would reveal a more subtle pattern.

10. William R. Catton, *Overshoot: The Ecological Basis of Revolutionary Change* (Urbana: University of Illinois Press, 1980).

11. Allan Schnaiberg, *The Environment: From Surplus to Scarcity* (New York: Oxford University Press, 1980).

12. Reasonably balanced sociological accounts occur in Stan L. Albrecht, "The Environment as a Social Problem," in Armand Mauss (Ed.), *Social Problems as Social Movements* (Philadelphia: Lippincott, 1975); and David L. Sills, "The Environmental Movement and Its Critics," *Human Ecology,* Vol. 3, No. 1 (Jan. 1975), pp. 1–42.

13. Critiques of the environmental movement have also been made by representatives of the scientific establishment. Among the many works of this kind are John Maddox, *The Doomsday Syndrome: An Attack on Pessimism* (New York: McGraw-Hill, 1972); and George Claus and Karen Bolander, *Ecological Sanity: A Critical Examination of Bad Science, Good Intentions and Premature Doomsday Announcements of the Ecology Lobby* (New York: McKay, 1977). Maddox is editor of *Nature,* one of the most influential scientific periodicals in the world. See also David Vogel, "A Big Agenda," *Wilson Quarterly,* Autumn 1987, pp. 51–68.

14. In France there is a regular intellectual industry for psychoanalysis of the population's "groundless fears" of nuclear power. If nuclear power were as critical to the American economy as it is to the French economy, a similar effort would undoubtedly achieve success here.

15. Allan Schnaiberg is an important exception.

16. Wiebe Bijker, Thomas P. Hughes, and Trevor Pinch (Eds.), *The Social Construction of Technological Systems* (Cambridge, Mass.: MIT Press, 1987).

17. Michel Callon, Jean-Pierre Courtial, William A. Turner, and Serge Bauin, "From Translations to Problematic Networks: An Introduction to Co-Word Analysis," *Social Science Information,* Vol. 22, No. 2 (1983), p. 193.

18. Adapted and abridged from Michel Callon, John Law, and Arie Rip (Eds.), *Mapping the Dynamics of Science and Technology* (Dobbs Ferry, N.Y.: Sheridan House, 1986), pp. xvi–xvii. Copyright © 1986 Michel Callon, John Law, and Arie Rip. Adaptation approved and permission to reprint by the authors.

19. Bruno Latour, *Science in Action: How to Follow Scientists and Engineers Through Society* (Cambridge, Mass.: Harvard University Press, 1987).

20. See Christiaan Barnard, "Medicine Negated," in A. Asimov (Ed.), *Living in the Future* (New York: Beaufort Books, 1985), pp. 224–231.

21. For instance, see Ivan Illich, *Tools for Conviviality* (New York: Harper & Row, 1973); and Ivan Illich, *Medical Nemesis* (New York: Bantam Books, 1977).

22. Wiebe Bijker and Trevor Pinch, "The Social Construction of Facts and Artifacts: Or How the Sociology of Science and the Sociology of Technology Might Benefit Each Other," *Social Studies of Science,* Vol. 14, No. 3 (Aug. 1984), pp. 399–442; reprinted in Bijker et al. (Eds.), *Social Construction,* pp. 17–50.

23. Wiebe Bijker, "The Social Construction of Bakelite: Toward a Theory of Invention," in Bijker et al. (Eds.), *Social Construction,* pp. 159–190.

24. Technologist and philosopher Amrom Katz, for many years a consultant for the Rand Corporation, was once confronted with a particularly large and complicated piece of photographic apparatus. He remembers wondering, "What is the problem if *this* is the solution?"

25. Thomas Parke Hughes, *Networks of Power: Electrification in Western Society, 1880–1930* (Baltimore: Johns Hopkins University Press, 1983).

26. Luther Gerlach, an anthropologist, has done work along very similar lines to those of Hughes, in the study of energy networks; see Luther P. Gerlach and Gary B. Palmer, "Adaptation Through Evolving Interdependence," in P. C. Nystrom and W. H. Starbuck (Eds.), *Handbook of Organizational Design,* Vol. 1 (New York: Oxford University Press, 1978), pp. 323–381. Surprisingly, this important article is cited neither in *Networks of Power* nor in the Bijker et al. collection.

27. These arguments are reiterated and extended in Hughes's later writings, notably in his book *American Genesis: A Century of Invention and Technological Enthusiasm 1870–1970* (New York: Viking Penguin, 1989), pp. 184–248.

28. See Thomas P. Hughes, "The Evolution of Large Systems," in Bijker et al. (Eds.), *Social Construction,* pp. 51–82.

29. Hughes, in a personal communication of March 10, 1989, states that he thinks "the social construction of technology" is most evident when a technological field is just being opened up. When the industry becomes mature, "momentum" tends to prevail.

30. Hughes, *Networks of Power,* p. 121.

31. Again see Hughes, "The Evolution of Large Systems," in Bijker et al. (Eds.), *Social Construction,* pp. 51–82.

32. Langdon Winner, *Autonomous Technology: Technics-out-of-Control as a Theme in Political Thought* (Cambridge, Mass.: MIT Press, 1977).

33. David F. Noble, *America by Design: Science, Technology, and the Rise of Corporate Capitalism* (New York: Oxford University Press, 1977).

34. Edward Clinton Ezell, *The Great Rifle Controversy* (Harrisburg, Pa.: Stackpole Books, 1984).

35. John U. Nef, *Industry and Government in France and England 1540–1640* (Ithaca, N.Y.: Cornell University Press, 1964).

36. Arnold Pacey, *The Culture of Technology* (Cambridge, Mass.: MIT Press, 1983).

37. Donald MacKenzie and Graham Spinardi, "The Shaping of Nuclear Weapon System Technology: U.S. Fleet Ballistic Missile Guidance and Navigation," *Social Studies of Science,* Vol. 18, Nos. 3 & 4 (Aug. and Nov. 1988), pp. 419–464, 581–624; also see Hughes, *Networks of Power,* pp. 51–82.

38. Victor C. Ferkiss, *Technological Man: The Myth and the Reality* (New York: New American Library, 1969).

39. Weinstein, *Sociology/Technology.* Nor is he alone. For instance, Chris Argyris, Robert Putnam, and Diana McLain Smith in *Action Science: Concepts, Methods, and Skills for Research and Invention* (San Francisco: Jossey-Bass, 1985) contend that "rigorous research" is often less effective in producing change than is action- and implementation-oriented research.

PART THREE

Originators & Managers of Technology

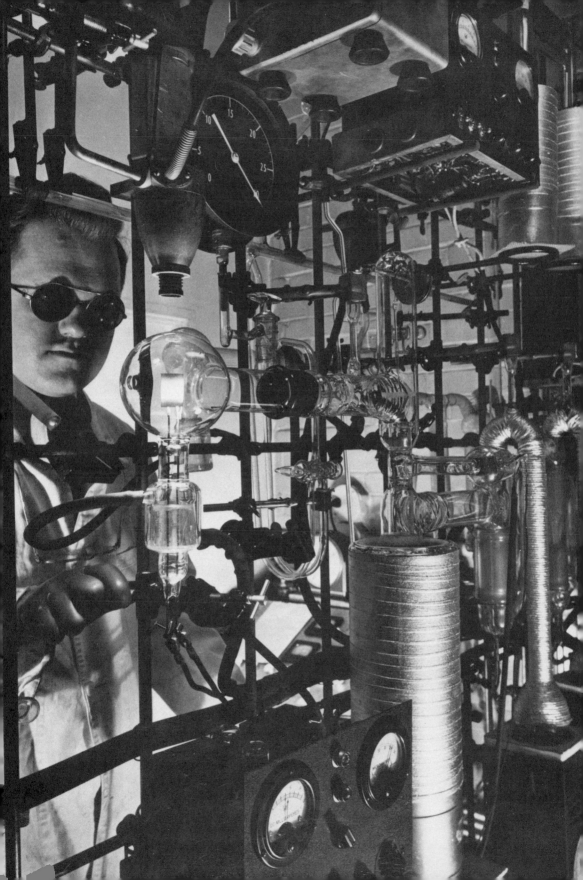

CHAPTER 5

Inventors & Inventions

Among the key questions regarding technology are "How are new technologies developed?" and "How do they become part of society?" The general process by which such changes take place is *technological innovation,* but in this chapter we will be interested in a particular part of the process: *invention.* Every society has a set of institutions by which it recognizes, trains, and encourages inventors, including its "knowledge" institutions such as schools and universities, as well as reward systems such as patents and prizes for inventors. Also important are social norms regarding the value of mechanical creativity and the honor bestowed on technological experts and heroes. Societies that are better at encouraging inventors are likely to experience a wider range of technological choices. But before we consider such institutions, we need to study the people they are designed to foster. We need to look at inventors.

Inventors are people who develop inventions. We call a device an invention when it is a genuinely novel artifact based on new principles, new materials, or a new combination of parts. For the sake of convenience, it is helpful to think of an invention as a device that could, in principle, receive a *patent*.[1] The underlying idea of a patentable invention is a device that is a "nonobvious" improvement over existing practice; that is, one that requires a "breakthrough" of some sort.[2] Although, as we will see, inventions are often developed by teams and are often incomplete, usually one person—the "inventor"—is primarily responsible for their "reduction to practical use." In this chapter we will consider the personality and activities of the inventor.

What Do We Know About Invention?

Because the development of new devices is a social activity with powerful economic consequences, a considerable amount of study has been devoted to it. Nonetheless, there are enormous gaps in what we know. Here we will examine the nature of inventors—their thought processes and their activities. The biographies of inventors may reveal some of the psychological features that make people particularly inventive; indeed, by tracing the relation between inventions and the inventor's life, such studies may show how inventions are influenced by background, training, and other life experiences. Biographical studies, however, are enormously time-consuming, and generalizing from individual studies is not always appropriate. Furthermore, studies of individuals may give society as a whole little guidance about ways to increase inventive activity, for inventors are notoriously idiosyncratic. And finally, comparative biographical studies of inventors, though valuable, are rare. Nonetheless, studies of individuals do offer some important insights, and in the absence of more systematic information, they provide a useful beginning.

In the next chapter, we will consider other sources of information about the inventive process to examine how inventions relate to the society that produced them. Among the studies we will consider are those that probe the origins of inventions; such studies can show us much about the relation between an invention and its social context and can answer some of the questions we have regarding how important inventions come about. A number of studies have been conducted to determine either the relative importance of single inventors versus corporate teams or the role of the type of organization in relation to creativity. Other studies have investigated the role of demand in stimulating invention. We will see that studies of "independent multiple invention" are particularly valuable in showing how invention responds both to social demands and to the "state of the art."

The Nature of Inventors

Inventors are creative people with a penchant for mechanical things. Their preoccupation with technology begins early in life. As children, inventors of all kinds tend to take apart things—the family clock, for instance—to see how they work.[3] They may also have an ability to fix family appliances; many earn a reputation as a "Mr. Fixit" while still quite young. Consider what the father of Burt Rutan, an inventor of lightweight airplanes, has said about his son:

> When he was nine, he refused ready-made model kits. "I know just what I want to build," he'd say. "Just get me a big sheet of paper to draw on and some balsa wood." So I'd get him the paper from a butcher shop and the materials from a hobby shop, and he'd construct these models that won first-place trophies in the state and national meets. Models were his whole life, just as flying was [his brother] Dick's.[4]

This early preoccupation became a lifetime pursuit crowned by impressive accomplishments. In December 1986, the *Voyager,* a lightweight plane built by Burt Rutan and piloted by Jeanna Yeager and Burt's brother Dick, made the first around-the-world flight on a single tank of aviation fuel.[5]

Now consider this description of Steven Wozniak, engineer and inventor of the Apple II computer:

> Steve Wozniak was a precocious youngster and a tinkerer extraordinaire almost as soon as he learned to read. In fact, a lot of his early reading was in the engineering literature his father brought home from work. He became especially interested in ham radio, and built his own transmitter and receiver from a kit. He was only a sixth-grader when he earned his ham radio operator's license. A ham license is not that easy to obtain. An applicant has to take a fairly difficult test in which he must demonstrate an overall knowledge of electronic theory and also send and receive telegraphically a minimum number of words per minute in international code. It's a lot like learning a second language. Wozniak mastered it when he was only eleven years old.
>
> Right from the start, Wozniak understood spatial relationships. "When Steve got his Christmas toys, he'd start reading the instructions and be lost," Margaret Wozniak [his mother] says. "Instead of reading the instructions, he'd look at the toy, and whoosh, put it together, just like that."[6]

With proper encouragement and training these childhood leanings develop into professional competence. Parents' occupations provide one source of technical knowledge and expertise. College education (usually in engineering) and sometimes also graduate training provide more extensive technical knowledge. Inventors, especially major inventors, tend to be better educated than the rest of the population.[7] Education does not create the ability to invent; but it provides a good deal of scientific knowledge, it refines the ability to develop inventions, and it gives inventors tools for communicating the value of an invention to others.[8]

Inventors typically find positions in technological firms or strike off on their own to continue their inventive activities. Those who were loners as children tend to become independent inventors as adults. Future corporate inventors, although still strongly inclined toward mechanical things, tend to have greater social skills than independent inventors,[9] many of whom seem to resemble the eccentrics of the stereotypical "mad inventor."[10] Even the less flamboyant independent inventors may have a difficult time persuading the world of their ideas' value.[11]

Independent and corporate inventors tend to share high energy levels, probing curiosity, and restlessness. Like successful scientists, successful inventors tend to be self-starters. One of Thomas Edison's assistants, Francis Jehl, wrote that "putting off a thing until tomorrow was a practice unknown to Edison. He kept going forward relentlessly and when an obstacle came into his path either passed round it or turned it to his advantage."[12] Similarly, Orville Wright (of the Wright brothers) remarked to a friend that he could "remember when Wilbur and I could hardly wait for morning to come to get at something that interested us. *That's* happiness."[13] Rene Dubos says of Louis Pasteur that "experiments were usually carried out as soon as they were discussed and prepared. This rapid passage from conception to execution accounts, in part, for Pasteur's phenomenal scientific productivity."[14]

Similar comments were made to me by Dean Wilson regarding the dynamism of the student scientific experiments on nutrition in a Colombian village; as soon as the results of one experiment were in, the investigators immediately suggested what the next experiment should be.[15] William McLean, a self-designated "gadgeteer" and chief designer of the Sidewinder air-to-air missile, often stayed up late at night working on the missile, as did the other members of the Sidewinder team. Interest and enthusiasm in the project, as well as loyalty to other team members, kept people working in their laboratories long after hours.[16] Howard Wilcox, McLean's assistant on the missile project, recalls that

> it was not unusual to go back to the lab after supper. When you went down to the lab, you'd find a little knot of cars in the parking lot. Those were the Sidewinder guys.[17]

Another engineer on the project, Warren Legler, said that interest was so intense that the project seemed to change from hour to hour, rather than from day to day.[18] We will examine such a "skunk works" arrangement—a small, high-performance innovation team—in the next chapter.

Inventors are never at rest; from the minute they encounter a new device they are at work trying to figure out how to make it better.[19] Sometimes they seem "programmed" to detect an inventive opportunity. Jacob Rabinow, an inventor with over 225 U.S. patents, describes inventing as "a disease"; he also describes it as an "art form" that can properly be appreciated only by those who do it.[20]

Steve Wozniak makes a very interesting comment about the artistry of inventing. Lamenting the lack of time (due to business pressures) for personal inventive work since he invented the Apple II computer, he speaks of the special joy he gets from elegant design:

> I remember once that I designed a PC board for our disk interface. I did a rare thing for an engineer. I laid out the board myself. At Apple, we had departments that usually did that. But I came in many nights in a row, working very, very late. I laid out the whole board, and then I got an idea to save one feed-through. So I took the board apart, I trashed maybe a week's worth of work, and then I started over.
>
> And I did it another way that saved another feed-through. No big deal. Nobody in the world would ever know that I laid it out to have very few feed-throughs—three instead of maybe fifty. None of this would ever be seen, but for some reason it seemed important to me in an artistic sense. You can have a feeling that all these things are important, but you can't necessarily justify them logically. The effort comes from being so close to your art.[21]

For inventors, constant improvement is often irresistible. One of the early indications that William McLean was a "gadgeteer" occurred during his days as a physics graduate student at Cal Tech. His professors observed that he often seemed to be more interested in the apparatus used to do the experiments than in the experiments themselves.[22] Inventors like McLean are constantly trying to improve the technical arrangements found in society. They want change, and the fuel for their desire for change is a continuous series of thoughts about how things could be made better, thoughts stirred by their own work experiences and frustrations, by technical advances made by others, and often by a sudden intuition that an improvement is possible.

Consider, for instance, the story of the inventor of the digital computer, Charles Babbage (1792–1871). Babbage's desire to invent a computer derived from his frustrations with tables of numbers that contained inaccuracies due to human errors in computation and copying. Babbage felt that he could eliminate human error if he could find a way to carry out the calculations—and the printing—by machine. Thus was born the idea of the "Difference Engine," the first of Babbage's computer concepts.

Babbage applied for and received a grant from the British government to work on such a machine, which, although started, was never completed. Once Babbage had developed the idea of a special-purpose computer, he wondered why he should not also invent a general-purpose one. Babbage then conceived a general-purpose machine he called the "Analytical Engine"; today we would call it a digital computer. This machine was to be built of cams and cogwheels and powered by steam; electronics was still far in the future.

Unfortunately, the mechanical resources available to Babbage in the early nineteenth century could not keep pace with his thought; the technology of the time could not supply in large quantity the kind of precision machining Babbage needed to complete the "Analytical Engine." He could design a computer, but was never able to build it. In fact, due to his restlessness, he never completed any of his machines. Lord Moulton said of him:

One of the sad memories of my life is a visit to the celebrated mathematician and inventor, Mr. Babbage. He was far advanced in age, but his mind was still as vigorous as ever. He took me through his workrooms. In the first rooms I saw the parts of the original Calculating Machine, which had been shown in an incomplete state many years before and had even been put to some use. I asked him about its present form. "I have not finished it because in working at it I came on the idea of my Analytical Machine, which would do all that it was capable of doing and much more. Indeed, the idea was so much simpler that it would have taken more work to complete the Calculating Machine than to design and construct the other in its entirety, so I turned my attention to the Analytical Machine." After a few minutes talk we went into the next workroom, where he showed and explained to me the working of the elements of the Analytical Machine. I asked if I could see it. "I have never completed it," he said, "because I hit upon an idea of doing the same thing by a different and more effective method, and this rendered it useless to proceed along the old lines." Then we went into the third room. There lay scattered bits of mechanism, but I saw no trace of any working machine. Very cautiously I approached the subject, and received the dreaded answer. "It is not constructed yet, but I am working at it and it will take less time to construct it altogether than it would have taken to complete the Analytical Machine from the stage at which I left it." I took leave of the old man with a heavy heart. When he died a few years later, not only had he constructed no machine, but the verdict of a jury of kind and sympathetic scientific men who were deputed to pronounce upon what he had left behind him, either in papers or mechanism, was that everything was too incomplete to be capable of being put to any useful purpose.[23]

Obviously, this jury lacked the imagination to see the possibilities. The lion's share of his legacy would be only concepts, for the kind of labor required at the time to build an analytical engine was more than Babbage or probably any private individual could command. As a result Babbage died a frustrated, bitter old man who conceived one of the most important inventions of all time.

The Components of Invention

The Stream of Ideas and the Notion of Originality. Invention does not occur in a vacuum; the inventor's social and intellectual environment is very important. The Wright brothers, for instance, did not start from scratch. They were excited by the early steps toward aviation and read extensively about the glider experiments carried out by engineers Otto Lilienthal and Octave Chanute.[24] Thus the Wrights' invention was based on an already extensive cultural knowledge made available to them in the form of books and articles. Far from occurring in a void, their invention took place in a society that was fascinated by the prospects of aviation. Their competitor at the Smithsonian Institution, Samuel Pierpont Langley, was trying—with little success—to invent a workable airplane, as indeed were others.[25] Similarly, Edison's idea of an electrical utility network was not a "bolt from the blue," but was derived from extensive experience with gas lighting.

Thus, very little that is invented is completely new. Even the basic idea for the invention has usually been "in the air" for a considerable time, often for decades or even for hundreds of years.[26] The components of the invention have frequently been available for some time. And the inventor seldom works alone, but is often assisted by others who may get relatively little credit in the history books, even though their ideas are often important. Others may be working on the same innovation but have chosen a less fruitful or more difficult path; they rarely get credit in the history books either. The inventions themselves arise in a context teeming with helpful ideas and alternative solutions.[27] Thus inventors may be likened to people who, after walking into a packed toolshed or workshop, emerge with parts that they then assemble into something new. Because the new device is so impressive, we tend to forget about the toolshed.

Inventors, furthermore, are often busy with many different problems and products. We find that ideas that lead to important inventions are usually part of a continuous stream of ideas that occur to the inventor. As inventors attempt to solve the problems of interest to them, a stream of solutions comes to mind; some are committed to laboratory notebooks, and others are developed further through experimentation or construction of prototypes.[28] Each of these ideas becomes a stepping stone to still newer ideas. Sometimes inventors do not completely develop ideas fully themselves, but instead communicate them to others who then go on to complete them. The English electrical engineer Sir John Ambrose Fleming explained the process of invention nicely:

> Invention consists in overcoming the practical difficulties of the new advance, not merely talking or writing about the new thing, but in *doing* it, and doing it so that those who come after have had real obstacles cleared out of their way, and have a process or appliance at their disposal which was not there before the inventor entered the field. In most cases, however, the removal of the obstacles which block the way is not entirely the work of one person. The fort is captured only after a series of attacks, each conducted under a different leader. In these cases the inventor who breaks down the last obstruction or leads the final assault is more particularly associated in the public mind with the victory, though his intrinsic contribution may not actually be of great importance.[29]

Thus it is most useful to view invention as a contribution to an ongoing process of which the inventor is only a part. One way of visualizing this is to imagine a stream of events like this:

$$A \rightarrow B \rightarrow C \rightarrow D \rightarrow E$$

Now let's suppose that the act of invention in which we are interested is D in the diagram above—the invention, let's say, of the electric light bulb by Thomas Edison. Although unquestionably Edison was a genius, a number of previous inventive acts made important contributions to the invention of the light bulb, and that invention in turn led to other inventions. By choosing to look only at D we are isolating from the stream of events only one of its elements. But there may be other, parallel streams of actions occurring at the same time, for instance:

$$I \rightarrow J \rightarrow K \rightarrow L \rightarrow M$$

These parallel streams may be quite important, for two of them may merge to form a new stream of events:

$$K \rightarrow L \rightarrow M \searrow$$
$$\rightarrow Q \rightarrow R \rightarrow S$$
$$C \rightarrow D \rightarrow E \nearrow$$

Thus, the development of airplanes and of steam turbines contributed to the invention of the jet plane.[30] The jet's flight surfaces had been developed through experience with propeller planes, and its engine resulted from an application of steam turbine technology.

An invention, then, is not an isolated event, but rather one that occurs as part of a long chain of actions that takes place in a context in which other chains are also in the process of development. The inventor is more a member of a relay team who finds a state of development X and moves it forward to state X + 1. In some cases the leap is a great one, and sometimes it is breathtaking. When Edison invented the electric light, he also invented the entire electrical utility system that went with it. Edison was not merely an inventor, but a systems engineer as well.[31]

Similarly, the development of the airplane by the Wright brothers included a whole series of inventions related to lift, steering, and propulsion, each of which was carefully considered for its effects on the whole system. Wilbur Wright did not merely invent a propeller; he tested a number of different designs to find the best one.[32] Inventions involving these kinds of leaps in imagination are rightly seen as works of genius, for they involve work on several fronts at once. Still, even though a major aim of the inventor is the cultivation of clusters of ideas, these ideas may be so numerous that they overwhelm the inventor's ability to test or develop them, as happened with Edison's near-discovery of radio transmission. Edison came very close to discovering radio transmission, but in regarding the "Edison effect" as a mere curiosity rather than a potential real point for research, he missed a key opportunity.[33]

The "Right Question": The Importance of the Prepared Mind. The stream of ideas is driven partly from within and partly from without. It is striking how often something from the outside—a notice in a technical journal, a chance observation, a challenge—will set the inventor going. Percy Scott, the British inventor of continuous-aim firing on naval ships, noticed one day that one of his gunners was using the gun crank to compensate for the roll of the ship, thus keeping the gun on target. Scott realized that what one man could do by improvising, others could do through training. This was the beginning of a virtual naval revolution, led by Scott, that was to change forever the nature of gunnery training.[34] Scott developed a highly structured training program with rewards for performance, as well as some of the first gunnery simulators, and eventually he redesigned the entire gunnery system for the Royal Navy—all from a single observation. Still, the observation that causes such changes must fall on fertile soil. During his inaugural lecture at the University of Lille, Louis Pasteur remarked that Oersted's discovery of electromagnetism resulted from a chance observation. But, he said, "chance only favors the mind which is prepared."[35]

Of course, "prepared" means many things. First, the inventor must be prepared by technical competence. This does not mean that he or she must be an expert in the field in question, but rather that technical competence, when not already possessed, must be acquired. Second, the inventor must have appropriate motivation, whether in the form of financial reward, organizational need, or intrinsic interest. And third, the mind must be prepared in the sense that the chance encounter must seem to hold the promise of a solution; it must be approached as the piece that could potentially complete the puzzle.[36]

Where do such puzzles come from? What provides, in other words, the problem to be solved in the first place? Here we must differentiate the *problem* that stimulates the invention from the *internal processes* in the inventor's mind that lead to the solution. The problem may arise from some external group, as for instance in the case of the farmers who approached Louis Pasteur because their cows were dying of anthrax. In other cases it may come from within—from the inventor's frustrations with current technology. For Percy Scott, the problem was the shamefully poor performance of naval gunnery in his time (during one of the battles of the Spanish-American War only about 3 percent of the rounds fired hit their targets). Scott's question, then, was how this abysmal performance could be improved.

Another way of identifying problems is looking for critical shortcomings in current technologies. Thomas Hughes points out that many of the heroic inventors at the turn of the century spent considerable time developing an awareness of such critical problems in fields in which they were working. He calls these problems, borrowing from military terminology, "reverse salients," because they were problems that hold up the further development of whole systems.[37] Elmer Sperry, inventor of the gyroscope, tried to identify these reverse salients by reading patents and recognizing instances in which invention would generate maximum (financial) payoff.[38]

Yet the key to successful invention is learning not only to choose problems, but also to ask questions about them in the right way.[39] Knowing the method of developing key questions is the touchstone of genius, for finding the answers to the questions may be easier than figuring out what one should be seeking in the first place. The inventor's striking originality is often an ability to pose questions in a way no one else has thought of. Paul MacCready, whose human-powered aircraft have won several important prizes, is said to have this ability to pose key questions:

> What MacCready likes to do best is to solve "annoyingly complex" problems. He does it, he says, by ignoring ordinary thinking and by asking the kinds of "stupid questions" that make more conventional engineers roll their eyes.

Specific questions: won't strong wire work as well as fancy wing supports? And general questions: why does a plane have to go fast, as long as it goes fast enough? Perhaps, he continually reminds you, it is an approach nobody else bothers with, but nearly every aviator in the world knows his name because his simple answers work.[40]

Paul MacCready comments:

The most important aspect of inventing is choosing the right challenge, hopefully one that can just barely be achieved with the latest technology. Asking "stupid questions" is important, and is easy to do when you do not have a reputation in that field to worry about. The key questions are basic ones: What is the real, underlying goal? Are the key issues resolvable? What size and weight are needed? Ignore what anyone else has done and start with fundamentals. With the big picture clear (or at least becoming clearer), the selection of design details becomes straightforward (and in many cases not especially critical). It is often hugely important to do building and testing quickly, so as to clarify your thinking—this is the case in the pioneering stage of a field, not the advanced stage such as designing a new airliner. It is vital to be able to draw on outside expertise whenever required, without feeling threatened by those more competent than you. You need to know enough about the expert's field to fit it into the context of your challenge.[41]

It is often useful to consider this kind of inventive question as taking the form, "What if?"[42] The basic kernel of an inventive idea is the notion of some task that might be done and something that might accomplish it; the associated question might be "What if there were something like an X; would it be able to do Y?" Usually, of course, the purpose comes first and the device second, but not always; in some cases the inventor may hit upon an answer and look for a problem to which it can be applied. In fact, the problem is important for invention only so far as it provides a motive for searching for a solution.

Chronic discontent with current arrangements can provide such a motive for searching, but discontent alone is not enough. The inventor's essential trait is a predisposition not only to look for, but also to work toward, a solution to the problem. Without this positive tendency to seek solutions, the discontent becomes pure negativism. Only when discontent results in an active search for improvement does it lead to invention.

Robert Goddard, the rocketry pioneer, described the origins of his interest in rocketry at age 17 thusly:

This was the situation when, on the afternoon of October 19, 1899, I climbed a tall cherry tree at the back of the barn, on a plot where I had visions of some kind of frog-hatching experiments, and armed with a saw which I still have, and a hatchet, started to trim the dead limbs off the cherry tree. It was one of the quiet, colorful afternoons of sheer beauty which we have in October in New England, and I as I looked toward the fields at the

East, I imagined how wonderful it would be to make some device which had even the *possibility* of ascending to Mars, and how it would look on a small scale, if sent up from the meadow. I have several photographs of the tree, taken since, with the little ladder I made to climb it, leaning against it.

It seemed to me then that a weight whirling around a horizontal shaft, moving more rapidly above than below, could furnish lift by virtue of the greater centrifugal force at the top of the path. In any event, I was a different boy when I descended the tree from when I ascended, for existence ever afterward seemed very purposive."[43]

Goddard ever afterward referred to this as his "Anniversary Day." For the inventor this early intimation may constitute only the first step on a long path that must be traveled. Goddard was later to spend most of his life developing rockets, although the great achievements in rocketry (such as the space program) would occur after his death in 1945. None of Goddard's designs, by the way, used the form of inertial propulsion he originally conceived.[44]

Persistent Inventors. Ideas require development. The inventor's success derives not only from framing original concepts, but also from being willing to pursue them relentlessly.[45] Thomas Edison once said that "invention is one percent inspiration and ninety-nine percent perspiration." In questioning inventors about what they saw as the key characteristics of successful inventors, Joseph Rossman found that they picked "perseverance" twice as often as any other characteristic, including even originality.[46] Getting the basic idea is only the beginning; extensive further development may be required, for when the idea comes it is frequently incomplete.[47]

The device or process as originally conceived by the inventor is thus often a mere intuition, a sketchy outline with only some of the details filled in. Indeed, some of the parts of the invention may not yet exist, or may even be considered impossible to create. The invention before its completion is an imagined thing, and in proposing it the inventor has brought into view an imagined world in which the invention exists. He or she will then try to make this imagined world a reality. The inventor stands with one foot solidly planted in the world of known devices and with the other foot resting in the imaginary world. To build a solid bridge to the imaginary world is what drives the inventor on; invention is thus a form of self-transcendence.

The incomplete nature of the bridge to the imagined world must be rectified, for after the inventor has conceived the device the task of filling in the details, testing them in operation, and actually bringing the concept to fruition remains; the invention has not been invented until these steps are carried out. And the desire to carry them out is important for the process of invention, for without the drive to complete the process, the invention remains merely an interesting concept, as was the case with Babbage's computers. How far from the beginning the ultimate goal can be is often staggering.

When Howard Head first thought of the possibility of a metal ski in 1946, he was working for Martin Aircraft. He had experienced frustration at his lack of skill in his first attempts at skiing and had blamed it on his wooden skis. Certain he could produce a better ski out of lightweight aircraft materials like aluminum, Head started working on the project in his spare time. By 1948 he had produced the first model, the Mark I, which broke almost immediately when a ski instructor tried to flex it. Head quit his job to devote all his time to developing the ski. He began sending the models off for testing; they all came back broken, until one day in 1950 he produced a ski that was sufficiently flexible. It was the fortieth attempt. About his efforts, Head said this:

> If I had known then that it would take 40 versions before the ski was any good, I might have given it up. . . . But fortunately you get trapped into thinking the next design will be it.[48]

Even an initial success may meet with daunting obstacles. Consider the following description, by the nineteenth-century inventor Sir Henry Bessemer, of some of the early difficulties in the development of the steel manufacturing process which bears his name. Bessemer, having announced the discovery that steel might be mass-produced, was suddenly confounded by impurities in iron ore that made his process unreliable.

> I . . . worked steadily on. Six months more of anxious toil glided away, and things were in very much the same state, except that many thousands of pounds had been uselessly expended, and I was much worn by hard work and mental anxiety. The large fortune that had seemed almost within my grasp was now far off; my name as an engineer and inventor had suffered much by the defeat of my plans. Those who had most feared the change with which my invention had threatened their long-vested interests felt perfectly reassured, and could now safely sneer at my unavailing efforts; and, what was far worse, my best friends tried, first by gentle hints, then by stronger arguments, to make me desist from a pursuit that all the world had proclaimed to be utterly futile. It was, indeed, a hard struggle; I had well-nigh learned to distrust myself, and was fain at times to surrender my own convictions to the mere opinion of others. Those most near and dear to me grieved over my obstinate persistence. But what could I do? I had had the most irrefragable evidence of the absolute truth and soundness of the principle upon which my invention was based, and with this knowledge I could not persuade myself to fling away the promise of fame and wealth and lose entirely the results of years of labor and mental anxiety, and at the same time confess myself beaten and defeated. Happily for me, the end was in sight.[49]

(Note particularly Bessemer's use of the words "fame and wealth." Some scholars have claimed that the urge to create is greatest when driven by such intrinsic rewards as the joy of creation, as opposed to money and recognition.[50] However, there are quite a number of instances in which inventors have frankly confessed that this or that particular innovation was driven by the spur of economic need or desire to gain recognition by others. Paul MacCready, for instance, in explaining his motive for creating the *Gossamer Condor,* the first successful human-powered aircraft, has stated that a huge debt incurred by a brother-in-law was a powerful stimulus to his imagination. MacCready invented the *Gossamer Condor* to win the Kremer Prize—£50,000, just the size of the debt; the juxtaposition of the debt and the prize proved to be the necessary spark.[51])

It may appear that for the inventor merely to set foot on the path to invention is the critical accomplishment. But many begin and few finish. The key point is that for the inventor, the as yet imaginary world of the successful invention seems attainable; to others, however, the distance yet to be traveled may seem far more significant. For the inventor, the important thing is to have glimpsed this new world; for others, the salient feature is that the new world remains imaginary. What overcomes doubt in the end is the finished product—the working invention. But the road to this point is long, rocky, and treacherous, and not all inventors reach the end.

"Getting It Out the Door": Marketing the Invention. Inventors often develop worthwhile things, only to find they can't make any money on them. Inventors sometimes naively believe that taking out a patent is all they need to do to get the invention produced.[52] But even when the inventor has patented the device, the problems have in fact only begun. It is very common for inventors to go broke in the process of getting their devices "out the door." There are two basic reasons for this problem: The first is that many inventors have little business ability; the second is that they often face overwhelming commercial opposition. Even inventors who work for corporations must convince others in the corporate structure to provide funds for them. Similar problems often hamper inventors inside government, who often are unable to "sell" the product to the bureaucracy and are opposed by the traditions and vested interests of long-established institutions.

Ralph Waldo Emerson once said that if you build a better mousetrap, the world will beat a path to your door. In real life, however, it seldom works this way. In two senses inventions must be sold to the public: The public must be sold on the idea that the invention is worthwhile, and the invention must be sold at a price that the public will pay for it. Marketing is a very difficult activity for many, as is evidenced by the high rate of failure in marketed products generally. Thus it is not particularly surprising that inventors are not very good at it either, lacking, as they often do, both training and inclination. Fortunate inventors find talented entrepreneurs to help them get the product to the marketplace. Often, though, the most difficult thing to accomplish is the change in public attitudes that makes the innovation desirable to consumers.

When we look at many of the products around us, we see that some of them have well-defined uses. However, we often fail to realize that it took time to establish these uses. There are, of course, cases in which an invention is so indispensable to an industry that it is immediately taken up, as was the case with the cotton gin. Or a group of consumer enthusiasts may breathlessly await the product, as was the case with the first personal computer, the Altair. Interest in the Altair among computer hobbyists was so great that one enthusiast boarded a plane and rented a car to get to the factory to pick up his unit.[53] The first ballpoint pens apparently elicited this kind of fanatic enthusiasm.[54] Such inventions need little selling, and do not even have to work very well (the Altair didn't, for instance). They are automatically absorbed into the market, and competitors rapidly appear. But this is not the usual case.

More commonly inventors must invent not only the invention, but the use for it, too; then they must persuade potential users that the invention will meet their needs. Surprisingly, inventors sometimes do not understand the most important uses of their inventions. Edison thought that the primary use for the phonograph would be office dictation, not mass entertainment.[55] But the public became so interested that the resulting "phonograph craze" ensured powerful commercial interest in and a rapid emergence of the phonograph as a working consumer product. When holograms and lasers were invented, both were considered to have narrow ranges of application; they were considered to be "solutions without problems."[56] The real scope of application for both became clear only with time.

Even more frequently inventors are unable to convince others of the uses that they envision. Inventors are experts in *things;* to sell their inventions, they often require the services of experts in *people.* Furthermore, many firms are not interested in the work of outside inventors—the "not invented here" syndrome.[57] Thus often for a considerable period the invention's real capabilities may lie fallow. This unwillingness to take new inventions seriously accounts for the reluctance of many armies to use machine guns effectively for many years.[58]

But vision is only part of the problem. A greater difficulty is making the invention pay off. If the invention is to be sold commercially it must be manufactured, and it will not be manufactured unless someone thinks a profit can be made by doing so. In other words, invention is only the first phase; the second phase, commercialization, is equally important. And for the individual inventor commercialization of inventions is difficult. Unless the invention becomes available in quantity—unless it *diffuses*—it cannot influence society very much.[59] Diffusion demands the development of a business, and usually—because few inventors are good businesspeople—it requires the involvement of an *entrepreneur.*

We can define an entrepreneur as a person who brings together the elements required to start a new business. Typically these are funds, personnel, a business plan, and the management expertise needed to run the new firm. Entrepreneurs, like inventors, have distinctive orientations and characteristics.[60] Like inventors they are highly creative, but they are creative with organizations rather than with things.[61] They also tend to be different in background and psychological makeup from ordinary business executives.[62] Frequently the entrepreneur is indispensable in getting an inventor's idea into production because the business expertise of the entrepreneur is needed to complement the technological skill of the inventor. Usually the funds to start up the firm come from local sources, such as personal savings, bank loans, and loans from family and friends, although in rare cases professional investors—"venture capitalists"—provide them.[63]

The work of the entrepreneur is many-faceted. Developing a new business involves the difficult task of maintaining a sensitive balance between the requirements of investors, inventors, and staff and the marketing plan. Horrendous, often unforeseen problems may occur in manufacturing, even as money continues to be spent without any sales receipts. Underfinancing is a common problem, making the new enterprise vulnerable; the entrepreneur must persuade the project's backers to keep extending funds even though no product has yet appeared. Relationships among personnel are often somewhat tense, for regular employment cannot be assured until the product becomes successful. The inventor may want to keep improving the product, whereas the key issue may be to get it out the factory door. The entrepreneur must be able to keep all these forces in balance. The willingness to work long hours and the energy to do so are often required.

But even when the new firm gets off the ground, another problem becomes important. New inventions threaten established businesses, and these businesses often go to great lengths to neutralize such threats.[64] Thus the new business is in danger of being bought out, commercially overwhelmed by competitors' products, or in some cases driven out of business by lawsuits over patents.[65] Alexander Graham Bell's patent for the telephone and Henry Ford's patent on his automobile were both hotly contested. Patent infringers have threatened to bury inventors "in a ton of paper," often no idle threat.[66] The current average cost of a patent suit has been estimated to be around $250,000, no mean sum for a fledgling firm.[67] Nor are courts likely to be helpful in defending a patent's validity against firms that want to see it invalidated.[68] Today it is not unusual for legal fees in major patent lawsuits to run into the millions of dollars. It is easy to see that few small business firms could afford this kind of financial threat.

Inventors, then, face serious problems in getting their inventions "out the door." Not only may they lack experience and ability in business, but even if they find an entrepreneur willing to help them, their firm may be unable to hold its own. We will examine further in a later chapter the kinds of problems they face in the marketplace, even when competition is fair. In the United States today, the inventor and the small firm face major hurdles in getting the invention produced and making a profit from it. Although federal and state governments have tried to assist such firms, the amounts of assistance are small and inadequate.

Conclusion

Although we know a great deal about individual inventors, our society seems to possess relatively little systematic information about these individuals who are so important to us. This lack of information reflects our neglect of the inventor as a social resource. In American culture inventors are not as honored as scientists, just as synthesis is less honored than analysis. We also tend to believe, wrongly, that individual inventors are no longer important to us and that invention is now largely done by scientific teams in large companies. Although such teams are important, they are not the whole story, and even teams require inventors. How inventors are recognized, encouraged, and trained is important, then, not just for individual inventiveness, but also for industrial teams. Love for and skill in inventing does not originate the moment someone is hired for a position in an R&D department, but is prepared by all the life experiences of each team member.

Invention is a highly creative process. It is an art form distinct from other kinds of creativity. Our society, so dependent on inventions, would do well to cultivate this kind of creativity. Its practitioners, the inventors, are valuable to us for their tireless desire and ability to improve the things with which they come into contact. But by itself, invention is not enough. To make objects useful for society, the devices must be developed and mass-produced; this often means that inventors must depend on other people for money and managerial talent.

Invention is often an individual process, but it affects and is in turn affected by society. In this chapter we have discussed the role of entrepreneurs in helping inventors produce their inventions. But there are other ways in which society is important to the inventive process. In the next chapter we will explore the ways in which these other social forces affect invention.

Notes

1. Exactly what constitutes an invention is a matter of some debate; see, for instance, Lynn White, "The Act of Invention: Causes, Context, Continuities and Consequences," *Technology and Culture,* Vol. 3, No. 4 (1962), pp. 486–500; and Abbott Payson Usher, *A History of Mechanical Inventions* (New York: Dover, 1988). We are on firmer ground with patents; see Christine MacLeod, *Inventing the Industrial Revolution: The English Patent System 1660–1800* (New York: Cambridge University Press, 1988).

2. Again, exactly what is "nonobvious" can be the subject of some debate; see Richard L. Gausewitz, *Patent Pending: Today's Inventors and Their Inventions* (Old Greenwich, Ct.: Devin-Adair, 1983).

3. See Joseph Rossman, "A Study of the Childhood, Education and Age of 710 Inventors," *Journal of the Patent Office Society,* Vol. XVII, No. 5 (May 1935), pp. 411–421.

4. Berton Bernstein, "The Reporter at Large: The Last Plum," *The New Yorker,* Aug. 4, 1986, p. 51. It is quite interesting to note that Paul MacCready, another inventor of lightweight planes, also refused ready-made kits at age 13; see Patrick Cooke, "The Man Who Launched a Dinosaur," *Science 86,* Apr. 1986, p. 30.

5. William D. Marbach and Peter McAlevey, "Up, Down, and Around," *Newsweek,* Dec. 29, 1986, pp. 34–44.

6. Doug Garr, *Woz: Prodigal Son of Silicon Valley* (New York: Avon Books, 1984), pp. 28–29. © 1984 by Doug Garr. Reprinted by permission of the author.

7. Rossman, "Study," pp. 411–421; and Colleen Lyons and Donald Beaver, "Heroic Inventors and the American Dream," paper presented to the Midwest Junto of the History of Science Society, Apr. 7, 1978. These findings suggest that education may be important to successful future inventiveness. Although the inventors may create their first inventions while in their teens, their first patents may not come until their late twenties. Formal education may well provide the confidence and formal skills necessary to transform creativity into productivity.

8. Having lots of ideas is important, but knowing how to choose among them is equally important. Jacob Rabinow, a prolific inventor, argues that the difference between the amateur and the professional inventor is that the professional will discard much more quickly those ideas unlikely to work out (personal interview).

9. John Stuteville, *A Study of the Life History Patterns of Highly Creative Inventors.* Doctoral Dissertation, University of Southern California, 1966.

10. Adrian Hope, "It's a Wonderful Idea, but . . . ," *New Scientist,* June 1, 1978, pp. 576–581.

11. Irving Siegel, "Independent Inventors: Six Moral Tales," *Idea,* Vol. 9, No. 4 (Winter 1965), pp. 643–655.

12. Francis Jehl, *Menlo Park Reminiscences,* (Dearborn, Mich.: Edison Institute, n.d.) Vol. I, p. 233.

13. Quoted in Alfred Gollin, *No Longer an Island: Britain and the Wright Brothers* (London: Heinemann, 1984), p. 17.

14. Rene Dubos, *Louis Pasteur: Free Lance of Science,* (Boston: Little, Brown, 1950), p. 63.

15. Personal interview with Dean Wilson; these experiments are also mentioned in Donald Schon, *The Reflective Practitioner: How Professionals Think in Action* (New York: Basic Books, 1983).

16. Unpublished reminiscences of Howard Wilcox, and personal communication with Charles P. Smith, Dec. 29, 1986. See also John Fialka, "Weapon of Choice," *Wall Street Journal,* Feb. 15, 1985, pp. 1, 30. It is not true, as this article claims, that McLean carried out most of the work in his garage, although he did some work there. He had a small space in the back of the Michelson Laboratory at the Naval Ordnance Test Station, and he could often be found there until two or three in the morning. Of course, if important ideas occurred to him late at night or on the weekend, then the garage might well be used, according to his deputy, Haskell G. Wilson (personal interview). See also H. Pickering, "William B. McLean, 1914–1976," *National Academy of Sciences Biographical Memoirs,* Vol. 55 (1985), pp. 398–409.

17. Howard Wilcox, personal interview.

18. Warren Legler, telephone interview.

19. This trait is clear in Peter Padfield's *Aim Straight: A Biography of Admiral Sir Percy Scott* (London: Hodder and Stoughton, 1966). The first sentence of his text says it all: "Percy Scott was a projectile." See also Joseph Rossman, *Industrial Creativity: The Psychology of the Inventor* (New York: University Books, 1964), p. 152; "desire to improve" is a close second to "love of inventing" as motivations cited by inventors for their work.

20. Interview with Jacob Rabinow, June 19, 1987. See also the portrait of Rabinow by Robert Kanigel, "One Man's Mousetraps," *New York Times Magazine,* May 17, 1987, p. 48 et seq., and Jacob Rabinow, "Invention Is an Art Form: It Should Be Supported as Such," *Industrial Research and Development,* Nov. 1980, pp. 108–112 and "Is Invention an Art? Since It Is Fun, Should Inventors Be Paid?" *Industrial Research and Development,* Dec. 1980, pp. 88–91.

21. From Kenneth A. Brown, *Inventors at Work: Interviews with Sixteen Notable American Inventors* (Redmond, Wash.: Microsoft Press, 1988), p. 231. Reprinted by permission of Microsoft Press. Copyright © 1988 by Microsoft Press. All rights reserved.

22. Interview with Haskell Wilson, in Mar. 1987; Wilson got this observation from Prof. Charles C. Lauritsen, formerly professor at the California Institute of Technology.

23. Hermann Goldstine, *The Computer from Pascal to Von Neumann* (Princeton, N.J.: Princeton University Press, 1980), p. 24.

24. John Evangelist Walsh, *One Day at Kitty Hawk: The Untold Story of the Wright Brothers* (New York: Crowell, 1975).

25. Gustave Whitehead, a German-American inventor, was also trying to produce a successful powered-airplane flight at the same time as the Wright brothers. There is some evidence that he may have achieved powered flight earlier, but these claims are open to dispute; see James Brooke, "Bat-Winged Plane to Challenge Wrights' Claim," *New York Times,* Feb. 20, 1986.

26. S. C. Gilfillan, *The Sociology of Invention* (Chicago: Follett, 1935).

27. The existence of these alternatives and competitors often leads to protracted court battles over who the original inventor was, and who therefore has the right to file a patent. Exactly who, for instance, first developed the laser is a matter of serious dispute; was the inventor Nobel Prize-winner Charles Townes or his then graduate student Gordon Gould? According to Eliot Marshall, in "Gould Advances Inventor's Claim on the Laser," *Science,* Vol. 216 (Apr. 23, 1982), pp. 392–395, at p. 395, "one of the few points that is clear in this long, contested record is that several people almost simultaneously hit on the laser concept, and that a few in quick succession built lasers that proved to be well engineered."

28. Thomas G. Lang, the inventor of the first SWATH ship, *SSP Kaimalino,* told me that the SWATH design had emerged after literally years of filling notebooks with ideas about ship flotation. The SWATH (small waterplane area twin hull) ship is an unusual catamaran with torpedolike tubes along the bottoms of the twin hulls. He had had experience in inventing and selling a small hydroplane conversion kit he developed with the help of his father. But he explored hydroplanes, surface effect vehicles, and various SWATH designs before arriving at the design leading to the *Kaimalino.* See also his account in "*SSP Kaimalino:* Conception, Developmental History, Hurdles and Success," American Society of Mechanical Engineers Paper 86-WA/HH-4 (presented to the ASME at Anaheim, California, Dec., 1986).

29. Quoted in W. Rupert McLaurin, *Invention and Innovation in the Radio Industry* (New York: Macmillan, 1949), p. 56.

30. Edward Constant II, *The Origins of the Turbojet Revolution* (Baltimore: Johns Hopkins University Press, 1980).

31. Robert Friedel and Paul Israel with Bernard S. Finn, *Edison's Electric Light: Biography of an Invention* (New Brunswick, N.J.: Rutgers University Press, 1986). A good popular treatment of systems engineering is C. West Churchman, *The Systems Approach* (New York: Dell, 1979); for a more advanced approach, see Arthur D. Hall, *A Methodology for Systems Engineering* (Princeton, N.J.: Van Nostrand, 1962).

32. Walsh, *One Day at Kitty Hawk.*

33. Matthew Josephson, *Edison: A Biography* (New York: McGraw-Hill, 1959).

34. Elting Morison, "Gunfire at Sea: A Case Study in Innovation," in his *Men, Machines, and Modern Times* (Cambridge, Mass.: MIT Press, 1966), pp. 17–44.

35. Rene Vallery-Radot, *The Life of Pasteur* (New York: Dover, 1960), p. 76.

36. See Abbott Payton Usher, *A History of Mechanical Inventions,* 2nd ed. (New York: Dover, 1988), pp. 60–72.

37. Thomas Hughes, "How Did the Heroic Inventors Do It?" *American Heritage of Invention and Technology,* Vol. 1, No. 2 (Fall 1985), pp. 18–25. In this essay Hughes distinguishes between "reverse salient," which represents a state of technical imbalance, and "critical problem," which is the inventor's perception that the benefits of correcting such an imbalance are worth the efforts involved.

38. Hughes, "Heroic Inventors," p. 20.

39. Jacob Getzels, "The Problem of the Problem," in R. Hogarth (Ed.), *New Directions for Methodology of Social and Behavioral Science: Question Framing and Response Consistency,* No. 11 (San Francisco: Jossey-Bass, 1982).

40. Patrick Cooke, "The Man Who Launched a Dinosaur," *Science 86,* Apr. 1986, p. 30.

41. Quoted with permission of Paul B. MacCready from a letter written to the author dated Aug. 2, 1987.

42. For instance, in an interview Nobel Prize-winner Linus Pauling spoke of the genesis of his most important discovery, the theory of covalent chemical bonds (Sandra Tessler, "The Solver of Puzzles," *Detroit News,* June 12, 1986, D-1):

    He described his discovery as a "simple equation" which began with the question "what if?" . . . As Pauling recalled, "in December of 1930 or January of '31" he asked some basic questions about molecular structure and "within a few minutes" had figured out a hypothesis of bonding between atoms. It took much longer to prove the theory and win worldwide recognition.

43. Esther Goddard (Ed.), *The Papers of Robert H. Goddard,* Vol. 1 (New York: McGraw-Hill, 1970), p. 9.

44. Milton Lehman, *Robert Goddard: Pioneer of Space Research* (New York: Plenum, 1988).

45. The restlessness of some individuals may conflict with persistence. This was true of the radio inventor Lee De Forest, who would hop to another problem if difficulties loomed. See McLaurin, *Invention and Innovation,* p. 79.

46. Joseph Rossman, *Industrial Creativity: The Psychology of the Inventor* (New York: University Books, 1964).

47. John Enos, in reviewing six innovations in petroleum processing, states that "perhaps original ideas came easily, but in the petroleum industry they were all implemented only by an amazing amount of work." See John Enos, *Petroleum Progress and Profits: A History of Process Innovation* (Cambridge, Mass.: MIT Press, 1962), p. 228.

48. "Howard Head Says I'm Giving Up the Thing World," *Sports Illustrated,* Sept. 29, 1980, p. 68.

49. Henry Bessemer, *An Autobiography* (London: Engineering, 1905), pp. 173–174. Copyright *Engineering* 1905. Reprinted with permission of the publisher. Similar problems plagued Rudolph Diesel, whose engine required a decade of development before it found eventual success; see Donald E. Thomas, Jr., *Diesel: Technology and Society in Industrial Germany* (Tuscaloosa: University of Alabama Press, 1987).

50. See Teresa M. Amabile, "The Social Psychology of Creativity: A Componential Conceptualization," *Journal of Personality and Social Psychology,* Vol. 45, No. 2 (1983), pp. 357–376; and Teresa M. Amabile, Beth Ann Hennessey, and Barbara S. Grossman, "Social Influences on Creativity: The Effects of Contracted-For Reward," *Journal of Personality and Social Psychology,* Vol. 50, No. 1 (1986), pp. 14–23.

51. This statement was made by MacCready several times in the course of lectures he delivered in the Ann Arbor, Michigan, area on May 11, 1988.

52. Stuart Macdonald, *The Individual Inventor in Australia,* a report published by the Department of Economics of the University of Queensland, 1982. A published excerpt from this report is "The Individual Inventor in Australia," *Australian Director,* Feb. 1983, pp. 44–51.

53. Paul Freiberger and Michael Swaine, *Fire in the Valley: The Making of the Personal Computer* (Berkeley, Calif.: Osborne/McGraw-Hill, 1984), pp. 37–40.

54. Victor Papanek, *Design for the Real World* (New York: Random House, 1971), p. 27.

55. Matthew Josephson, *Edison: A Biography* (New York: McGraw-Hill, 1959), pp. 172–173.

56. John D. Anderson, "Holograms: Popping Up All Over," *New York Times* Aug. 2, 1987, Section 4, p. 4; and Winston E. Kock, *The Creative Engineer: The Art of Inventing* (New York: Plenum, 1978), pp. 99–101.

57. Adrian Hope, "It's a Wonderful Idea, but . . . ," *New Scientist,* June 1, 1978, pp. 576–581; and Adrian Hope "The Death of an Idea," *New Scientist,* Sept. 13, 1979, pp. 794–797.

58. John Ellis, *The Social History of the Machine Gun* (Baltimore: Johns Hopkins University Press, 1986), pp. 47–78.

59. At least not in a direct physical sense. Inventions in the prototype or even the blueprint stage, however, may have very powerful effects on the minds of other inventors. Charles Babbage never finished any of his computing engines, but his work had enormous impacts on later thinkers.

60. See Orvis Collins and David G. Moore, *The Organization Makers: A Behavioral Study of Independent Entrepreneurs* (New York: Appleton-Century-Crofts, 1970).

61. A good popular account, with a variety of interesting "success stories," is A. David Silver, *Entrepreneurial Megabucks: The 100 Greatest Entrepreneurs of the Last 25 Years* (New York: Wiley, 1985); see also Clinton Woods, *Ideas That Became Big Business* (Baltimore: Founders, 1959).

62. Collins and Moore, *The Organization Makers,* pp. 15–48.

63. See the testimony of Albert Shapero, Ohio State University, in *Small, High Technology Firms, Inventors, and Innovation,* hearings before the Subcommittee on Investigations and Oversight of the Committee on Science and Technology, U.S. House of Representatives, 97th Congress (Washington, D.C.: U.S. Government Printing Office, 1981), pp. 24–27.

64. Richard Dunford, "The Suppression of Technology as a Strategy for Controlling Resource Dependence," *Administrative Science Quarterly,* Vol. 32, No. 4 (Dec. 1987), pp. 512–525.

65. Bernhard J. Stern, "Restraints upon the Utilization of Inventions," in his *Historical Sociology* (New York: Citadel, 1959), pp. 75–101; see also Edwin Sutherland, *White Collar Crime* (New York: Holt, Rinehart & Winston, 1961), pp. 95–110.

66. Tom Wolfe, "Land of Wizards," *Popular Mechanics,* Vol. 163, No. 7 (July 1986), pp. 127–139.

67. House hearings on *Small, High Technology Firms, Inventors, and Innovation,* p. 115.

68. *Ibid.,* p. 125.

CHAPTER 6

Invention as a Social Process

Invention does not take place in a vacuum; it is shaped by larger social institutions. One portrait of inventors depicts them as free spirits who pursue apparently randomly chosen problems through equally random methods. This portrait is just as false as the opposite view, which portrays inventors as merely practical extenders of basic science. The act of invention is intensely creative, but the creativity involved is constrained by social processes that need study. Invention is social in two notable ways.

The first social impact on invention involves the collaborative nature of the inventive process. Some inventors work alone, but many function as parts of teams of various sizes. Inventions are produced by teams of a broad range of sizes, from the Apple II computer, built by Stephen Wozniak and Steven Jobs in Jobs's garage, to the massive Apollo space program, run by the National Aeronautics and Space Administration and costing tens of billions of dollars.[1] A second influence on invention exists because society provides a field of problems and solutions in which the inventive process is immersed. Inventors are functioning parts of their societies, and their creations arise out of this broader social context. From their own social experiences inventors gain knowledge of both techniques and society's needs. Howard Head's experiences in skiing made him think about better skis. His experiences in the aircraft industry made him think about making such skis out of aluminum. The existence of a skiing industry made him think that an invention would be worthwhile. Because the inventor clearly is a member of society, we need to consider how this membership affects the process of invention.

The Origins of Inventions

How did we get the inventions we have? Rather than starting with inventors, we might start with inventions and try to reconstruct the events leading up to them. Presumably, by careful study of how major inventions get produced, we can understand the social processes that lead to inventions. Three major approaches are taken in these studies:

1. *Individual genius.* One favorite explanation of the origins of major inventions is that they were developed only because some genius thought them up. This explanation is correct in some cases, but it is not sufficient because some important inventions were not conceived by geniuses, and many are thought up by more than one inventor.

2. *The state of the art.* This explanation, a favorite of the Ogburn group, holds that inventions are inevitable and largely depend on the level of development the technology has reached. When there is a great deal of technological activity in a field, more people are likely to be thinking about ways to improve products and processes. And it is certainly true that the state of the art is important, as the phenomenon of multiple independent invention (see below) shows. Again, this explanation is insufficient by itself.

3. *Social demand or explicit choice.* The central argument in this explanation is that inventions are made in response to social demand for them. The invention is desired by some set of people who are willing to pay handsomely for its use. Although many inventions do not seem to result from any preexisting demand, in other cases social demand leads to directed research or mandates expensive technological development.

Interest in these social factors in invention directs our attention toward broader social issues. Even if all inventors worked alone, they would still exist in a society with a certain level of knowledge, with certain kinds of technologies, certain industrial problems, and certain demands for products of a particular kind. Even if inventors worked alone, these social contexts would be important, for better mouse-traps depend not only on mousetrap inventors, but on mousetrap technology and mousetrap customers as well.

The Role of Scientific Knowledge in Invention

One of the most important factors in inventiveness is scientific and technological knowledge. When inventors invent, they draw on an immense cultural fund of knowledge, as Ogburn and others have pointed out. Technologists refer to the level that knowledge has reached as "the state of the art"; it is the current level at which technology is best able to invent, transform, and manufacture. In many respects it represents a kind of starting point for the individual's stream of ideas.

The state of the art is made up of a variety of different kinds of knowledge. Some of this knowledge is basic science resulting from studies on how physical or biological systems work. Other knowledge is essentially technological and concerns practical methods for attaining certain ends. Some part of this technological knowledge consists of knowing the properties of current devices and systems, many of which are likely to go into a new invention. Bruno Latour calls these devices and systems the "black boxes" of technology[2]—anything that can be taken for granted and can be put into a new device without too much further development.

A common view is that "science discovers and technology applies." Many people take the view that technological advances depend on advances in science because technology borrows basic ideas from science. Indeed, technology is sometimes described as "applied science." But is it really true that ideas are developed first in "pure" science and then transferred somehow to engineering and technology? Whereas there is some truth to this idea, it is an extremely crude first approximation that is false at least some of the time,[3] and it glosses over a variety of ways in which science and technology interact.[4]

One way science and technology can affect each other is through the *transfer of ideas*. This relationship, usually conceived of as the transfer of ideas from science to technology, is often what people mean by the term *applied*. But how do the ideas get from science to technology?[5] Do ideas migrate through scientific journals and conferences or largely through the movements of people?[6] Engineering education in the United States, for instance, includes a very strong dose of science; some new scientific ideas are thus transferred as engineering students go from university to workplace.

But not all the movement is from science to technology; ideas may originate in technological areas and then move to science when scientists are given the task of explaining why a technology works. Thus, for example, difficulties in understanding the dynamics of steam engines led to the scientific development of thermodynamics.[7] Similarly, coping with the behavior of telephone networks led to an interesting branch of mathematics called queueing theory.[8] "Computer science" obviously grew out of the technology of computing machines, just as "aerodynamics" grew out of technological problems in building airplanes. And finally, metaphors developed in one field may transfer to the other, increasing the ability to understand and model the behavior of complex systems.[9]

Another form of influence is the *transfer of people*. Again, scientists often transfer to technological settings, where they carry out research for "applied" purposes. Many examples of this occur in military research; during the world wars, for instance, chemists, physicists, and biologists devoted themselves to designing, improving, and developing a variety of weapons.[10] An exchange from technology to science occurs when engineers and technicians join scientific laboratories to collaborate, bringing with them technological knowledge, skills, and patterns of thought. There is also the occasional case in which an engineer engages in "pure" scientific research, as when Karl Jansky and Grote Reber, both radio engineers, did their pioneering work in radio astronomy.[11]

Finally, there is the *transfer of devices*. A device developed for scientific measurements may have strong technological applications, or vice versa. Sometimes the device comes complete with a small community of knowledgeable practitioners required to make the thing work. New measurement devices are just as likely to be developed by scientists as they are by engineers.

The same individual, furthermore, may be active in both areas. Scientists may invent and patent their inventions, and those who have gone into industry may be engaged both in fundamental research and practical product improvement. In engineering, particularly with such top practitioners as Thomas Edison or Charles Kettering, scientific innovation may occur as spin-offs from the practical work of inventing and design. However, for such spin-offs to become scientifically fruitful, they must be adopted by scientific laboratories per se.

Sometimes technologists discover that a process that is mysterious to them nonetheless allows a complex system to work. For many technologists, this mysterious behavior, although intriguing, is simply used unquestioningly.[12] For instance, in developing the Sidewinder missile, magnetic amplifiers were used. Although some of the reasons they worked were unknown, they functioned well, and thus they were incorporated into the missile in spite of the mysteries associated with their dynamics.[13] A scientist is much more likely than a technologist to find such anomalous behavior intriguing enough to make it the special object of investigation.

Some technologists may also be more interested in the more "basic" aspects of the technology than are others. An interesting example of this difference in attitude occurred while I was researching quality control in some plants in Michigan. At one small factory, which made automobile fans, much of the organization's tiny R&D operation spent its time testing their fans. "Here," said one of the engineers, "we don't have time to figure out why or how the fan falls apart; we just want to know *whether* it falls apart." He then went on to describe a large, complicated scientific project taken on by one of the larger companies in which the disintegration of the fan was carefully recorded on film. This was taken as an example of the kind of "useless" research that people get into when their research operation is too well funded and staffed.

To sum up, then, science does contribute to the stock of knowledge from which inventions—and technological advances generally—come. However, the flow between scientists and technologists is certainly not one-way, and a variety of interactions are important. It is also important to note that a basic research advance does not automatically become a commercial product; this transformation requires entrepreneurship and hard work. In recent years Americans, with one of the best scientific communities in the world, have assumed that scientific advances would automatically transform themselves into commercially viable products. This, however, has not happened. Other countries, especially Japan, have proved to be much better at this "technology transfer" process.[14]

Roles in the Inventive Process

Problem-Finders and Problem-Solvers. Many important inventions are created by groups.[15]

In such cases the invention emerges through the interaction of several individuals. Obviously a technical team may require people with different kinds of competences: researchers, designers, developers, prototype builders, testers, and so forth. But even if we look at the purely mental activities, it may be possible to see distinct inventive roles. For instance, some people may be good in the role of question-posers, whereas others may be good answer-providers; some may be good in both. Thomas Hughes observed in his *Networks of Power* that Edison developed with members of his research team a symbiotic relationship that often led to the development of distinctive roles in the inventive dialogue:

Edison also impressed [physicist Francis] Upton with his talent for asking original questions. "I can answer questions very easily after they are asked," Upton lamented, "but I find great trouble in framing any to answer." Edison posed questions that could be translated into hypotheses, which in turn established the strategy and tactics of experimentation. His questions were often drawn from his doubts about accepted explanations and procedures. According to Upton, Edison never took anything for granted; he always doubted what others thought possible. Sometimes the result was that he found a new way.[16]

Even stronger support for the idea of distinct roles in the inventive process comes from a study by Gerald Gordon.[17] In studying three sets of innovation teams—sociologists, chemists, and engineers—Gordon identified distinct traits that differentiated those who were good at finding problems from those who were good at solving them. The good problem-solvers scored high on the Remote Associates Test (RAT), which assesses how well a person can find a link between seemingly unrelated items. High-RAT scorers individually had a higher rate of patentable ideas, patent proposals accepted, and actual patents. Clearly, these are good people to have on an innovation team!

But Gordon's study also found that team leadership by someone with problem-*finding* skills made a big difference as well. Gordon asked the leaders of research and R&D teams to rate the people in the teams. He found that some leaders differentiated among their team members on these assessments, whereas others didn't; these "high-differentiators" (high-DIFFs) were more sensitive to individual differences in creativity. Gordon theorized that those who are good at making distinctions among people are also good at finding problems, for spotting problems requires an ability to make fine discriminations. More significantly, these leaders had much more innovative teams.[18] Teams of medical sociologists led by high-DIFF chiefs were generally rated higher than teams with low-DIFF chiefs; this was true regardless of the quality of the interactions within the teams. Similarly teams of chemists were more valued by their companies then they were led by high-DIFF leaders. The leaders who did not perceive these differences had teams that were more successful at extending traditional knowledge than they were at innovating.

Thus creative teams led by the right sort of people may be necessary for high levels of innovation. Edison, obviously, had exceptional problem-finding and problem-solving skills.

Technological Gatekeepers. Getting the right technical information is important to invention. The value of institutions and publications that distribute technical information should not be underestimated. Many inventors—the Wright brothers are a good example—draw directly on the contributions of others who have brought the concept to a certain point but are unable to go further. The Wright brothers did not design the airplane in a technical void; they found Otto Lilienthal's glider experiments essential in designing their own gliders. In some cases earlier researchers may have made a critical observation that is the key concept for the invention but had been unable to do anything with it. In multiple invention it is striking how many times earlier discoveries remained undeveloped until later discoverers with more drive or imagination, or better circumstances, pushed the invention further. Once this information becomes widely available, others often can use it to advance their own efforts.

Thus another important role in innovation is that of the so-called technological gatekeeper. In a study of R&D firms, Thomas J. Allen found that certain individuals in technical organizations tend to act as brokers of information that originates outside the organization.[19] These brokers, the "technological gatekeepers," typically read more technical journals and had more outside contacts than did other workers in these laboratories. They usually worked in the first supervisory tier of the organizations, just above the "bench" technologists, and they tended to have more formal education and more academic contacts than the other workers as well.

A prime example of the importance of the technological gatekeeper is the story of John H. Dessauer, at one time the director of research of the Haloid Corporation. As head of R&D for this small Rochester, New York, company, Dessauer was looking for new fields that Haloid could enter. He chanced across an article about a dry photocopying process being developed by Chester Carlson, and he realized that this might be the beginning of a new industry.[20] He contacted Carlson, whose process had been turned down by some 20 industrial giants, including RCA, Remington Rand, Kodak, and IBM. Development of Carlson's process was undertaken by Haloid, which changed its name to Xerox, and the company began a period of growth through which it became a corporate giant.[21] In his role Dessauer was not only required to recognize the possibility for innovation, but also to possess the imagination that enabled him to see the innovation's market potential.

Product Adversaries. Even adversaries may have an important role in this continuous inventive dialogue. Simply by opposing the invention, adversaries may force the inventor to strengthen it. In some cases they may be vital for discovering the invention's flaws, which can perhaps be corrected during the development process. In the previous chapter we saw how Bessemer's opponents forced him to identify the exact limits within which his process would work; once these parameters were determined, Bessemer's steel-making process was far more secure than it had been before their opposition. In other cases, of course, opposition may be extremely destructive, as is sometimes the case during patent fights in which the inventor's scarce financial resources are expended in protecting the ownership of creative products. Many inventors have lost their fortunes trying in vain to protect inventions against powerful and sometimes dishonest opponents.

Product Champions. Another important role in corporate invention is played by product champions.[22] These people are often the actual inventors of the product, but not always. Their role is to act as internal promoter of the product—to convince others in the corporation that it is worth developing.[23] Gifford Pinchot III has labeled such people "intrapreneurs"; they are analogous to entrepreneurs who start new companies,[24] but they start projects *inside* firms.[25] Not only users, but also other members of the firm that spawned the innovation, must be convinced of its value. This job of selling the innovation inside the firm requires high personal energy, for it is frequently a long, tough fight. Although some firms encourage product champions, others discourage those who develop projects that "no one has asked for."

This point brings up one of the major problems of innovation in large organizations. Worthwhile ideas for products frequently occur to people who have not been given the responsibility for thinking them up, even though many firms expect innovations to come mostly from people in designated "thinking" roles, such as those in the R&D department. Thus innovations are expected to come from those in "staff" positions, whereas "line" personnel are expected to be "doers" rather than "thinkers." But what happens when someone in a "doing" role comes up with a new product idea? If the idea is valuable, the person faces a double hurdle. The first hurdle is that any new idea, even if it occurs to the right person, is hard to sell. People are already committed to the ideas they have; why should they listen to new ones? The second hurdle is that the person who has the idea also has to convince others in the firm that it is all right for him or her to have gotten the idea in the first place, even though "thinking" is not his or her organizational role. Highly creative organizations, of course, encourage ideas from everybody, but such organizations are in the minority.

Percy Scott, who developed continuous-aim firing in the Royal Navy (among many other innovations), found his ideas hampered because he was a line officer:

[I]n government offices they do not like suggestions coming from the outside which could have originated in the office itself. It was the same with all my proposals. They were boycotted, because the people—mostly my juniors in age, and with far less experience—dealt with these matters at the Admiralty, and felt aggrieved that the suggestions had not emanated from themselves.[26]

Because the Admiralty deemed themselves the "designated idea people," line officers such as Scott were considered to be insufficiently qualified. When Scott later became Inspector of Gunnery, however, he opened the system up to ideas from line officers. In one telegram he stated:

Heretofore it has been the custom to boycott all ideas from sea-going ships I am trying to change that and make use of the talent of the Navy please distribute enclosed to any officers who have ideas and explain that the boycott is off as far as Xlent [the gunnery training center] is concerned.[27]

There are many classic accounts of product championship in business firms. One of the most famous is the development of Post-It Notes by the 3-M Corporation, which is well known for development of new products.[28] Spencer Silver, a chemist at 3-M, discovered a new monomer, and he began to experiment with it "just to see what it would do." Unlike many adhesives that 3-M has discovered, this monomer was not particularly sticky, but Silver was convinced that something could be done with it. He tried for five years to interest others at 3-M in it, without success. Eventually he was given a small "venture team" of helpers to develop a product for the market. One of the members of this team, Arthur Fry, had the insight that the only slightly sticky chemical might be used to make notes stick to things, yet still be detachable. This insight occurred when Fry, a church choir member, was frustrated in having the place markers in his hymnal fall out between services.

Once Fry had discovered this use, then ways had to be developed to produce the "notes" with the sticky monomer in a single line on the back. This proved to be very difficult. Then the idea of the product had to be sold to the corporation. After a long fight to convince 3-M executives that it would sell, it was finally market tested, but it flopped. The key to marketing Post-It Notes turned out to be repeated use: Only after people had used the notes did they become addicted to their use. Then, but only after the marketing strategy had been redesigned to include large giveaways, did the notes' true commercial strength become apparent. It has now become a multi-million-dollar product. In this case one person discovered the chemical, but it took a team to figure out how to manufacture and market it.

Much of the important innovation in the world is done not on the products themselves, but on the processes of their production. Successful process improvements depend on an organizational climate of learning and a willingness to share credit among departments. Unhappily, in the United States this kind of organizational climate seems more the exception than the rule.[29] As Charles "Boss" Kettering put it, "The greatest durability contest in the world is trying to get a new idea into a factory."[30] Other countries seem more willing than the United States to encourage these small, incremental improvements in making things.[31] One of the problems with American firms is that thinking is assigned to a specialized department, and little effort is made to educate and enlist the rest of the work force in the firm's creative activities.[32] This often provides a serious barrier to innovations once they have been conceived, and it discourages workers from conceiving of and developing innovations themselves.

Spectacular success can come to a firm willing to nurture ideas developed from within. A prime example is the development of float glass by Pilkington Brothers. The float glass process provides large sheets of plate glass that do not require buffing after forming; float glass is manufactured by continuous extrusion of a glass sheet over a bath of molten tin. The process was conceived by Alistair Pilkington in 1952 while he was washing dishes one evening. At that time he was an assistant, without any firm assignment, to the production director of the Pilkington Brothers Glass Company. His proposal, which offered the chance for important future profits, was adopted by the company, which decided to develop it under the most stringent commercial secrecy. The project was progressively "scaled up" until a full-fledged production operation was under way. Development took seven years and £7 million.[33]

Had this company, at that time privately owned, not been prepared to put up such a huge amount of money, Pilkington's idea might have remained just that. But because the company was a family firm and Pilkington was a relative of the owners, strong financial support went hand in hand with a willingness (actually, a necessity!) to get the idea into production. Some have suggested that had the company been publicly owned, this kind of expensive development never would have been funded because many shareholders would have seen it as a long shot and insisted on a quicker return on their investments. The development of the Xerox machine took place under similar, supportive circumstances.[34] Thus if ideas are to become viable products, they must arise within a social context that is favorable to their exploitation. Without such a context, they may remain solutions without problems.

Settings in the Inventive Process

The R&D Department. Special departments carry out most of the invention in private industrial firms. Although in principle all employees of a firm could be considered responsible for inventing new things, in fact most large firms have special departments that are expected to produce the inventions the firm uses.[35] These research and development (R&D) departments are usually staffed with engineers and applied scientists. Numerous studies of these departments have been carried out because engineering creativity is so important for such firms.[36]

One of the best accounts of a highly creative R&D team is Tracy Kidder's *The Soul of a New Machine*. *Soul* is a portrait of computer engineers struggling to produce a usable product in record time.[37] A computer is so complicated that two teams are required, one to design the hardware and another to design the software. Kidder shows the enormous amount of effort that sometimes goes into a sophisticated commercial product like a computer; a considerable amount of effort must be spent just on getting the "bugs" out of the product before it can be sold. Although strictly speaking a new computer is not an invention per se, a good deal of invention goes into making it, and various parts of such a new computer are likely to receive patents. As we will see in the next chapter, the line between invention and design is not a sharp one.

A large volume of R&D is also carried out by the government, much of it concerning military systems. The U.S. Navy alone has roughly a dozen "laboratories" that carry out research and development work, from the most basic research to the most pedestrian improvements to already-used systems. Some of these "labs" have 5,000 people working in them—engineers, technicians, and supporting personnel.

Sometimes invention is necessarily a group matter because of the size and scale of the project in question. The first atomic bomb was designed by a very large group of people in the so-called Manhattan Project.[38] J. Robert Oppenheimer provided charismatic leadership for this group of people, but most of the invention was done by teams of scientists.[39] Getting such groups to be productive is a major problem because the results are considered so important and research on this scale is very expensive.[40]

In other instances a group may not be necessary in principle, but the ideas of others are helpful in perfecting the design. Others' input may be extremely important in spotting problems, developing novel aspects, and providing critiques. This kind of group creativity is apparent in industrial design groups set up to develop product innovations. The basic method for running such groups was developed by Alex Osborn, who called it "brainstorming."[41] Osborn found that a group that encouraged people to produce ideas with relatively few constraints would help them discover novel methods and products. Discredit was thrown on brainstorming during the 1940s when some studies suggested that individuals could do just as well alone as they could in a group. But later studies have clarified the situations in which group strategies may be quite valuable.[42] Other popular approaches are "Synectics," developed by William J. J. Gordon, and "Lateral Thinking," made popular by Edward de Bono.[43] Many of these group problem-solving techniques have been used in industry for decades.

"Skunk Works." Of course, much R&D work is routine—it's "just a job."[44] But it may be useful for us to consider some operations that have a very different spirit, that are much more like highly creative teams of scientists who make basic discoveries. Some of these operations are large, like the Manhattan Project, but some are relatively small, like the "skunk works" concept.

A "skunk works" can be defined as a small (often less than 50 people), high-performance innovation team. Its members act in ways that are very different from those common in routine engineering. The phrase originally comes from a comic strip, Al Capp's *Li'l Abner*. The skunk works was a particularly smelly factory in the comic strip, a stereotype for many industrial and chemical "process" plants. The term was first applied to a real operation at Lockheed Aircraft, where an unusual department had been set up to produce the first operational U.S. jet fighter in 1943. This department, under the direction of Clarence "Kelly" Johnson, was developing the jet in a makeshift structure with walls made of packing crates and a roof made from a circus tent. The tent was next to the Lockheed wind tunnel in literally the only space available for this important effort. Irving Culver, one of the 23 engineers Johnson "stole" from other parts of Lockheed, used the phrase during a telephone call, and it stuck.

The Skunk Works was destined for great things. The first prototype of the P-80 "Shooting Star" was produced in 143 days, a record time. And it was an excellent jet fighter. Other Skunk Works projects also became famous: the U-2 and SR-71 spy planes, the C-135 Hercules transport, and the F-104 Starfighter. What is more significant is that many of these airplanes were produced in record time, with cost *under-runs* (a rarity in military development) and small numbers of personnel.[45] The Skunk Works became a model for small, lean, high-performance teams in research and development.[46] Today many industrial "skunk works" exist in addition to those in military laboratories.

One of the things that Kelly Johnson had hoped to accomplish in creating the Skunk Works was a close relationship between designers and users (see the next chapter). He felt he could streamline the inventive process so that it could be done in shorter time and at less cost. Other engineers have also used this approach. One of them was William B. McLean, a physicist at the Naval Ordnance Test Station (NOTS) in California's Mojave Desert. NOTS had been developed after World War II as a government laboratory to test naval weapons.[47] It was no ordinary laboratory, however. It was designed to encourage maximum creativity in the same way that the Office of Scientific Research and Development had done during World War II.[48] When William McLean, a young scientist from the National Bureau of Standards, was posted there in 1945, he rapidly began to develop a project that became a technological star—the Sidewinder missile. As we saw in the previous chapter, the Sidewinder was developed with a team of about 30 engineers in a "skunk works" atmosphere under McLean's direction. As weapons systems go, it is a remarkably simple, inexpensive, and reliable mechanism, and it has achieved something of a legendary status.

Although McLean had been trained as a physicist, he showed a remarkable aptitude for "gadgeteering." McLean's intelligence and gift for creative engineering was evident to all who worked with him.[49] The Sidewinder had been born in McLean's mind when he worked for the National Bureau of Standards during World War II. His work on a large, slow missile, called the Bat, generated a philosophy further refined by his experiences concerning what a missile ought to do. When McLean was sent to NOTS, he set about developing the small team that worked on the problem. McLean noted that

> communications were facilitated by the fact that the working group was isolated in a small community in the desert about 150 miles from the nearest large city. People could and did communicate with each other all day, through the cocktail hour, and for as long as the parties lasted at night. This isolation

in a location where the job could be performed provided large measures of the intimate communication which is so essential for getting any major job completed.[50]

Another factor was the congeniality of McLean's small team. Work after normal working hours was common on the Sidewinder project. Even though McLean was not a natural leader, he inspired admiration from most of his subordinates because of his genius and technical excellence, very important features in a technical leader.[51] Mutual respect, trust, and confidence inspired the continuous effort needed to complete the project in the face of technical difficulties and external resistance. The resistance came from a military establishment that had already decided it wanted something else: a larger, more expensive, radar-guided missile, instead of the small, infrared-guided one that McLean was trying to produce.

McLean's team persevered in the face of this official resistance. At one point, the Sidewinder project was cancelled by Washington because "studies" showed that only an all-weather missile would do. The project, however, was continued in a clandestine manner and eventually emerged again as an open project. In 1953 the first successful test of the missile took place, and in 1956 it was put into general service by the U.S. Navy. Shortly thereafter, it was adopted by the Air Force. Versions of Sidewinder were still operative and considered superior in 1988, and it is still the least expensive air-to-air missile in the U.S. inventory. Countries all over the world have designed their own "Sidewinders" or copied the United States's version. During the Falklands conflict between Great Britain and Argentina, the British fired some 26 Sidewinders; 19 Argentinian planes were shot down, a very high success rate for such weapons.[52] Thirty years after the Sidewinder's introduction, it is still a high-performing military system.

A number of important products have been produced in such a "skunk works" manner. The IBM Personal Computer was produced for IBM at a "skunk works" in Boca Raton, Florida. The novel SWATH ship design, which uses a special tubular hull system, was developed by a "skunk works" at the Naval Ocean Systems Center.[53] The 3-M Corporation is well known for encouraging the development of "skunk works" by telling employees to "steal" 15 percent of their time to work on self-initiated projects.

Aspects of the Social Process of Invention

The role of the inventor is to advance the state of the art, to use current knowledge to develop new things, or to develop new knowledge. Some inventors take items in the experimental stage and move them into the black box stage. Very original inventors, of course, not only do this but dream up totally new systems. In any case, the role of the inventor is to advance the state of the art.

To a large extent, it is this characteristic that differentiates the inventor from the architect or the creative engineer. Architects are highly creative people, and in a sense each new building is an invention.[54] But generally a new building does not advance the state of the art. Only buildings that represent truly novel principles make an architect an inventor. Similarly, engineers can often be quite creative in their designs. But only if they advance the state of the art do they become considered inventors. Admittedly, the dividing line is a fine one, but the distinction is important. We will say more about this point in the next chapter.

Multiple Independent Invention. We have seen that inventors' primary activity is to generate a stream of inventive ideas. These streams are tested by experiments and, increasingly today, by simulations. What is surprising is that these inventive streams seem, on occasion, to arrive at the same result independently. This is shown in the interesting phenomenon called "multiple independent invention."

William F. Ogburn first called attention to this simultaneous occurrence of apparently identical inventions. He pointed out that according to the history textbooks of different countries, a citizen of each country seems to have invented the steam engine, the airplane, the motor car, and so forth. He compiled a list of some 148 inventions and discoveries that had been made independently by two or more people.[55] What is involved, of course, is that often several countries claim that *their* inventor created the invention in question. This kind of claim is possible because frequently several people do seem to have gotten the same idea at the same time and to have developed it into a workable device. To Ogburn, these multiple occurrences meant that inventions came into being when "the time was ripe," rather than resulting from the random appearance of a "great man" or genius.

One example of such a "multiple invention" is the miner's safety lamp. Carrying a lighted candle in a coal mine is very dangerous because if the candle contacts a pocket of methane gas, a very serious explosion may be caused. In 1815 both Humphrey Davy, an eminent scientist, and George Stephenson, a talented inventor with little formal education, worked on and invented serviceable safety lamps for miners. As we might expect, because both lamps used similar principles, a considerable dispute arose over who had invented first. Public opinion naturally saw Davy, the more eminent man, as the inventor, but the evidence suggests that each person developed the invention independently.[56] How is such multiple, and ostensibly independent, invention possible?

Before we answer this question we must make an elementary but important distinction between *devices* and the *principles* upon which they are based. Consider the electrical telegraphs invented by the team of Wheatstone and Cooke in England and by Morse in the United States. In the sense that they developed telegraphs that transmitted signals by electricity, they invented the same thing. But in actual configuration, the devices they invented were quite different.[57] The signals a Wheatstone device transmitted moved a pointer; those the Morse device transmitted came out in bursts of sound we call dots and dashes. The telegraphs were thus similar, but not identical. Most "simultaneous independent inventions" are of this type: the same principle, but different details.

There are a few inventions in which identical, or at least astonishingly similar, devices have been independently invented. An example is the Pelton water wheel, a prime mover that uses a jet of water to turn distinctive cups mounted around a wheel.[58] In this case the device was relatively simple; when the device is complex, the chance that an identical device will be independently invented is smaller. In fact, we might advance as a general hypothesis the idea that *the more complex the device, the smaller the probability of an identical independent invention of the same device.*

Consider, for instance, the digital computer, invented independently in Germany, England, and the United States.[59] The computer was invented in Germany in 1938 by Konrad Zuse, whose Z-1 used electromechanical relays. It was invented in England by Alan Turing to decode German codes in 1943 and was named the Colossus (5,000 electronic tubes). In the United States, an electromechanical version (the Mark I) was built by Howard Aiken in 1943, and an electronic version was produced by Eckert and Mauchly (ENIAC, 18,000 tubes) in 1946. The British and German versions were kept secret until after World War II, and in fact Colossus's existence was kept secret until long after it was obsolete. Early histories of the computer omitted mention of Colossus, even though Turing's role in computer science otherwise was well known.

What is the likelihood that two independently invented digital computers would be identical? Virtually zero, most would think. And in fact the details of these four computers were very different. Yet we are still faced with the question of why they independently came into existence at about the same time. The answer to the question involves three factors:

1. *A common base of knowledge.* All three countries shared a common base of knowledge in mathematics, computing, and machinery. All four teams therefore could draw upon the same overall knowledge available in technical journals, books, and other sources. At least before the war, publications describing principles of these machines were freely available to the mathematical community.

2. *A similar competitive situation.* The three countries were involved in the same highly competitive situation in regard to military capability.[60] It is no accident that in almost every case computer development was funded by the military and that its primary use was to be in such military fields as aeronautics, ballistics, and cryptography.

Together, factors 1 and 2 explain the similarity in cultural contexts that generated the inventions: Similar abilities and similar needs led to similar devices.

3. *Different orientations and resources.* The computers were different, however, because the inventors had different guiding ideas and different financial and material resources. Konrad Zuse would have used electronic tubes for his Z-1, had he been able to afford them. The British Colossus was basically a single-purpose device designed largely for its primary role: deciphering codes. The Mark I, a huge dinosaur of a computer put together with aid from IBM, did not

use tubes because they were considered unreliable. As a result, it was hundreds of times slower than ENIAC, which did use tubes and was finished only three years later.

Multiple inventions, then, are likely to have both the similarities that make them multiple and significant differences in detail. But the large premium that inventors place on originality and the importance of invention to national pride mean that disputes over priority are likely to be long and fiercely fought. Although these disputes are interesting as sociological phenomena, they need not further detain us here.[61]

The existence of multiple inventions indicates that many times inventors' streams of ideas seem to arrive at a similar point independently. This does not make an invention inevitable in the strictest sense of the word. But the appearance of an invention that uses an important principle—such as steam power, jet propulsion, or ultra-high-speed centrifugation—does appear to be reasonably predictable. It is not that inventors are unimportant; in fact, the brighter ones are particularly important, for bright inventors discover things earlier.[62] But it is true that most ways to do things will be discovered in the long run.

It is also true that an inventor can make a unique choice that is not at all inevitable and that affects the technology for decades. This is true of Sholes's QWERTY keyboard for the typewriter.[63] The choice to use alternating instead of direct current was also an important, noninevitable decision.[64] Computer designers, because the things they invent are so complex, are likely to leave their particular stamp on the equipment they invent.[65] For software, the uniqueness factor is even more important. The invention of Fortran by John Backus was a unique decision that shaped programming for over a quarter-century.[66] Ironically, Konrad Zuse had already developed a method superior to Fortran during World War II, but he did not reveal it to the Allies after he was captured because he did not like the treatment he received at the hands of his captors.[67] The results of his withholding this method and the subsequent use of the less-efficient Fortran instead illustrate that individual inventors are indeed important.

Directed Technological Advances. During the 1950s, when I was growing up, I got the impression that technology just automatically advanced—that one invention led to another in a seemingly endless stream that no one needed to push faster. This general process was called "progress," and although it might occasionally require an Edison or a Ford somewhere along the line, it was unstoppable. This philosophy was taught, as we have seen in a previous chapter, by William Fielding Ogburn and his school. Close study of the history of technology, however, suggests that the process is neither automatic nor fixed in sequence. It can be directed and focused by market conditions and by the decisions of government bureaucracies, and it can flow at very different rates. And finally, once a particular configuration is decided upon, this decision may have a strong influence on everything that follows.

Social Demands. One factor influencing the flow of technological advancement is social demand.[68] Groups in society with strong needs may provide incentives to inventors to develop desired inventions. These incentives may be implicit in the situation or may be consciously promoted. They may be implicit, for instance, when some technical problem—such as food spoilage—reduces profits. The inventor is sure to have an interested audience if he or she can figure out a way to keep the food from spoiling—for example, by the invention of a refrigerated railroad car. In other cases the problem may have more far-reaching public implications. Louis Pasteur's invention of the vaccines for anthrax and rabies is an example of research directed at the solution of social problems at the urgings of interested social groups.[69]

Fame and financial gain are thus powerful incentives for some inventors, especially those who can create inventions for which there is already a brisk demand. But demand can also be stimulated. Many inventors are likely to tackle problems because their solutions appear to be urgently wanted by others. This was true of Elmer Sperry, for instance, who scanned the *Official Gazette* of the U.S. Patent Office to identify potentially lucrative inventions.[70]

Pasteur's inventions vividly illustrate how social needs can stimulate intense efforts to develop solutions.[71] A great many of Pasteur's most famous scientific discoveries took place under the pressure of a clearly identified social need to which Pasteur was personally directed by those with particular concerns. For instance, his investigations into problems concerning the silkworm industry, the beer industry, and anthrax in cattle were all direct responses to pressing social needs. In some of these cases, such as the silkworm problem, Pasteur engaged in work that today we would call "operations research." Although such research is really applied science or engineering, rather than fundamental scientific work, many of Pasteur's basic discoveries also came from these investigations. Pasteur not only helped the industry in each case, he also developed new scientific principles. A good many medical discoveries have been made under similar pressures. For instance, the Salk vaccine for polio was invented under similar circumstances.[72]

Yet another impressive example of this search for a solution was the struggle to find an alternative to the Houdry catalytic cracking process in the petroleum industry. Houdry had sought help from the large petroleum companies to develop his catalytic cracking process. Few companies were interested in helping him out, so when he succeeded in developing it, he avenged himself by charging high license fees for the use of his patents. The companies did not wish to pay these fees, so they decided to try to find a method that would work without using his ideas. The research staffs of six corporations spent six years and $15 million before they invented an alternative, the fluid catalytic cracking process.[73] The intensity of this research effort has been compared to the effort to invent the atomic bomb. Today, the United States is using a similar research drive to invent space weapons as part of its Strategic Defense Initiative, commonly known as "Star Wars."

Rewards. Sometimes rewards have been offered to inventors in hopes that a required device will be invented. In the seventeenth century, for example, one of the most serious problems of navigation was an inability to determine a ship's longitude at sea. The longitude of a place is the number of degrees it lies east or west of Greenwich, England. Many a ship was lost on the rocks or steered hundreds of miles off course because only the latitude, which is easily determined by the position of the sun, could be known with certainty.[74] Determining the longitude required an accurate clock, so that a navigator could compare the local time (determined from the sun's position in the sky) with the time at some point of reference (in this case, Greenwich, England), which would then allow the navigator to determine how far away his ship was from this point of reference. The British government was so concerned about this problem that it offered a large reward to anyone who could invent an accurate clock, or chronometer. This reward, made law in 1714, included a sliding scale to encourage accuracy:

£10,000 for any method of determining a ship's longitude within one degree

£15,000 if it determined longitude within two-thirds of a degree (40')

£20,000 if it determined longitude within half a degree (30').[75]

It is telling that in spite of the enormous sums offered—consider the date—it required 45 years for even the least of these awards to be won. The state of the art did not permit more rapid progress. In fact, some were convinced that the answer was not a clock at all, but better astronomy. Isaac Newton was convinced (in 1721) that "this improvement must be made at land, not by watchmakers or teachers of Navigation or people that know not how to find the Longitude at land, but by the ablest Astronomers."

It must have been disturbing to many of these ablest astronomers, then, when the solution appeared from the hands of a mere Yorkshire clockmaker, John Harrison. After a considerable period of development, Harrison was finally given the reward in 1759.[76] This is a splendid example of an explicit demand for technology driving the progress of innovation. Similarly, Paul MacCready has credited the incentive of the £50,000 Kremer prize with being a major motivation for his invention of the first successful human-powered aircraft, the *Gossamer Condor*.[77]

Patents. Progress in innovation may also be encouraged by the protection afforded by patents. Governments offer inventors patents to provide a temporary period—17 years in the United States—of monopoly on the use of their patented inventions. Getting a patent not only confirms the inventor's originality, but confers a potentially important financial reward on the winner of the patent. Unhappily, much of the inventor's resources may be taken up fighting to have the patent enforced against unscrupulous competitors. Such a battle, for instance, is still taking place over the invention of the laser.[78] During World War II, when the National Inventors Council sought from inventors ideas that might aid in the war effort, Bert Adams submitted a design for a battery that used water instead of acid as its electrolyte. He hoped that the agency interested, the Army, would recognize the value of his invention and compensate him for it. The Army accepted and used Adams's ideas but never told him about it. After 25 years of struggle and $200,000 in legal costs, the U.S. Supreme Court ruled that Bert Adams had invented the water battery and that the Army had not been fair in its dealings with him. A year before his death at the age of 66, he received a $2.5 million settlement for his claims.[79]

That patenting confers certain advantages is, however, not the whole story. Many experts claim that patents are not particularly important in getting inventors to invent.[80] Many people have pointed out that inventors would often invent in the absence of patents or in situations in which patents do not apply. This argument should be tempered, however, by the recognition that having some kind of monopoly is of great importance to *firms* developing the invention, whether this monopoly comes about because of a patent, a well-kept industrial secret, difficulty of entry into the market, or an arrangement between firms (whether legal or not). A great advantage that Japan has had in the development of its innovations is a highly negotiated economy in which companies are designated by the Ministry of International Trade and Industry to develop particular products. The de facto monopoly may be an important incentive for firms to undergo the expensive research and development needed to develop these new products. In the United States, by contrast, such developments are left to the initiative of private firms. In a variety of fields—high-definition television, photovoltaics, and superconductors—these initiatives have been insufficient, and the market is being lost to Japan.

Reverse Salients. Just as the efforts of inventors are often focused from the outside, the inventors themselves sometimes choose problems for which the solutions will be socially and financially important.[81] Inventors' attention is thus directed toward problems that Thomas Hughes calls "reverse salients"—those parts of a productive process that provide problems for the entire system, the sticking points that seem to hold everything up.[82] The phrase comes from military practice and refers to a certain area that has remained behind the lines and under enemy control after a general advance. John Hobson described such problems in the cotton industry:

> The pressure of industrial circumstances direct the intelligence of many minds towards the comprehension of some single central point of difficulty, the common knowledge of the age induces many to reach similar solutions: that solution which is slightly better adapted to the facts or "grasps the skirts of happy chance" comes out victorious. . . .
>
> Thus in a given trade where there are several important processes, an improvement in one process which places it in front of the others stimulates invention in the latter, and each in its turn draws such inventive intelligence as is required to bring it into line with the most highly-developed process. Since the later inventions, with new knowledge and new power behind them, often overshoot the earlier ones, we have a certain law of oscillation in the several processes which maintains progress by means of the stimulus constantly applied by the most advanced process which "makes the pace." . . . As the invention of the fly-shuttle gave weaving the advantage, more and more attention was concentrated upon the spinning processes and the jenny was evolved; the deficiency of the jenny in spinning warp evolved the water-frame, which for the first time liberated the cotton industry from dependence on linen warp: the demand for finer and more uniform yarns stimulated the invention of the mule. These notable improvements in spinning machinery, with their minor appendages, placed spinning

ahead of weaving, and stimulated the series of inventions embodied in the power loom. . . .[83]

Because solving the technical imbalances posed by such reverse salients is likely to be well rewarded, inventors naturally focus their attention on such points, which they define as critical problems. We have already seen a major example of such focusing in the invention of the fluid catalytic cracking process noted above.

A good deal of inventive activity, then, is focused by explicit or implicit bottlenecks in productive processes and other social activities. Whereas in some cases formal rewards are offered for solutions to these problems, in other cases inventors realize that a solution would be extremely valuable. "Better mousetraps" do pay off, sometimes handsomely, but they can also be unexploited by a society that is indifferent to them. Just as prepared minds are necessary to seize on the important observation, a prepared society is necessary to seize on a proffered invention, particularly if it is complicated.

Technology "Push" or Demand "Pull?"

A long debate has taken place over the respective roles that technology's internal evolution and social demand play in the genesis of invention: Is technology pushed from within or pulled from without? We have already discussed a number of instances in which a preexisting demand focused social efforts on invention. It is equally easy to find inventions for which uses were not immediately obvious. For instance, Col. John T. Thompson, who invented a submachine gun (the "tommy gun"), had a great deal of difficulty finding ready customers for it.[84] Similarly, when Edison invented the phonograph, he had no idea that it would become widely used for entertainment.[85] We have already noted the difficulties Spencer Silver had in finding a use for the monomer that eventually stuck Post-It Notes firmly into the modern office. So there is no question that in some cases inventions predate knowledge of their major uses.

The big question, though, is this: What do the overall patterns show? Does technological change come about because the processes of research lead to unforeseen results, or because market or similar forces cause people to search for certain effects? This question was addressed most strongly by Jacob Schmookler, who showed that patents seemed to increase concomitantly with heavy investment in a given industry.[86] This is how Schmookler visualized the process: As money flowed to a particular industrial sector, inventive creativity was focused on that sector's technological problems. As successes came from this research, they then became evident in the form of patents. Thus increasing *demand* for the sector's products led to increased inventiveness.

Critics have noted several problems with Schmookler's argument.[87] The first and most basic difficulty is using patents as a measure of invention. There is almost always a much larger number of patentable ideas than there are patents. And there is considerable evidence that the percentage of these inventions that are in fact patented is not constant.[88] If this is true, it may well be that during massive investment surges inventions of a certain kind are more likely to be patented because they are worth more money. Further, if Schmookler's theory is correct, there should be a time lag between investment and patenting, but instead patenting and investment seem to increase simultaneously in Schmookler's data. This may mean that investment is a response to patenting, rather than the reverse. It is certainly true today that getting a patent may be a first step in making a major investment and that "getting a patent position" may be a key to successfully bringing a product to market. Finally, some evidence suggests that companies often exploit an invention months before the patent is taken out.

Unquestionably, demand is important for inventions; but development of basic knowledge is often critical, too. An important theory of the roles of basic research and market forces was developed by William Abernathy and James Utterback.[89] They suggest that in the early phases of technology, basic research is more important. As the technology gets going, however, demand-induced (applied) research for refining the product becomes more important. Finally, as the product nears its limits of improvement, the processes that go into manufacturing it become the subject of further inventiveness. Thus science-push may be more important in the beginning, but demand-pull may become important later on.

Some careful empirical studies have emphasized that scientific and technological developments are just as important as the demand for invention. For instance, in their examination of the commercial aircraft industry between 1925 and 1975, David Mowery and Nathan Rosenberg showed that supply also played an important role.[90] Government policies in particular played a strong role in promoting the industry, both in providing important technological expertise (via research for warplanes) and in providing an environment in which demand was high. A study of the chemical industry by Vivien Walsh showed that whereas demand played an important role, so did basic research, especially in stimulating "breakthrough" inventions.[91] Similarly, a study of the synthetic rubber industry by Noreen Cooray showed that the accumulation of knowledge that led to inventions was a product both of the industry's demand and of basic scientific research.[92]

Research and theory, then, suggest that both the progress of science and pressures of the market lead to inventions. Although many of the more fundamental inventions come out of scientific research, not all do. It is also true that although relatively little market demand may exist before truly novel inventions appear, demand may develop rapidly after the introduction of the inventions into use. But market pressures may focus inventive attention on certain kinds of technological "sticking points," and such pressures can be very valuable for increasing the efficiency of the processes by which products are made. Thus both supply and demand are key forces in stimulating invention.

Conclusion

Invention is a social process. Although the independent inventor may be personally isolated, he or she is not culturally isolated, but instead participates in rich surroundings teeming with ideas and opportunities. Much invention, furthermore, takes place in a group setting and is often consciously aimed at achieving certain results. Whereas some inventors resemble "hackers," others are extremely systematic, and in fact some of the most interesting inventions (say the lunar landing system) are the products of immense R&D teams. Society, then, is constantly involved in the inventive process to such an extent that sometimes separated inventors develop very similar yet distinct inventions that differ only in the details. And the particular configuration of parts the inventor develops is important in the way the system works, how efficient it is, how it operates with other systems, and so on. In the next chapter on design we will examine yet more influences of creative style on the shape of technological products.

Notes

1. Doug Garr, *Woz: Prodigal Son of Silicon Valley* (New York: Avon, 1984); and Courtney G. Brooks, James Brooks, and Loyd Swenson, Jr., *Chariots for Apollo: A History of Manned Lunar Spacecraft* (Washington, D.C.: NASA, 1979).

2. Bruno Latour, *Science in Action* (Milton Keynes: Open Universities Press, 1986).

3. Some technological advances occur without any scientific breakthroughs; consider Walter Vincenti, "Technological Knowledge Without Science: The Invention of Flush Riveting in American Airplanes, ca. 1930—ca. 1950," *Technology and Culture,* Vol. 25, No. 3 (July 1984), pp. 540–576.

4. See Melvin Kranzberg, "The Unity of Science-Technology," *American Scientist,* Vol. 55, No. 1 (1967), pp. 48–68; and "The Disunity of Science-Technology," *American Scientist,* Vol. 56, No. 1 (1968), pp. 21–34.

5. Thomas J. Allen, *Managing the Flow of Technology: Technology Transfer and the Dissemination of Technological Information Within the Research and Development Organization* (Cambridge, Mass.: MIT Press, 1978), pp. 141–181; and Lars Hoglund and Olle Persson, "Communication Within a National R&D System: A Study of Iron and Steel in Sweden," *Research Policy,* Vol. 16 (1987), pp. 29–37.

6. R. G. A. Dolby, "The Transmission of Science," *History of Science,* Vol. 15 (1977), pp. 1–43.

7. D. S. L. Cardwell, *From Watt to Clausius: The Rise of Thermodynamics in the Early Industrial Age* (Ithaca, N.Y.: Cornell University Press, 1971).

8. Philip M. Morse, *Queues, Inventories and Maintenance* (New York: Wiley, 1958).

9. David Edge, "Technological Metaphor," in D. O. Edge and J. N. Wolfe (Eds.), *Meaning and Control: Social Aspects of Science and Technology* (London: Tavistock, 1973).

10. Daniel Kevles, *The Physicists: The History of a Scientific Community in Modern America* (Cambridge, Mass.: Harvard University Press, 1987); and Ronald Clark, *The Rise of the Boffins* (London: Phoenix House, 1962).

11. David Edge and Michael Mulkay, *Astronomy Transformed: The Emergence of Radio Astronomy in Britain* (New York: Wiley, 1976), pp. 10–11.

12. Alberto Cambrosio and Peter Keating, "Going Monoclonal: Art, Science, and Magic in the Day-to-Day Use of Hybridoma Technology," *Social Problems,* Vol. 35, No. 3 (June 1988), pp. 244–260.

13. Howard Wilcox, personal communication.

14. Often U.S. firms will develop the first products or prototypes, but other countries will refine and improve the product much more successfully; see Edwin Mansfield, "Industrial Innovation in Japan and the United States," *Science,* Vol. 241 (Sept. 30, 1988), pp. 1769–1774.

15. An outstanding and exhaustive review of this topic is provided by Edward B. Roberts, in "Managing Invention and Innovation," *Research Technology Management,* Vol. 31, No. 1 (Jan.-Feb. 1988), pp. 11–29.

16. Thomas Parke Hughes, *Networks of Power* (Baltimore: Johns Hopkins University Press, 1983), p. 26.

17. Gerald Gordon, "The Identification and Use of Creative Talent in Scientific Organizations," in Calvin Taylor (Ed.), *Climate for Creativity* (New York: Pergamon Press, 1972), pp. 109–124.

18. Among the chemists, Gordon found that the most innovative individuals were those who scored high on both the Remote Associates Test and on differentiating the work of others. These people, Gordon suggested, were good at both finding problems and solving them. They had nearly double the rate of patent disclosures, applications, and actual patents than that of the next highest category—those with high RAT but not high-DIFF scores. Those who scored low on both RAT and DIFF had few disclosures and no patents.

19. Allen, *Managing the Flow,* pp. 141–181.

20. John H. Dessauer, *My Years with Xerox: The Billions Nobody Wanted* (New York: Manor Books, 1971), pp. 41–42.

21. Gary Jacobson and John Hillkirk, *Xerox: American Samurai* (New York: Macmillan, 1986).

22. The classic article on product champions is Donald A. Schon, "Champions for Radical New Inventions," *Harvard Business Review,* Mar.-Apr. 1963, pp. 77–86.

23. See William A. Fischer, Willard Hamilton, Curtis P. McLaughlin, and Robert Zmud, "The Elusive Product Champion," *Research Management,* May-June 1986, pp. 13–16.

24. Gifford Pinchot III, *Intrapreneuring: Why You Don't Have to Leave the Corporation to Become an Entrepreneur* (New York: Harper & Row, 1985).

25. And also, of course, inside government; see Eugene Lewis, "Admiral Hyman Rickover: Technological Entrepreneurship in the U.S. Navy," in Jameson W. Doig and Erwin C. Hargrove (Eds.), *Leadership and Innovation: Biographical Perspective on Entrepreneurs in Government* (Baltimore: Johns Hopkins University Press, 1987), pp. 96–124.

26. Admiral Sir Percy Scott, *Fifty Years in the Royal Navy* (New York: Doran, 1919), p. 95.

27. Peter Padfield, *Aim Straight: A Biography of Admiral Sir Percy Scott* (London: Hodder and Stoughton, 1966), p. 136.

28. P. Ranganath Nayak and John M. Ketteringham, *Breakthroughs: How the Vision and Drive of Innovators in Sixteen Companies Created Commercial Breakthroughs that Swept the World* (New York: Rawson, 1986), pp. 50–73.

29. Jordan D. Lewis, "Technology, Enterprise, and American Economic Growth," *Science,* Vol. 215 (Mar. 5, 1982), pp. 1204–1211. This is a very valuable article, and it deserves more attention than it has gotten. I recommend it to anyone interested in this topic.

30. Quoted by T. A. Boyd, *Professional Amateur: The Biography of Charles Franklin Kettering* (New York: Dutton, 1957), p. 134. See another version of this saying in Michael E. Wolff, "What to Do About 'NIH'," *Research Management,* Jan.-Feb. 1987, pp. 9–11.

31. Lewis, "Technology," p. 1204–1211.

32. Lester Thurow, "A Weakness in Process Technology," *Science,* Vol. 238 (Dec. 18, 1987), pp. 1659–1663.

33. Eric N. Barbour, "Pilkington Float Glass (A)," Harvard Business School Case Study Number 9-672-069 (Boston: Harvard Business School, 1971); see also Gregory H. Wierzynski, "The Eccentric Lords of Float Glass," *Fortune,* Vol. 78, No. 1 (July 1968), pp. 91, 92, 121–124.

34. Dessauer, *My Years with Xerox,* p. 63.

35. Joseph Rossman, "Stimulating Employees to Invent," *Industrial and Engineering Chemistry,* Vol. 27, No. 11 (Nov. 1935), pp. 1380–1386, and Vol. 27, No. 12 (Dec. 1935), pp. 1510–1515; see also Mark Shepherd, Jr. and J. Fred Bucy, "Innovation at Texas Instruments," *Computer,* Sept. 1979, pp. 82–90.

36. A useful review of this topic is Frederick Betz, *Managing Technology: Competing Through New Ventures, Innovation, and Corporate Research* (Englewood Cliffs, N.J.: Prentice-Hall, 1987).

37. Tracy Kidder, *The Soul of a New Machine* (Boston: Little, Brown, 1981). For follow-ups on this case, see Mitchell Lynch, "At Data General, a Subdued Soul," *New York Times,* June 26, 1983, Business Section; and William M. Bulkeley, "Venturing Out: Computer Engineers Memorialized in Book Seek New Challenges," *Wall Street Journal,* Sept. 20, 1985, pp. 1, 16.

38. Richard Rhodes, *The Making of the Atomic Bomb* (New York: Simon & Schuster, 1986).

39. Nuel Pharr Davis, *Lawrence and Oppenheimer* (New York: Simon & Schuster, 1968); see also Richard Feynman, "Los Alamos from Below," in Lawrence Badash (Ed.), *Reminiscences of Los Alamos 1943–1945* (Dordrecht, Holland: Reidel, 1980), p. 106.

40. Hans Mark and Arnold Levine, *The Management of Research Institutions: Looking at Government Laboratories,* NASA SP-481 (Washington, D.C.: U.S. Government Printing Office, 1984).

41. Alex F. Osborn, *Applied Imagination: Principles and Procedures of Creative Thinking,* 2nd ed. (New York: Scribner, 1957).

42. For a complete review of group creativity, see Morris Stein, *Stimulating Creativity, Vol. II: Group Procedures* (New York: Academic Press, 1975).

43. William J. J. Gordon, *Synectics: The Development of Creative Capacity* (London: Collier Books, 1961); and Edward de Bono, *Lateral Thinking: Creativity Step by Step* (New York: Harper & Row, 1970).

44. An excellent series of portraits of ordinary R&D operations appears in six installments in *IEEE Spectrum:* Tekla S. Perry, "How Does Your Office Affect Your Work?" Oct. 1983; "Company A: A Friendly Place, But . . . ," Dec. 1983; "Company B: The Competitive Edge," Feb. 1984; "Company C: Quiescent and Secure," Apr. 1984; and "Company D: A Fortress Undisturbed," May 1984; and Douglas Carmichael, "Do EE Careers Suffer from Poor Management?" July 1984.

45. How far this is from usual military practices can be appreciated by knowing that in many cases the documentation submitted to get an airplane contract is roughly the weight of the airplane itself; see Norman Augustine, "Paper Airplanes," in *Augustine's Laws* (New York: Viking Press, 1986), pp. 245–252.

46. Clarence L. Johnson with Maggie Smith, *Kelly: More Than My Share Of It All* (Washington, D.C.: Smithsonian, 1985); see also John Tierney, "The Real Stuff," *Science 85,* Sept. 1985, pp. 24–35.

47. See two volumes on the history of the Naval Weapons Center, China Lake: Albert Christman, *Sailors, Scientists, and Rockets* (Washington, D.C.: U.S. Government Printing Office, 1971); and J. D. Gerrard-Gough and Albert Christman, *The Grand Experiment at Inyokern* (Washington, D.C.: U.S. Government Printing Office, 1978).

48. For a well-written popular history of OSRD, see William P. Baxter III, *Scientists Against Time* (Boston: Little, Brown, 1946).

49. For instance, Jacob Rabinow, a prolific inventor himself, describes McLean, who had a Ph.D. but no engineering degree, as the "best engineer I ever knew."

50. William B. McLean, "The Sidewinder Missile Program," in F. Kast and J. E. Rosenzweig (Eds.), *Science, Technology, and Management* (New York: McGraw-Hill, 1962), pp. 166–176.

51. The importance of a "hands-on" style of technical leadership is emphasized by Arthur Squires in *The Tender Ship: Governmental Management of Technological Change* (Boston: Birkhäuser Boston, 1987); see also Gerald M. Weinberg, *Becoming a Technical Leader* (New York: Dorset House, 1986).

52. Jeffrey Ethell and Alfred Price, *Air War South Atlantic* (London: Sidgwick and Jackson, 1983), p. 215.

53. Thomas Lang, "*SSP Kaimalino:* Conception, Developmental History, Hurdles and Success," A.S.M.E. Document 86-WA/HH-4 (New York: American Society of Mechanical Engineers, 1986).

54. Henry Petroski, "Engineering as Hypothesis," in his *To Engineer Is Human* (New York: St. Martin's Press, 1985).

55. William Fielding Ogburn and Dorothy S. Thomas, "Are Inventions Inevitable?" *Political Science Quarterly,* Vol. 37, No. 1 (Mar. 1922), pp. 83–98.

56. Samuel Smiles, *The Life of George Stephenson* (Boston: Ticknor and Fields, 1858), pp. 99–131.

57. Waldemar Kaempffert, *A Popular History of American Invention,* Vol. I (New York: Scribner, 1924), pp. 286–299.

58. Edward Constant, "On the Diversity and Co-Evolution of Technological Multiples," *Social Studies of Science,* Vol. 8, No. 2 (May 1978), pp. 183–210.

59. Christopher Evans, *The Making of the Micro: A History of the Computer* (New York: Van Nostrand, 1981), pp. 64–79.

60. Jet planes were also being developed quasi-independently in Germany and England; see

Edward Constant, *The Origins of the Turbojet Revolution* (Baltimore: Johns Hopkins University Press, 1980).

61. See, for instance, Robert Merton, "Multiple Discoveries as a Strategic Research Site," in Norman Storer (Ed.), *The Sociology of Science* (Chicago: University of Chicago Press, 1973), pp. 371–382.

62. And sometimes, much earlier, as we have seen in the case of Charles Babbage.

63. Jan Noyes, "The QWERTY Keyboard: A Review," *International Journal of Man-Machine Studies,* Vol. 18 (1983), pp. 265–281. I owe knowledge of this article to the kindness of Paul Green. See also Stephen J. Gould, "The Panda's Thumb of Technology," *Natural History,* Jan. 1987, pp. 14–23, brought to my attention by Reinaldo Perez.

64. Hughes, *Networks of Power,* pp. 106–139.

65. See for instance the portrait of Seymour Cray by William D. Metz, "Midwest Computer Architect Struggles with Speed of Light," *Science,* Vol. 199 (Jan. 27, 1978), pp. 404–407.

66. Anonymous, "Reinventing Computer Science," *High Technology,* June 1983, p. 75.

67. P. E. Ceruzzi, "The Early Computers of Konrad Zuse, 1935–1945," *Annals of the History of Computing,* Vol. 3, No. 3 (July 1981), pp. 241–262.

68. Nathan Rosenberg, "The Direction of Technology: Inducing Mechanisms and Focussing Devices," in his *Perspectives on Technology* (Cambridge: Cambridge University Press, 1976).

69. Rene Vallery-Radot, *The Life of Pasteur* (New York: Dover, 1948).

70. Thomas Hughes, "How Did the Heroic Inventors Do It?" *American Heritage of Invention and Technology,* Vol. 1, No. 2 (Fall 1985), pp. 18–25.

71. For instance, see the biography of Pasteur by his son-in-law: Rene Vallery-Radot, *The Life of Pasteur* (New York: Dover, 1960); and Rene Dubos, *Louis Pasteur: Free Lance of Science* (Boston: Little, Brown, 1950). I am grateful to Bruno Latour for introducing me to Pasteur's work.

72. Richard Carter, *Breakthrough: The Saga of Jonas Salk* (New York: Trident Press, 1966).

73. John L. Enos, *Petroleum Progress and Profits* (Cambridge, Mass., M.I.T. Press, 1962), pp. 163–186.

74. Rupert Gould, *The Marine Chronometer* (London: Holland Press, 1959).

75. Gould, *The Marine Chronometer,* pp. 13–14. As Gould's book shows, this was only one of many attempts to hasten progress in navigation through state-sponsored cash rewards.

76. The Board of Longitude took its time paying the reward money, and at age 80, John Harrison had to petition King George III to get his money. Even then it was paid only after deducting certain questionable expenses; see David Landes, *Revolution in Time: Clocks and the Making of the Modern World* (Cambridge, Mass.: Harvard University Press, 1983), pp. 145–157. The quotation from Isaac Newton occurs on page 146.

77. See Kenneth A. Brown, *Inventors at Work: Interviews with Sixteen Notable American Inventors* (Redmond, Wash.: Microsoft, 1988), p. 2. MacCready credited Kremer's prize with being the incentive for his major accomplishment on several occasions when he visited Ann Arbor, Michigan, in May 1988.

78. Eliot Marshall, "Gould Advances Inventor's Claim on the Laser," *Science,* Vol. 216 (Apr. 23, 1982), pp. 392–395.

79. Richard L. Gausewitz, *Patent Pending: Today's Inventors and Their Inventions* (Old Greenwich, Conn.: Devin-Adair, 1983), pp. 54–66. The behavior of the government in this

case gives some substance to inventors' common paranoia about larger firms and bureaucracies stealing their inventions. It also provides one reason why more inventors don't submit their ideas to national inventors' councils! See James N. Mosel, Barkev Sanders, and Irving Siegel, "Incentives and Deterrents to Inventing for National Defense," *Patent, Copyright, and Trademark Journal of Research and Education,* Vol. 1, No. 2 (Dec. 1957), pp. 185–215.

80. For instance, in a study of Italian inventors it was found that in 69.2 percent of the cases, the inventors claimed they would have invented the device for which they obtained the patent even if no patent had been available; see Giorgio Sirilli, "Patents and Inventors: An Empirical Study," *Research Policy,* Vol. 16 (1987), pp. 157–174.

81. Although financial incentives may be important for some inventors, for the majority it appears that the problem-solving process itself is the real stimulus.

82. Hughes, *Networks of Power,* p. 22; see also his "How Did the Great Inventors Do It?" *American Heritage of Invention and Technology,* Vol. 1, No. 2 (Fall 1985), pp. 18–25. Hughes had mentioned the concept of "reverse salient" in the *Proceedings of the 175th Anniversary of the U.S. Patent System 1790–1965;* Vol. I (Washington, D.C.: Patent Office Society, 1966); p. 360. Nathan Rosenberg has also referred to the same phenomenon as "technical imbalance"; see his "Direction of Technology," pp. 108–125.

83. John A. Hobson, *The Evolution of Modern Capitalism* (New York: Scribner, 1916), pp. 79–81. Reprinted with permission of Macmillan Publishing Company.

84. John Ellis, *The Social History of the Machine Gun* (Baltimore: Johns Hopkins University Press, 1986), pp. 149–151.

85. Matthew Josephson, *Edison: A Biography* (New York: McGraw-Hill, 1959), pp. 172–173.

86. Jacob Schmookler, *Invention and Economic Growth* (Cambridge, Mass.: Harvard University Press, 1966).

87. David C. Mowery and Nathan Rosenberg, "The Influence of Market Demand upon Innovation: A Critical Review of Some Recent Empirical Studies," in N. Rosenberg (Ed.), *Inside the Black Box: Technology and Economics* (Cambridge: Cambridge University Press, 1982), pp. 193–241. This article, though not easy to read, is a valuable review of a considerable body of research.

88. Barkev Sanders, "Commentary on *Invention and Economic Growth,*" *Idea,* Vol. 10 (1967), pp. 487–508. Sanders's comments on this book must be taken very seriously, for he has conducted more research on patenting activity than almost anyone in the world.

89. William J. Abernathy and James M. Utterback, "Patterns of Industrial Innovation," in Michael L. Tushman and W. L. Moore (Eds.), *Readings in the Management of Innovation* (Cambridge, Mass.: Ballinger, 1982), pp. 97–108.

90. David C. Mowery and Nathan Rosenberg, "Technical Change in the Commercial Aircraft Industry 1925–1975," in Rosenberg, *Inside the Black Box,* pp. 163–177.

91. Vivien Walsh, "Invention and Innovation in the Chemical Industry: Demand-Pull or Discovery-Push?" *Research Policy,* Vol. 13 (1984), pp. 211–234.

92. Noreen Cooray, "Knowledge Accumulation and Technological Advance," *Research Policy,* Vol. 14, No. 1 (1985), pp. 83–95.

The Role of Designers in Technology

Seldom do inventors give devices their final form. Although in rare cases a device can be used just as it was invented, usually some further refinement of it must take place. This is the role of *designers*. It is designers who give devices their particular form. As we have seen, the configuration of a device is important, since the device's social impact depends on it. Most of us are aware of the designer's role in creating fashion merchandise ("designer clothes"). Yet the concept is a more general one; virtually all products are designed by somebody.[1] We need to inquire into the characteristics of these form givers, to explore the nature of designers and the design process. Who are these designers and what do they do?

The Nature of Design
and Designers

In a variety of fields, from toothpicks to space shuttles, the forms of things that we use are shaped by designers.[2] Sometimes designers develop a basic invention into something that is workable. In other cases they recombine subsystems to produce a different configuration. Whether designers are also called engineers, architects, or stylists, their role is very important. A new design for a pencil may produce fewer cramped fingers. A new design for a building may enable people to be less stressed by noise, vibration, and so on. A different aircraft design may have very different performance characteristics in terms of speed, safety, maneuverability. Novel designs for appliances may reduce energy usage, and so forth.

Sensitivity to the consequences of design is important. The mechanical functioning of a device is not a guarantee of its effectiveness and side effects. Each element that goes into shaping the final form brings the possibility of new positive and negative consequences.

Where the precise distinction between inventor and designer rests is difficult to say, but it is partly a legal matter. If something can be patented, it is legally considered an invention. The inventor must show that what he or she has come up with is not "obvious" to other skilled practitioners and represents a "real" innovation. If the inventor has merely made a variation that, in principle, anyone "skilled in the art" could have made, then this may disqualify the new design from being an invention.[3]

Inventors might thus be thought of as people who develop new principles for devices, and designers as those who develop new forms for them. This distinction is probably as good as any, but there are lots of exceptions. For instance, whereas architects are definitely designers, in certain cases they are also inventors. Consider William Paxton's design of the Crystal Palace.[4] Built in London's Hyde Park to hold the International Exhibition of 1851, the Crystal Palace was one of the first "glass-wall" structures in the world, certainly the first of such a massive size. And William LeBaron Jenney, in developing the skyscraper, certainly invented the idea in addition to designing the building.[5]

Thus, the difference between inventor and designer is a relative rather than an absolute one. We might think of a continuum like this:

Figure 7-1

Inventor	Designer	Stylist
■------------------------■------------------------■		

At the left of this continuum are people who develop new principles and totally new kinds of things. In the middle are people who make substantial changes in configuration (the relationships among subsystems). And at the right we have people who only change the way things look, not how they work. Thus the person who thinks up an electric toaster in the first place is an inventor, the person who decides to make it toast two pieces instead of one is a designer, and the person who decides that the outside gets a floral design instead of a plain surface is a stylist. Stylists produce little more than good looks.[6] In real life, of course, these distinctions often blur considerably.[7] A continuum also exists for the background and training of the people involved: Toward the left, the innovators are more likely to have training in science or engineering, whereas toward the right they tend to be trained in the fine arts.

It is important to note the range of competences that go into designs. Designers of things that are invisible—industrial processes or mechanisms that transmit motion or power—are typically engineers, who are usually trained in engineering colleges. In the United States today, there is a strong emphasis on science in engineering, as opposed to practical skills, which means that we have excellent engineering researchers and often (by world standards) mediocre products.[8]

Sometimes the design work that is most prestigious is the least "hands-on" of all, involving simulation or truly invisible electronic components.[9] Work on products for ordinary consumers—as opposed to those for space, military, or other high-tech clients—is not considered to be nearly as exciting as many of the more exotic areas related to high technology. Status hierarchies in engineering start with the most abstract and scientific specialties—aerospace and electronic engineering—and end with civil engineering and forestry at the bottom. Thus engineers tend to see the most prestigious engineering work as that most closely related to science and pure research.

Another reason for status hierarchies is that in defense and space research, for instance, government funds allow experimentation that would be prohibitively expensive in private engineering. The kinds of test equipment, prototypes, and materials available for development of weapons and space systems are often looked upon with envy by engineers in more "humble callings."[10] Salaries are often proportionately higher in these high-tech fields as well. Engineers like to experiment, and experimentation is often more exciting in the defense and space fields than it is in the design of consumer products, for which budgets for engineering are usually more limited.

On the other end of the spectrum are designers trained by design institutes or schools of fine art. They design most of the things that are visible and make decisions about surfaces, colors, and configurations. Designers ideally want products of genuine aesthetic value, but practical considerations dictate that they design products that sell well because most of them work for commercial operations. Although the best practitioners of, for instance, industrial design, take users into consideration, their real reference group in most cases is the design profession itself. Highly respected designers achieve professional recognition through the appearance of their design in design publications; how well their products work for their users is often unmeasured, except by sales figures.

Training and professional orientation thus may have very profound effects on who designers design *for*. It is significant that both engineers and industrial designers, for instance, tend to view consumers as a relatively low-status (and therefore a relatively unimportant) group as compared to fellow professionals. Because in each case the highest status is accorded to designers who engage in custom work for special clients, the ordinary product for the ordinary person receives much less attention, although it may be widely disseminated and have significant social impacts.[11] But whether the result is visible or invisible to the user, the product's performance is strongly shaped by what the designer puts into it.

Factors That Affect
Design Criteria

Utility. Designs may embody values. For some objects utility may be the prime consideration. However, designers often find their designs judged by other designers and by design critics, who are a key to understanding why designers do some of the things they do.[12] For instance, architects are usually given awards for designing buildings that look good from the outside or from within; they often give too little consideration to the needs of the people who will occupy the buildings. Often the photographs of buildings in architectural magazines show the buildings without people in them. But because people will occupy them eventually, and some of these people may not agree with the aesthetics, what architects win awards for may sometimes be very uncomfortable to use.[13] Thus, Reyner Banham, in his book *The Well-Tempered Environment,* finds that architects have traditionally paid little attention to heating, cooling, and air circulation systems.[14]

Safety. Often designers seem oblivious to the consequences that small design changes could have on the users of the products.[15] For instance, by the 1960s automobile safety experts had noticed that cars could be made much safer in collisions if fewer knobs and protrusions jutted out from the dashboard. But neither manufacturers nor consumers showed much interest in these safety improvements; safety features did not help "sell" a product. After all, some of these potentially dangerous features were attractions to the car buyer.[16] The current emphasis on safety came later, as a result of campaigns by Ralph Nader and others.[17] (Nader was present when a little girl was decapitated by a glove-compartment door that sprung open while the car was in motion; this experience got him interested in the relationship between auto design and safety.)

Design may thus have fatal consequences in systems that must function in a life-and-death environment. One study of airplanes showed that some models were much more prone than others to induce pilot errors and thus cause accidents.[18] In designing hospital tools and appliances, this kind of consideration should be paramount, and aesthetics should be kept to a minimum. In the design of household things, however, aesthetics may play a larger role. This leads us to an important question: Where do the goals of the designer come from?

Political Considerations. Sometimes the designer's goals are essentially political. During the Vietnam War, for instance, many American soldiers found that their M-16 automatic rifles constantly jammed in combat. The original form of the M-16, the Armalite AR-15, was designed by Eugene Stoner, an employee of the Armalite Corporation. Stoner had been asked to develop the weapon at the request of the Infantry Board, after the "Salvo" studies that suggested that high-velocity, small-caliber weapons would be superior for combat.[19] But another part of the army rifle establishment, the Ordnance Board, did not like it, claiming it was too inaccurate to be the marksman's weapon they desired. For instance, a marksman would require a rifle accurate up to, say, 800 yards, whereas the AR-15 was designed to be accurate only up to 500 yards. Thus when pressures built up for the Army to adopt the weapon, it was "improved" by an interservice committee, and this "improvement" changed the characteristics of the rifle and made it more prone to jamming. (Further, the kind of ammunition Stoner had designed it for was not used.) Although the Special Forces demanded and got the original version, the weapon issued to most GIs needed frequent cleaning and was notoriously unreliable. This redesign was disastrous for many infantrymen, who died under hostile fire while trying to clean or unjam their weapons.[20]

Management Considerations. The M-16 case raises another issue. Those who make decisions about designs are often not designers, but are instead managers who are often remote both from the design process and from the scene of operations. The officers (mostly colonels and generals) who made decisions about the M-16 were not part of combat units; they were military bureaucrats. They did not have to face the consequences of their bumbling; nor did they have as much motivation to make the right choices as did the foot soldier. Similarly, engineers play a relatively insignificant role in many manufacturing organizations in the United States, as compared to many foreign countries where the heads of firms have technical training.[21] Thus decisions about design will often be made by those without knowledge either of the technology or of the users.

Design Issues

The "Appropriate Technology" Philosophy. Technologies of production and transport frequently have enormous impacts on people's lives. In developing countries particularly, energy is expensive, and capital equipment is difficult to repair and service. Importing the latest high-tech farming and production equipment from developed countries has produced serious problems.[22] Often the users are better off with small-scale, labor-intensive forms of production,[23] which utilize local skills and are more appropriate to local conditions. Similar remarks can be made about transport. Western automobiles are ill-designed, for instance, to respond to the needs of tropical countries; they are too heavy and use too much gasoline, and because they are built low to the ground, they are easily disabled by ruts and rocks.

Such concerns have led to a movement and a design philosophy known as "appropriate technology."[24] Even though the achievements of appropriate technology have been impressive, the sponsors of the more high-tech and energy-intensive forms of technology have shown little interest in it. Because the inner logic and supporting social structure of appropriate technology is in some respects antithetical to "modern" high-tech approaches, sponsors of high-tech machinery have periodically mounted attacks on it.[25] It is easy to dismiss the appropriate technology movement as a fad,[26] for the movement has seldom been able to get research funds and massive governmental support equal to those of its more conventional competitors. So the attacks have been, for the most part, unnecessary. Appropriate technology has gained little headway in Western countries because incentives to produce simple, energy-saving designs are as rare in engineering education as they are in actual design practice.[27] As long as energy is easily available, designs that conserve it are not likely to become popular.

Product Longevity. Another issue important for products is how long they are designed to last.[28] A great many things we use—from automobiles to paper bags—are designed for a life span that is considerably less than it might be. Not only does a poorly designed product break down, often at an inconvenient time, but it then becomes trash that must be discarded or recycled;[29] sometimes this trash is hazardous to incinerate or bury. This is true, for instance, of radioactive wastes from nuclear power operations.[30] Even nuclear isotopes used in comparatively small quantities for medical purposes can become dangerous. In Goiania, Brazil, a single nuclear isotope machine discarded from a clinic led to a number of deaths and contaminated hundreds of people.[31]

One of the ways in which product life can be extended involves designing for durability. However, durability may entail significant costs above what the consumer is willing to pay. Many Americans are willing to take out a 20- or 30-year mortgage on their house, but even if one could design cars that would last this long (and represent a significant saving), it is difficult to know whether consumers would buy them. Life-cycle cost is often neglected by consumers and industrial firms alike. In industrial products, for instance, the life-cycle cost of a new machine may be accounted for by two different entries, one for the initial cost and another for maintenance. A low initial cost may accompany a short factory life and an overall high life-cycle cost; a more durable product may have a much higher initial cost, but last perhaps three times as long. Even though the latter may represent a lower life-cycle cost, it may be refused by management. The initial cost is part of the capital budget, and the maintenance cost is part of the maintenance budget, and thus these two items may be considered separately. Ultimately, money saved by buying a lower-priced machine may be lost as the machine breaks down more often or wears out faster. Government bids often fail to take life-cycle costs into account, even though vendors often do and as a result make huge profits on parts and repairs.

Product Recyclability. An important design issue in recent years has been designing food and beverage containers for recyclability. Once the consumer has used whatever product is sold in the container, the container is sent back to the distributor to be cleaned or melted down and used again. Many states have passed laws that force the sale of soft drinks, for instance, in recyclable containers. These laws have been fiercely resisted by the beverage industry, which is apparently indifferent to the litter created by discarded bottles and cans. The passage of such laws has created less solid waste and less litter as a result of the monetary incentive to return the bottles. It has also created a number of entrepreneurs who find that rifling trash bins and collecting litter can provide money for everything from candy, comic books, and skateboards to food, liquor, and drugs.

Energy Efficiency. Energy efficiency may one day become an important criterion in a variety of products; in recent years automobiles have been strongly affected.[32] The petroleum price increases of the 1970s led to innovations designed to save fuel. Two case histories of design illustrate how fuel economy considerations can affect large and small cars.

The first example is the downsized Cadillac Seville, announced in 1975. Cadillac chief engineer Robert Templin speculated that a fuel-efficient version of the Seville might be a very popular car. Accordingly, engineers rapidly made drawings and built prototypes for a smaller version. Very little innovative engineering was possible; to the maximum extent possible, the car used technology that was well tested by the firm (an exception was electronic fuel injection). But for a Cadillac, the car was small, weighing 2.17 tons as opposed to the 2.6 tons of the larger Sedan de Ville. General Motors president Edward Cole did not like the idea and did not think it would be successful. At the eleventh hour, however, the chairman of the board intervened, and the engineers were allowed to proceed. The car was a best-seller among Cadillacs.[33]

The Fiat Panda was designed by Giorgetto Giugiaro of the firm ItalDesign in collaboration with Fiat, the manufacturer. Fiat set relatively few constraints on what the Panda should look like but did restrict weight, size, and type of engine. The car was also expected to be relatively simple to manufacture. Further, the Panda was deliberately designed around the driver, from the inside out. The "no frills" result, launched in 1980, proved to be very successful.[34] (Incidentally, Giurgiaro was also capable of designing very fine sports cars—he designed the successful Lotus Esprit.[35])

Regrettably, energy efficiency ceases to be a goal for managers once fuel becomes cheap. Thus in the building, automobile, and consumer appliance industries, the end of the "energy crisis" was accompanied by the return of energy-wasteful designs. The lack of values other than commercial ones on the part of such companies probably reflects a general failure of society to provide ecological awareness.

As the greenhouse effect and ozone hole problems become recognized, however, environmental awareness may increase.

Manufacturing Costs. Simplicity in manufacture is an important goal because it is related to the cost of manufacture and thus the product's price. As a product is manufactured widely the cost tends to decrease due to the "learning curve."[36] Aircraft manufacturers noticed that the cost of an airframe tended to drop over time, due in part to economies of scale (that is, it's cheaper to make more of something) but also because shortcuts in manufacturing allowed the same product to be made more cheaply.[37] The IBM Selectric typewriter as originally designed had 2,300 parts; currently it has only 190, a result of "learning curve" considerations. Reliability is also said to have increased.[38] Thus organizations learn to manufacture their products ever more cheaply as a result of this constant learning process.

Gender Considerations in Design

The issue of gender in design has been highlighted in recent years as women's consciousness and their rights in society have broadened.[39] There are a variety of reasons why gender is important in design:

First, devices are often designed with males in mind. This is apparent in the difficulties women have in operating these devices, whether because of their body configuration or strength, or the skills required. In some cases the neglect of women's needs is accidental, but in other cases it is intentional. Many of the improvements in automobiles have helped women gain increased mobility. Among these are power steering and power brakes, electric starting, adjustable seats, and automatic transmission. Automobiles without these improvements were often unfriendly environments for women, requiring superior strength or mechanical skills they did not have. Sometimes devices have intentionally been made too heavy for women to use comfortably, as with some printing technology.[40] Certain kinds of designs lock women out of certain work or social areas.

Second, women have unique physiological needs—for instance, in technology related to reproduction. But technology and institutions related to giving birth are often designed and controlled by males.[41] Recently, for example, there has been considerable controversy about such subjects as monitoring the fetus by ultrasound and using drugs to relieve pain during childbirth. Some people contend that women should not take painkillers during the delivery process, but some women have found them extremely helpful.[42] Birth control devices and strategies often have disproportionately high physiological effects on women. The use of midwives—almost always women—to assist in the delivery process is also a salient issue to many feminists.

Third, technology itself often embodies male values. This is evidently true of much military technology, upon which huge amounts of money, intellectual labor, and time are spent.[43] Few females show the fascination with weapons that many males do.[44] But there are also less obvious male influences in the design of automobiles, public places, industrial devices, medical equipment and procedures, and consumer products generally. Because designers and decision-makers in design organizations are often male, technology may often emphasize power, destructiveness, manipulation, and speed, as opposed to security, comfort, nurturance, or beauty. Masculine values tend to support high industrial productivity to the exclusion of more feminine values, such as preservation of or harmony with the environment.[45]

Fourth, women disproportionately engage in many activities, such as housework and child care. The design of devices and procedures for these activities, then, tends to shape women's lives more than men's. Considerable research has cast doubt, for instance, that "labor-saving" devices in the home really save labor, or that the lives of homemakers have genuinely been improved by household technology.[46] In some cases the machine may actually save labor, but the effect may be nullified by changing social norms, such as the need for "cleaner houses."

Women work disproportionately in certain kinds of jobs.[47] Until quite recently, for example, many people felt that a "secretary" was always a woman and a "surgeon" was always a man. Jobs intended for women may be routinized to a greater extent than those for men and may be tailored to require less skill. This seems to be happening with computer programming, for instance.[48]

Finally, the amounts of training in technological skills people receive may differ by gender. Even though there has been some equalization over the last two decades, important differences between males' and females' abilities to repair cars, sew, and troubleshoot mechanical appliances still exist. These differences are cultivated early, are reinforced through the school curriculum and hobbies, and are further heightened in college and graduate schools, where male professors teach mostly male students engineering skills and female professors teach mostly female students in "home economics" or (increasingly) in "human ecology." Complicated "hard" systems like computers and spaceships are generally designed by males, whereas "soft" systems such as clothing and interiors are generally designed by females. Although many women are in the field of computing, the intense positive feeling toward the machines that is typical of male computer hackers and designers is more rare among women. It has been theorized that such differences are the result of early cultural conditioning, but it is also possible that they reflect genetic predispositions.[49]

Although technology is important to women, much of it is designed and controlled by men, often with men in mind. Women's control of technology may be an important battleground in the future, as women seek to exert more influence over the designed environment.

Linkages Between
Designer and User

One of the most important issues of design is how the designer and user communicate. Obviously, it would be desirable for designers to be able to observe the use of their products so that the design could better serve users' needs. Seldom, however, does this happen. Designer and user are often separated by a considerable social and physical distance, over which communication takes place only in very limited ways.

For instance, consider the typical way in which we get a new product. We seldom see or even hear about the designer of our car, our house, our washing machine, or the bottles in which we get beer. Designers assume that consumers make important decisions in the marketplace; that is, the purchase patterns of a product are considered signals to producers about what people want. So when Brand X pantyhose sells better than Brand Y, all pantyhose are likely to become more like Brand X. It is obvious, though, that this is a very crude system, especially with large, complicated products. Often, to cope with different kinds of customers, firms will separate consumers into different categories according to "life-style" or socioeconomic status.[50] Categories like these allow firms to figure out to which groups of customers they should send which catalogues, the bane of many of us who get a dozen catalogues a week. The value of trying to figure out what consumers want seems debatable in comparison to the value of asking them more directly.

Of major importance is how much the concerns of users are taken into account at all in design.[51] Some studies of architects and designers suggest that little attention is paid to "human factors" aspects of design.[52] Yet by comparison with the clients for other forms of technology, the clients of architects are relatively powerful: They can often set the design criteria and can even demand changes at the last minute.[53] But more often than not, customers get something that does not meet their needs.

One problem is that what is important to designers is rarely important to users.[54] One of the reasons for this lack of interest in and knowledge of users is the education of designers, which often has a "fine arts" emphasis that is deficient in psychology and sociology.[55] But another problem is the nature of the communication system itself, which has too many delays and too many links in the chain between designers and users. The principles involved in designing with use in mind are reasonably well known.[56]

Firms may try to get information from potential customers directly, via consumer surveys, "focus" discussion groups, and devices like "taste tests."[57] This kind of research, although more expensive than market research that simply notes the number of items purchased, at least tends to provide a more immediate form of feedback. The results of such "market research" improve when the designers speak directly with the customers; such direct communication usually does not occur.[58] The "focus" groups are conducted by market research people, who in turn eventually communicate their findings to designers in watered-down, written recommendations. This is pretty thin material compared to direct feedback from user to designer. But what else can be done? An automobile designer cannot communicate directly with the thousands and sometimes millions of people who will use the product. Thus sampling is a necessity. Some firms have hired anthropologists to observe users and have even filmed people using their products.[59]

Sometimes users have thought of a better use or have developed certain modifications that can be designed into the product.[60] We will look into this sort of activity in Chapter 12. Here it is important to note that certain products, like wind-surfing boards, were first developed by customers and were only subsequently commercially produced.[61] Through this kind of process producers learn that consumers are using products in certain unexpected ways, sometimes contrary to safety or their best interests. In any case, what producers learn from these customer modifications is often useful in redesigning the product.

Some of the most interesting examples of user-designer interaction come out of military experience.[62] Because weapon systems have long lead times and are often quite expensive, the process by which such systems are designed and produced has been carefully studied. One of the more interesting facts to emerge from such studies is that the United States has failed to do a very good job of getting designers and users to cooperate. Such observations, though, have done little to change the excessively rigid design process.

The reason for this failure is that the process by which weapons are requested by the armed services is extremely bureaucratic. A request for a weapons system usually weighs several pounds. As one expert designer put it, if you know this much about the weapon, you've already got it designed! Of course, those requesting the weapon do not have it designed; they are only proposing that it be done a certain way. Then firms proposing to develop the weapons system bid on it, hoping to get the even more lucrative follow-up contract to produce it. Firms often present a highly optimistic forecast about their ability to develop and produce the weapon in order to get the contract. Only when they in fact get a development contract must they suddenly confront reality. Selling the government on the idea is one thing; getting the weapon to work is something else again. Consequently, the results of the development process are often a poor compromise between what the armed services requested and what the contractor was capable of doing. Requirements take on an artificial quality when user and designer can't interact.

It may be much more effective to include the people who are going to use the system in the development process itself. This does happen, but only rarely. For instance, at the U.S. Navy's Naval Weapons Center at China Lake, California, an Experimental Officer from the fleet has the responsibility for placing "operational" personnel assigned to the center into actual weapon development teams. These "representatives of the user" are likely to be much more critical of a system's performance than are the nonmilitary scientists and engineers who design the technical aspects of the weapons. These naval personnel thus form a critical link between what is designed and what goes to the fleet. Another naval laboratory, the Naval Ocean Systems Center, sent representatives out into the fleet. In this program, senior engineers would report directly to fleet commanders and listen to the technological concerns of operational personnel in action.[63] This provided yet another interface connecting designers and users.

The "design team" concept, which links not so much designers and users but rather designers and producers, is becoming more common in the United States. In a design team, people involved in virtually every aspect of production get together in the design process. Often designers, once they get something designed, simply "throw it over the wall," washing their hands of further responsibility for the product. "Well, I designed it; now you build it," goes this sentiment. This "throwing it over the wall" often results, of course, in unbuildable designs.[64] Serious compromises must then be made in production so that the device can actually be manufactured. To avoid this kind of irresponsible design, representatives of the manufacturing plant participate in the design team to offer suggestions as design proceeds.[65]

Another possibility for the linkage of designers and users is *participative design*. In this strategy, users are direct participants in the design of technological changes that affect them. A classic industrial experiment by Coch and French showed that workers who were allowed to participate in designing changes in the work process were likely to be more productive.[66] Surprisingly, very little has been done in the United States to follow up this insight, even though the study is well known.[67] In the 1970s architects and environmental psychologists in the United States became more interested in what Louis Davis calls "joint optimization," which involves designing factories with full input from those who will manage and work in them. Admittedly factories are usually "custom-made"—designed one at a time—but the same principles can be applied to mass-produced products.

Collaborative design offers a way to give people who use technological systems a chance to participate in the design of the environment and the equipment they will use. Because it allows a closer fit between the user's requirements and the designer's activity, it promises not only to make designers more aware of users' needs, but also to make users aware of the full spectrum of possibilities.[68] It is interesting to note that "custom-made" often denotes exactly this quality: The product has been designed to fit the user exactly.

Conclusion

Many forces shape a product's ultimate design. Many of these forces are commercial or bureaucratic, and they may reveal more about the organization producing the product than about the users it is ostensibly designed for. There are "many fingers in the pie" in the design process. Marketing departments in commercial firms are often a major influence. Low budgets for engineering and design activities and time pressures also affect design: "There's never time to do it right." Engineers have their own particular biases, and may be affected by idiosyncratic influences. The materials engineers choose may be affected by their preference for particular vendors and their personal loyalties to them. The market-place and related competition also shape decisions about what to produce.

In this chapter we have discussed the role of designs and designers. It is important for students of technology to examine the role of designers because it often affects not only how products look but also how they function. Decisions made by designers are important because they shape not only products, but also the lives people lead. Every product has effects on the community to which it is introduced; we will examine some of these effects more closely in the chapter on technology assessment. So designers, in shaping products, are also shaping the world in which we live. How safe our dwellings and tools are, how comfortable life is, how much energy we use, and what kind of trash we produce are all aspects of our lives affected by the work of designers. Technology embodies values, and designers shape the technology to illustrate and implement these values.

It is also important to realize that products are shaped by many people who are not identified as designers. Most designs are shaped by organizations, and the designer may be only the "tip of the iceberg." Departments not directly involved in design, such as marketing or sales, may be influential. Interdepartmental conflict, the biases of higher managers and articulate vendors, and industry fashions all shape design. In all this the voice of the user may be a small one. It may be valuable for us to start thinking how this voice may be amplified.

Notes

1. See, for instance David Pye, *The Nature of Design* (London: Studio Vista, 1964). Some sense of the variety of fields of design can be grasped by looking through Stephen Bayley (Ed.), *Conran Directory of Design* (New York: Villard Books, 1985). But even this leaves out architects and city planners. "Design" in the context of this chapter usually refers to "industrial design," which usually means design for multiple or mass production, as opposed to the "one-off" work of architects or city planners; see Edward Lucie-Smith, *A History of Industrial Design* (New York: Van Nostrand, 1983); Tracy Kidder, *House* (Boston: Houghton Mifflin, 1985); and Ian Mitroff, "The Unexplored World of Engineering Design," *Technology Review*, Vol. 73, No. 7 (May 1971), pp. 29–33.

2. See, for instance, Laurence B. Holland (Ed.), *Who Designs America?* (Garden City, N.J.: Doubleday, 1966).

3. Richard L. Gausewitz, *Patent Pending: Modern Inventors and Their Work* (Old Greenwich, Conn.: Devin-Adair, 1983).

4. Henry Petroski, *To Engineer Is Human: The Role of Failure in Successful Design* (New York: St. Martin's Press, 1965), pp. 136–157.

5. Carl W. Condit, *The Rise of the Skyscraper* (Chicago: University of Chicago Press, 1952), pp. 112–139.

6. For an exposition of the "good looks" approach, see Harold Van Doren, *Industrial Design: A Practical Guide* (New York: McGraw-Hill, 1940). An effective critique of this ap-

proach is Victor Papanek, *Design for the Real World: Human Ecology and Social Change* (New York: Pantheon Books, 1971), pp. 152–184.

7. For a master practitioner's entertaining book on how a designer works, see Henry Dreyfuss, *Designing Things for People* (New York: Simon & Schuster, 1955); see also Ralph Caplan, *By Design: Why There Are No Locks on the Bathroom Doors in the Hotel Louis XIV and Other Object Lessons* (New York: St. Martin's Press, 1982).

8. Jonathan Rowe, "Why the Engineer Left the Shop Floor," *Washington Monthly,* June 1984.

9. Paul Wallich, "The Engineer's Job: It Moves Toward Abstraction," *I.E.E.E. Spectrum,* June 1984, pp. 32–37.

10. My cousin Keith Westrum, a designer of industrial equipment, notes that "engineers love to play, and you have bigger toys to play with in defense and space. The research and development environment is thus a lot more interesting."

11. In an older survey of engineers, the motivation "to help others" was reported by 39 percent of engineers with a B.S. degree, whereas only 26 percent noted that this motivation was characteristic of their jobs. By contrast, 82 percent and 72 percent rated "innovation" and "challenging work" as very important; Robert Perrucci, William K. LeBold, and Warren E. Howland, "The Engineer in Industry and Government," *Engineering Education,* Vol. 56, No. 7 (Mar. 1966), pp. 237–259.

12. Inventor Jacob Rabinow contends that inventors would benefit from invention critics. It is worthwhile to note that in engineering design, exceptional abilities are often widely known and appreciated, often throughout the whole professional network.

13. Robert Sommer, *Tight Spaces: Hard Architecture and How to Humanize It* (Englewood Cliffs, N.J.: Prentice-Hall, 1974), pp. 129–137; see also his *Social Design* (Englewood Cliffs, N.J.: Prentice-Hall, 1983).

14. Reviewed in John Kouwenhoven, "Architecture as Environmental Technology," *Technology and Culture,* Vol. 11, No. 1 (Jan. 1970), pp. 85–93.

15. Victor Papanek has done more than any other single designer to bring this point home; see his *Design for the Real World,* cited above, and his *Design for Human Scale* (New York: Van Nostrand, 1983).

16. Charles McCarry, *Citizen Nader* (New York: Saturday Review Press, 1972), p. 67.

17. Thomas Whiteside, *The Investigation of Ralph Nader* (New York: Pocket Books, 1972).

18. Bureau of Safety, Civil Aeronautics Board, *Aircraft Design-Induced Pilot Error* (Washington, D.C.: National Transportation Safety Board, July 1967).

19. Thomas L. McNaugher, *The M16 Controversies: Military Organization and Weapons Acquisition* (New York: Praeger, 1984), p. 59. The account given in this textbook is a very much abbreviated version of the actual complex interbureau struggles.

20. Arthur M. Squires, *The Tender Ship: Governmental Management of Technological Change* (Boston: Birkhäuser Boston, 1986), pp. 57–86.

21. Robert H. Hayes and William J. Abernathy, "Managing Our Way to Economic Decline," *Harvard Business Review,* July-Aug. 1980; and Cary Frey, "What Is Due the Engineer?" *I.E.E.E. Spectrum,* June 1984, pp. 64–68.

22. Arnold Pacey, *The Culture of Technology* (Cambridge, Mass.: MIT Press, 1984), pp. 35–54.

23. E. F. Schumacher, *Small Is Beautiful: Economics As If People Mattered* (New York: Harper & Row, 1973).

24. George McRobie, *Small Is Possible* (New York: Harper & Row, 1981).

25. For interesting arguments on both sides of this issue, see R. S. Eckaus, "Appropriate Technology: The Movement Has Only a Few Clothes On," *Issues in Science and Technology*, Vol. III, No. 2 (Winter 1987), pp. 62–71; Francis Stewart, "The Case for Appropriate Technology," *Issues in Science and Technology*, Vol. III, No. 4 (Summer 1987), pp. 101–109; and David Dickson, *The Politics of Alternative Technology* (New York: Universe Books, 1979).

26. This has also been true of the "soft paths" school of energy alternatives. Many of these criticisms deserve serious consideration. See Allan Schnaiberg, "Soft Energy and Hard Labor? Structural Restraints on the Transition to Appropriate Technology," in G. F. Summers (Ed.), *Technology and Social Change in Rural Areas* (Boulder, Colo.: Westview Press, 1983) and his "Did You Ever Meet a Payroll? Contradictions in the Structure of the Appropriate Technology Movement," *Humboldt Journal of Social Relations*, Vol. 9, No. 2 (Spring/Summer 1982), pp. 38–62. See also the special issue of *Sociological Inquiry*, Vol. 53, No. 2/3 (Spring 1983) on "The Sociology of the Environment."

27. Christopher Flavin and Alan Durning, "Raising Energy Efficiency," in Lester Brown (Ed.), *State of the World 1988* (New York: Norton, 1988).

28. Walter J. Stahel, "Product Life as a Variable," *Science and Public Policy*, Vol. 13, No. 4 (Aug. 1986), pp. 185–193.

29. Vance Packard, *The Waste Makers* (New York: Pocket Books, 1963). This book created a sensation when it was first published in 1960, and the situation it described has generally gotten worse: We are creating more trash. See Cynthia Pollock Shea, *Recycling Urban Wastes* (Washington, D.C.: Worldwatch Institute, 1987).

30. Fred C. Shapiro, *Radwaste: A Reporter's Investigation of a Growing Nuclear Menace* (New York: Random House, 1981).

31. Leslie Roberts, "Radiation Accident Grips Goiania," *Science*, Vol. 238 (Nov. 20, 1987), pp. 1028–1031.

32. For a very interesting article on how General Motors decided to use downsizing generally, see Joseph Kraft, "Annals of Industry: The Downsizing Decision," *New Yorker*, May 5, 1980, pp. 136–162. This article suggests that very important product decisions involving billions of dollars can be made in a haphazard manner.

33. David C. Smith, "The New Small Seville: Cadillac's King-Size Gamble," *Ward's Auto World*, May 1975, pp. 23–28.

34. Edward Lucie-Smith, *A History of Industrial Design* (New York: Van Nostrand, 1983), pp. 133–140.

35. Graham Robson, *The Third Generation Lotuses: Elite, Eclat, Esprit, Excel* (Chiswick, England: Motor Racing Publications, 1983), pp. 51–64.

36. Linda Argote and Dennis Epple, "Learning Curves in Manufacturing," *Science*, Vol. 247 (Feb. 23, 1990), pp. 920–924. Lowering costs through the use of less-expensive parts is now known as "value engineering."

37. T. P. Wright, "Factors Affecting the Cost of Airplanes," *Journal of the Aeronautical Sciences*, Vol. 3 (Feb. 1936), pp. 122–128.

38. Joel Kurtzmann, "For Better Products, Use Fewer Parts" (interview with Kailash C. Joshi of IBM), *New York Times*, June 26, 1988, Business Section, p. 2.

39. See, for example, Jan Zimmerman (Ed.), *The Technological Woman: Interfacing with Tomorrow* (New York: Praeger, 1983); and Joan Rothschild (Ed.), *Machina Ex Dea: Feminist Perspectives on Technology* (New York: Pergamon Press, 1983).

40. Cynthia Cockburn, "The Material of Male Power," *Feminist Review,* Vol. 9 (1981), pp. 41–58.

41. See, for instance, Diana Scully, *Men Who Control Women's Health: The Miseducation of Obstetrician-Gynecologists* (Boston: Houghton Mifflin, 1980).

42. Nadine Brozan, "Women Gain as Technology Becomes Part of Natural Birth," *New York Times,* Nov. 13, 1988, pp. 1, 14.

43. Seymour Melman, *The Permanent War Economy: American Capitalism in Decline* (New York: Simon & Schuster, 1974).

44. Keith Westrum points out that most engineers are males and "men like things that go boom." I have seldom found a similar excitement about planes, bombs, and guns on the part of women.

45. It is true that men have often been leaders of ecological movements or proponents of alternative life-styles. Nonetheless, consider the values implicit in the work of the Army Corps of Engineers as described by Gene Marine, *America the Raped: The Engineering Mentality and the Devastation of a Continent* (New York: Avon, 1969); and Carolyn Merchant, "Mining the Earth's Womb," in Rothschild, *Machina Ex Dea,* pp. 99–117.

46. Ruth Schwartz Cowan, *More Work for Mother: The Ironies of Technology from the Open Hearth to the Microwave* (New York: Basic Books, 1983); and Philip Bereano, Christine Bose, and Erik Arnold, "Kitchen Technology and the Liberation of Women from Housework," in W. Faulkner and E. Arnold (Eds.), *Smothered by Invention: Technology in Women's Lives* (London: Pluto Press, 1985).

47. Elizabeth Faulkner Baker, *Technology and Women's Work* (New York: Columbia University Press, 1964); and Iftikar Ahmed (Ed.), *Technology and Rural Women* (London: George Allen and Unwin, 1985).

48. Joan Greenbaum, *In the Name of Efficiency: Management Theory and Shopfloor Practice in Data-Processing Work* (Philadelphia: Temple University Press, 1979); and Heather Menzies, *Women and the Chip: Case Studies of the Effects of Informatics on Employment in Canada* (Montreal: Institute for Research on Public Policy, 1981).

49. John Markoff, "Computing in America: A Masculine Mystique," *New York Times,* Feb. 13, 1989, pp. A1, B10; see also the "Letters" column of the *Times* for Mar. 5, 1989, Section 4.

50. Brad Edmondson, "Who You Are Is What You Buy," *Washington Post,* Oct. 26, 1986, p. B-3.

51. See the conference proceedings by William B. Rouse and Kenneth R. Boff (Eds.), *System Design: Behavioral Perspectives on Designers, Tools, and Organizations* (New York: North-Holland, 1987).

52. See, for instance, David Meister and Donald E. Farr, "The Utilization of Human Factors Information by Designers," *Human Factors,* Feb. 1967, pp. 71–87.

53. One important problem in this situation is that the client is not necessarily the user. Thus, when an architect is contracted to design a house, the user is often present and influential; in designing public housing, however, the wishes and preferences of clients can often be ignored. See Devereux Bowley, Jr., "Architects and Residents of Public Housing," *Perspectives on the Professions,* Vol. 3, No. 4, (Chicago: Illinois Institute of Technology, 1983), pp. 1–2. Both this entire issue and the one following it are devoted to relationships between architects and their clients.

54. Charles Perrow, "The Organizational Context of Human Factors Engineering," *Administrative Science Quarterly,* Vol. 28, No. 4 (Dec. 1983), pp. 521–541; Lin Gingras, *The Psychology of the Design of Information Systems,* doctoral thesis, Graduate School of Business Administration, U.C.L.A., 1977.

55. Even very fine designers sometimes emphasize artistic considerations first and function second; see Owen Gingerich, "A Conversation with Charles Eames," *American Scholar,* Vol. 46 (1977), pp. 326–337. Yet both Eames and his employer, the Herman Miller Company, have taken users into account, not only through intention, but also through scientific study; see the sympathetic portrait of the firm by Ralph Caplan, *The Design of Herman Miller: Pioneered by Eames, Girard, Nelson, Propst, Rohde* (New York: Watson-Guptill, 1976). But if fine-arts designers are often insensitive to users' needs, engineering designers are often worse.

56. John D. Gould and Clayton Lewis, "Designing for Usability: Key Principles and What Designers Think," *Communications of the Association for Computing Machinery,* Vol. 28, No. 3 (Mar. 1985), pp. 300–311. Thanks to Paul Green for this article.

57. Claudia Deutsch, "What Do People Want, Anyway?" *New York Times,* Nov. 8, 1987, Section IV, p. 4.

58. It does occur, however, at Honda Motors in Japan, where the marketing department has been eliminated so that engineers speak directly with consumers.

59. Tamar Lewin, "Casting an Anthropological Eye on American Consumers," *New York Times,* May 11, 1986, Business Section, p. 6.

60. See Eric von Hippel, "Successful Industrial Products from Customer Ideas," *Journal of Marketing,* Jan. 1978, pp. 39–49.

61. Two of my students, Fred Seifert and David Chapman, conducted a study on the development of the Wind-Surfing Surfboard. Changes in recreational technologies such as wind-surfing often originate with the users.

62. On this subject generally, see James Fallows, *National Defense* (New York: Random House, 1981).

63. Interview with Captain Robert Gautier, U.S. Navy (Retired), former commanding officer of the Naval Ocean Systems Center.

64. Gerald Auerbach, formerly a vice-president of Aerojet General Corporation, told me about ordering a design engineer to spend an hour a day on the factory floor. This experience provided the engineer, who initially resisted the requirement, a powerful education in making things that were buildable.

65. The May 1987 issue of *I.E.E.E. Spectrum* is a special number devoted to "Good Design." In several of the articles in this special issue the design team concept is illustrated with actual industrial examples.

66. Lester Coch and John R. P. French, "Overcoming Resistance to Change," *Human Relations,* Vol. 1 (1948), pp. 512–532.

67. One exception was the work of Louis Davis; see his "Optimizing Organization-Plant Design: A Complementary Structure for Technical and Social Systems," *Organizational Dynamics,* Autumn 1979, pp. 3–15. A great deal of work has been done with "joint optimization" techniques in Great Britain, under the rubric "socio-technical systems theory." For a review of this subject, see Eric Trist, *The Evolution of Socio-Technical Systems* (Toronto: Ontario Quality of Working Life Center, 1981); also printed in Andrew Van de Ven (Ed.), *New Perspectives on Organizational Design and Behavior* (New York: Wiley, 1981).

68. See the proceedings of the conference on *Organizations Designs and the Future* (Ypsilanti: Eastern Michigan University, 1986), edited by Ron Westrum. The paper in this collection given by Robert Shibley and Lynda Schneekloth provides several examples of the way that collaborative design can produce impressive results.

CHAPTER 8

Innovation: Inventions & Institutions

It is a common practice to separate *invention,* the creation of something new, from *diffusion,* the distribution of the new thing to society at large. For a variety of reasons, however, it may be more useful to think of invention and diffusion as parts of a larger "innovation process."[1] Invention and dissemination are not as distinct as they might appear to be.

Consider the safety razor. King C. Gillette, it is true, thought up and patented the idea for a razor with disposable blades. But he would have been unable to turn it into a commercial product without the help of William E. Nickerson, a skilled mechanic who joined Gillette's organization. Over a period of six years Nickerson worked at and solved the very complicated problems involved in manufacturing the blades.[2] Gillette was thus the inventor of the safety razor, but Nickerson's practical activities were no less important in turning the invention into a widely diffused, commercially viable product. Thus making an idea workable may include figuring out how to manufacture it. But is the manufacturing process part of the invention process, or part of the diffusion process? It may be difficult to separate invention as a distinct act from other actions necessary to disseminate the invention.[3]

Another reason to consider invention and diffusion together is that continued tinkering—by the initial manufacturer and by others—takes place after the product is on the market.[4] The far-reaching social impact of computers would never have resulted from the machines as they were originally developed; continued tinkering has produced very different machines. The miniaturization of components has allowed computers to become smaller and thereby affect every sphere of modern existence. New designs ("architectures") have also made the machines more efficient. Thus there is a very strong possibility that a technology will change in the process of diffusion. Finally, very different conceptions of what the technology is reflect alternative ways of "framing" it. Boelie Elzen has shown how different conceptions of ultracentrifuges led not only to different devices, but also to different conceptions of what constituted an ultracentrifuge.[5] Because different people conceive of an invention differently, a number of different trajectories of development may coexist.

Innovation, then, consists not only of invention and diffusion, but includes many intermediate and bridging processes. Both inventiveness and innovativeness occur in an exceedingly complex social web. The innovation process, however, involves more than just activities; it involves emotions related to inventing, introducing new things, and thinking new thoughts. Inventions come about only partly to solve physical problems; their developers often have in mind broader purposes that may conflict with those of others in society who want to hold on to current institutions or perhaps introduce other new ones. Innovation, then, is bound to cause conflict as promoters of new institutions clash with defenders of old ones or with other promoters. These clashes are important for the fate of innovations, and we must consider them an intrinsic part of our subject. The subjects we will examine in this chapter are the nature of the enthusiasm that drives promoters and the resistance to change shown by defenders of current systems. We will examine enthusiasm and resistance first at the intellectual level, then in the context of social institutions.

Intellectual Resistance to Innovation

Inventions are often brought about by changes in the social fabric, and in turn inventions bring about other changes in society. Under these circumstances considerable resistance to invention and to its diffusion to a wider population is hardly surprising.[6] This resistance to innovation arises originally when the invention is first considered. Very few major inventions have not first been scorned by "experts" and social commentators.

Writers have compiled entire books of statements, later proved incorrect, regarding the impossibility of various inventions or discoveries.[7] Many of these statements seem absurd in the light of subsequent experience, but they were often delivered with great authority at the time, frequently by people whose opinions were very difficult to ignore. A great many of these statements, of course, were spoken casually to friends or associates; they often reveal little more than the speaker's lack of thought and research on the subject in question. Others, however, were the result of considerable reflection. In some cases the "expert" wrote an article or even a book presenting a well-developed brief against the possibility or usefulness of a given invention. These opinions cannot be dismissed as casual mistakes; rather, they require closer examination because they suggest some interesting features of thought about inventions.

An excellent example is Simon Newcomb's assessment, offered in 1901, of the impossibility of a heavier-than-air flying machine. Without "some new metal or some new force," Newcomb suggested, an airplane simply could not be invented.[8] Newcomb's statement was offered neither casually nor without considerable authority; at the time he was probably the most eminent scientist in the United States. Two years later the Wright brothers, without any such discoveries in physics or metallurgy, developed and flew the first powered airplane to carry a human being. Because Newcomb's statement is by no means unique, it is worth inquiring into this intellectual reluctance to consider new technological possibilities. Arthur C. Clarke has called the two major elements associated with this reluctance "failures of imagination" and "failure of nerve," respectively.[9] Let us examine them in turn.

Failures of Imagination. It is far easier to write an article explaining why something won't work than it is to invent it. Put another way, it is always easier to see the problems with a proposed invention than it is to see the solutions.[10] This is particularly a problem for experts, who know much better than laypersons how current systems work and who have encountered the barriers that inventors are trying to surmount. Whenever we cannot see how to do something, it is tempting to believe that others will fail at it as well. And even if others succeed, we may not attach much importance to their success. Such was the case of Elisha Gray, who nearly patented the telephone at the same time as Alexander Graham Bell but hesitated because his own arrogance and expertise blinded him to the importance of the invention.[11]

To invent something is to enter a new world, a world in which the invention is a reality. To consider willingly a new invention demands both a suspension of doubt and an extension of imagination. And in one sense people like Newcomb are always right: Invention demands that things change. But often they fail to understand exactly what must change. Often the changes they envision *are* impossible because they fail to see the possibilities in current knowledge. As a result they think that an invention of a certain kind would require much more radical change than experience proves is necessary.

One of the favorite ploys of skeptics is to invoke scientific laws to erect imaginary barriers to the invention. These barriers often prove surprisingly easy to penetrate. Thus Edison's skeptics suggested that he did not understand the laws of physics. One doubter, Dr. Paget Higgs, wrote that "much nonsense has been talked in relation to the incandescent electric light. A certain inventor has even claimed the power to divide the electric current indefinitely, not knowing or else forgetting that such a statement is incompatible with the well-proven laws of the conservation of energy."[12] Alas for the skeptics, Edison understood the principles better in certain particulars— the ones that counted—than his critics did. In such cases we might suspect that the doubters have not really examined carefully the known "facts" to determine what kind of possibilities are actually allowed by them. Thus one aspect of the failure of imagination is really a failure to *look,* an unwillingness to extend one's mind into the unfamiliar.[13]

Robert Fulton, who made a commercial success of the steamboat, recalled his reception during the experimental stage:

> When I was building my first steamboat, the project was viewed by the pub-
> lic either with indifference, or with contempt, or as a visionary scheme. My
> friends, indeed, were civil, but they were shy. They listened with patience to
> my explanations, but with a settled cast of incredulity on their countenances.
> As I had occasion daily to pass to and from the shipyard while my boat was
> in progress, I have often loitered unknown near the idle groups of strangers,
> gathering in little circles, and heard various inquiries as to the object of this
> new vehicle. The language was uniformly that of scorn, sneer, or ridicule.
> The loud laugh often rose at my expense; the dry jest; the wise calculation of
> losses and expenditures; the dull but endless repetition of "Fulton's Folly."
> Never did a single encouraging remark, a bright hope, a warm wish, cross
> my path. Silence itself was but politeness, veiling its doubts, or hiding its
> reproaches.[14]

It is fortunate that inventors, as we have seen, often possess the will to implement their ideas despite the lack of moral support. We might wonder, though, what happens to similar ideas in the hands of inventors whose drive to succeed is less strong, which brings us to the second failure associated with intellectual resistance, the failure of nerve.

Failures of Nerve. A failure of nerve is a fear of change. People fear change because it shakes up their world.[15] One of the virtues of adulthood is sufficient mastery of the world to have a reasonable grasp of how it is likely to operate. But inventions and discoveries force people to change these carefully developed perceptions, habits, and orientations.

For many people this kind of change is threatening due to insecurity or inflexibility of personality. Rather than envisioning a world in which things will be different, which requires personal adaptation, it is easier to deny that change is possible. This may be one of the reasons that a whole "pseudoscience" of imaginary dangers developed in relation to railroad travel.[16] The horrors a proposed innovation will bring about are often very much exaggerated, reflecting the feelings of a fearful person more than the likely outcome. Such fears, however, are not always without foundation; in some cases, as with the automobile, the negative effects anticipated were no worse than the reality, but they were different than those that actually occurred.

One striking example of almost willful ignorance is contained in a very interesting article by William Markowitz on whether UFOs could be real.[17] In discussing whether UFOs could be interplanetary spaceships or not, Markowitz states that UFOs could not be real because they do not look like the liquid-fueled rockets of the 1960s, do not leave behind burned ground, and so on.[18] "Hence," says Markowitz, "the published reports of landings and lift-offs of UFO's are not reports of spacecraft controlled by extraterrestrial beings *if the laws of physics are valid*" [my emphasis].[19] Note this powerful juxtaposition of the power of science and questionable reports of UFOs. Yet the laws of physics have very little to do with the issue. Markowitz must have known about the Lunar Excursion Module, even then being developed by NASA to land on the moon.[20] Although the LEM was not, strictly speaking, a spacecraft, it was certainly part of a spacecraft system—the relevant point here. The LEM could have easily fit the description of some of the reported UFO sightings—*and this was a device that was already in the test stage.* Thus Markowitz's argument about the impossibility of UFOs being spaceships could have been refuted by reference to the current technology of his own society. And of course we would expect real aliens to have even more impressive devices. Markowitz's failure to use current scientific knowledge in considering even the possibilities of spaceflight was not merely a failure of imagination; it suggests a genuine breakdown in reality-testing due to fear of the unknown.

Sometimes people manage the discomfort of change by maintaining that the innovation will change nothing of significance. A striking illustration of this blindness is offered by one of the great naval innovators of the early twentieth century. William S. Sims had brought the "gunnery revolution" of Percy Scott to the U.S. Navy and had been responsible for many reforms and improvements.[21] In 1909 Sims was present when New York City was holding its 300th anniversary celebration. One of the high points was a long flight by Wilbur Wright in an aeroplane over battleships anchored in the river. Sims was then commander of the USS *Minnesota,* a battleship. He watched Wright fly low over his ship, and

> later that day Sims was asked what would happen if such an event should take place in wartime. "At the height Mr. Wright was flying the ship would probably be able to get the range and destroy the aeroplane," Sims replied confidently. "At a greater altitude and going the speed Wright flew, the aviator's chance of dropping anything on a battleship would be small."[22]

Sims's attitude about airplanes and battleships remained unchanged by his extensive experience in World War I. It was not until 1921 that, as a result of simulations carried out at the Naval War College, he underwent a complete change of mind. He noted that one afternoon "we had a discussion with the entire staff over the whole matter, and it was easy to see that the question of the passing of the battleship was not an agreeable one to various members."[23] Indeed not. The Navy was moving into an uncomfortable world in which aircraft carriers would play a more important role than battleships. Subsequent study has shown that the most valuable weapon in World War I naval action, the torpedo, continued not to be taken seriously, even though mines and torpedoes accounted for 70 percent of combat losses among major warships in that conflict.[24]

Apparently the Navy still has a problem with such uncomfortable realities. In the Falklands War between Great Britain and Argentina the vulnerability of surface ships to airborne missiles was strikingly demonstrated.[25] Yet this lesson was not taken to heart by admirals who long before the 1980s had transferred their emphasis from battleships to aircraft carriers.[26] After the frigate USS *Stark* was put out of action by two Exocet missiles fired by an Iraqi plane, a reporter from the *New York Times*, John H. Cushman, Jr., had this exchange with the Vice Chief of Naval Operations for Surface Vessels, Vice Admiral Joseph Metcalf III:

> Cushman: "What does the attack on the *Stark* tell you about the vulnerability of surface ships in the face of increasingly sophisticated anti-ship missiles?"
>
> Adm. Metcalf: "The *Stark* took a hit in the worst possible place, and survived. Contrary to the statement that surface ships are vulnerable and can't take damage, I would suggest that the lesson from the *Stark* is the opposite. Our ship is there. It's floating [it was out of action]. The fire was right next to the ammunition magazines. Did they blow up? No, because we have developed insensitive munitions, that resist an explosion. They are pretty damn tough ships. This, a smaller ship, a ship which was not designed to go into heavy stress areas, survived."[27]

These and many similar remarks skim over an unpleasant fact: The ship *was* hit in the worst possible place. But there are probably no good places for a ship to be hit by a missile, and many antiship missiles are larger and more dangerous than the Exocets that hit the *Stark*.[28] This desire to avoid reality is worrisome. The Navy allows its admirals' affinity for the carrier to continue by providing an intellectual environment that does not admit the free flow of ideas and criticism generated by actual experience.[29]

One symptom of this affinity is the reluctance to allow sinking of aircraft carriers—large, vulnerable targets, but often admirals' flagships—during wargames or in computer simulations. One professor of the Naval War College, Thomas Etzold, stated bluntly that "in more than five years of experience as an umpire and adviser in high level naval war games, I have witnessed the unwillingness of senior naval officers to permit carriers to be sunk, even when taken under overwhelming attack."[30] Even when six hits were scored on a carrier during one simulation at the Naval War College, the officer responsible was given credit for reducing the carrier's efficiency by only 2 percent.[31] Thus even simulations of naval combat tend to preserve the importance of aircraft carriers by using protocols with unrealistic assumptions. The need for more submarines, obvious to many observers since World War I, is pushed aside in favor of giant carriers with huge price tags. The recently commissioned USS *Theodore Roosevelt* cost $2.46 billion to construct.[32] But many consider these huge, costly ships to be simply large, floating targets.[33]

People, then, develop skills, habits, and emotions associated with particular technological systems. Once a set of people has thus invested in a technology, resistance to changing it is natural because stability protects these investments in knowledge, skill, and emotion. Furthermore, we are often dealing not just with random individuals, but with a social institution with dynamics that may be intimately connected to particular technologies. It is not just admirals who do not like aircraft carriers to be sunk during wargames; the entire structure and functioning of the Navy protects such practices for many reasons, not the least of which is because it creates admirals in the first place. Technologies often support the social relations of the context that created them, and thus to change the technology is to change the technology's social system. It should not be surprising that those who occupy positions of importance in the current system resist overturning it in favor of a new one. *Intellectual resistance* to new technology is often the first line of defense in this struggle.

Calculative Rationality: The Rhetoric of Denial

Doubters of new technological possibilities have at their disposal a powerful *rhetoric of denial* that can be used to ward off concerns regarding technological change.[34] Such a rhetoric, or elements of it, can be brought into play whenever a novelty threatens to rearrange a social system associated with a technology. Here are some typical elements of such a rhetoric:

1. The proposed invention is impossible because it violates scientific laws.

2. The would-be inventors are not competent; true experts have already given up the possibility of the invention.

3. Even if the invention were developed, it would have only academic interest because routine use would cost too much.

4. If the invention were developed, it would have side effects so negative as to render it worthless or even dangerous.

5. Alternative lines of development that would meet the same needs seem more promising and are already being pursued by competent authorities; many of these improvements are made to current, well-understood systems.

Any inventor must face the possibility that doubters will use such rhetoric to discourage support for an invention. This intellectual resistance is important precisely because inventors seldom act alone. They may need technical assistance to aid in the development of the invention, they need backers to finance it, and they need institutions willing to adopt it. Without the interest and support of these groups, the invention may lie idle. So these doubts may have immediate negative impacts on the innovation process.

There is a flaw in this calculative rationality, however, and the flaw is the assumption that the rest of the world will remain the same, so that the conditions under which the technology works will never change much. But circumstances do change. An isolated culture may come in contact with other cultures, just as traditional Japanese culture was suddenly faced with the appearance of Commodore Perry in 1851. After 1851 the traditional strategies no longer worked; new ones had to be invented.

New technologies are also likely to create conditions such that doing things "the traditional way" no longer provides an efficient response. And much of social life involves conflict. In conflicts such as those involved in economic competition, law enforcement, and war, one can expect the other side to improve its capabilities; in doing so it may well develop new tools that are difficult to combat.[35] The existence of competition makes the use of calculative rationality questionable because competitors' innovations may change the effectiveness of one's own systems.

In their preparations for the America's Cup races of 1987, the contestants brought state-of-the-art capabilities into action in their attempts to develop an unbeatable yacht.[36] But no yacht will remain unbeatable forever; thus the need to anticipate change. The American automobile industry was buffeted sharply in the 1970s, first by oil crises in 1973 and 1978 and then by competition with high-quality automobiles from foreign countries. American firms, using calculative rationality, had vertically integrated[37] their production processes to such an extent that they could no longer adjust to changes rapidly. For instance, when Ford decided in the mid-1980s to produce its all-new Sable/Taurus line, the changeover was to cost $3.1 billion.[38] Clearly Ford was not "poised for change." Yet it was more poised for change than General Motors, which is even more vertically integrated.[39] Calculative rationality works less well when it must take into account the moves of opponents or a rapidly changing environment.

Generative Rationality:
The Rhetoric of Affirmation

For inventors, overcoming doubt is an important part of inventing, for doubts about an invention frequently disguise powerful emotions. The most important of these is "I don't want things to change!" The ability to cope with doubts and to infuse others with enthusiasm is essential to inventors, and willingness to persist in the face of doubt is necessary to keep those who are helping the inventor "on board."[40] It is no accident that one of the salient features of the creative personality is self-confidence; creators need this character attribute to continue the development of inventive ideas into physical form. Not only must inventors have courage in themselves; they must also infect others with the belief that the invention will work. Self-confidence is as essential as inventive ability because progress is often ambiguous.

Doubt is not irrational and is perfectly sound from the point of view of calculative rationality. People know what has worked for them before; it is simply not possible for some people to extend their minds beyond this experience. Inventors, however, are not so bound. They evoke a different, *generative rationality* that is based not on experience, but rather on imagined situations that the inventor proposes to create.[41] The inventor must use ingenuity to create these proposed situations and may not be able to specify in advance exactly how they can be brought about. This is the weak point upon which the doubters are likely to pounce and is a major target for the rhetoric of denial. What we have here, then, is a confrontation of *incompatible logics,* one oriented to the past, the other to the future. Each logic correctly predicts consequences from premises, but the premises are different.

Let us examine each of these rationalities a bit further. What we have called calculative rationality bases itself on a sure foundation: past experience. We know that something will work because we have tried it before and it worked. If we have tried it repeatedly and it has worked again and again, we are even more certain. If we tried alternatives and they proved to be less satisfactory, this adds more weight to our opinion. In short, to use the tried and true seems rational precisely because we know it well. There may be improvements, but we expect them to be marginal.

To avoid relying on the tried and true, the inventor or innovator may use a generative strategy. Along with it comes a rationale: the rhetoric of affirmation. Essentially the opposite of the rhetoric of denial, this series of guiding principles *supports* the innovation process. Consider a typical example of the rhetoric of affirmation:

1. Follow your hunches.[42]
2. If an answer to a problem exists, it will be found only by someone actively looking for it.
3. Today's impossibility will be tomorrow's off-the-shelf product.
4. If people did not experiment, *Homo sapiens* would still be living in caves.
5. Obstinate persistence is necessary to carry the project through.[43]
6. Even if serious bugs exist in the design, the best way to get them ironed out is to set a deadline by which the lack of solutions would create a disaster; this challenge assists the thinking process.[44]
7. The "experts" are usually wrong. Because all important innovations are opposed, the fact that opposition exists means nothing.

These ideas are not always articulated by the innovators themselves, but they are often implicit in much of what they do. The rationale provides aid and comfort to experimentation, risk-taking, and the pursuit of ostensibly wild ideas. The entire philosophy might be summed up in a single rhetorical question: "Why not try it?" Of course, the rhetoric of denial provides reasons why not. But the innovator surges on, eager to see what will happen.

Generative rationalities involve an imaginative leap from the known to the unknown. Whereas the calculative reasoner tends to think the way a spider weaves a web, carefully connecting all points in the progress of thought, the generative reasoner is more like a frog jumping from one lily pad to another. In the end, of course, the inventor must find some way of connecting the lily pads, a problem frogs do not have to face! The work of invention is thus dual: First comes the imaginative leap, then the often slow process of connecting the imagined possibility with current technological capabilities. We have already discussed how Goddard imagined a space vehicle in the flash of a promising possibility and how he spent the rest of his life developing rockets that could make the possibility a reality. It is significant to note that the "centrifugal drive" that Goddard mentioned in passing was not the means by which he later sought to reach out into space. Rather, it provided encouragement that his basic idea might be actualized and therefore ought to be pursued.

The point here is that believers are absolutely correct that in a large number of cases, *only those who diligently examine the technical possibilities will find the answers*. You will not find unless you look. Often we discover, after the fact, that inventors underestimated how much effort it would take to get the innovation going. Thus it is not unusual to hear statements similar to Howard Head's remark about inventing the metal ski: "If I had known it would take 40 versions before the ski was any good, I might have given it up." But fortunately inventors are exceedingly optimistic and keep plugging away until they succeed, while their critics believe that no matter how many times they try, the invention won't work.

William McLean, inventor of the Sidewinder missile (see Chapter 6), preferred to build a prototype of whatever system he was working on as soon as possible, instead of carrying out lengthy calculations and simulations in advance. McLean had a very high degree of "physical intuition" and often made striking forays into poorly understood areas. In these explorations, he wanted to think ahead, but he preferred not to address potential problems until he was sure they would make a difference.[45]

Toward the end of his life McLean was working on the Ocean Farm project, an attempt to farm the open oceans. In this project seaweed was grown on huge plastic meshes, which extended over considerable distances. One of the important aspects of the project was providing the seaweed a sufficient flow of mineral-rich water using upwelling pumps, in some cases 1,000 feet long, which brought water from below to the surface. The head of the project, Howard Wilcox, noted that McLean was developing a prototype for one of these pumps, and he was doing it without carrying out many of the calculations that would ordinarily be thought necessary with regard to the pump's performance. Wilcox then made some of the calculations on the pump himself. He showed them to McLean and asked why McLean did not do more calculations. McLean's reply is worth recording: "I prefer to let nature do the calculations. Then I know they are right."[46]

Calculative rationality, as we have seen, provides a whole series of "logical" reasons why things won't work.[47] Sometimes this skepticism is well founded, but sometimes it is simply an excuse for laziness. It is much easier to write an article that tries to prove that something is impossible than it is to invent the thing in question. It is therefore not surprising that for every major successful invention, there are a number of doubters who publicly voice their skepticism. These skeptics are often correct, and creativity does have limits. Enormous amounts of effort have been expended in the unsuccessful search for "perpetual motion" machines that would violate the Second Law of Thermodynamics. But it is ordinarily difficult to tell what will work until it is tried; only after the fact is it obvious what is and is not "possible." Hence those who try often disprove those who doubt.

Is Innovation Progress?

Technical progress occurs when better devices replace less adequate ones, and such replacement takes place through innovation—the invention, distribution, and use of new devices. However, it does not follow that innovation is always progress, that it is a transition to something better. It all depends, as we shall see, on what we wish to call progress.

To invent something is only to offer society a possibility; society can still decide whether the possibility offered is worth implementing or not. Thus the struggle to get something new into use is often more severe than that involved in developing it in the first place; it certainly involves more people. Decisions about adoption of the technology, furthermore, often hinge on a single question: Is the new technology better? Even from a purely technical point of view the answer to this question may be complex.[48] But adoption depends on more than this; it depends on what the technology will do to change the situation into which it is thrust, and often on the user's position in that situation.

Sometimes new devices are of dubious value. Many such devices are involved in modern medicine; one example is the fetal monitor. Fetal monitors are expensive ($10,000) and tricky to interpret, and they do not seem to produce any significant overall improvement in the health of babies. But thanks to habit and economic forces, they are still in use. "Technologies sort of appear abruptly, spread widely, and then don't fade away very fast even if later research doesn't show them to be very effective," commented one critic. Whereas many forces tend to maintain the use of fetal monitors, including commercial firms that sell them and legal threats (malpractice suits) if they are not used, few forces work to effect their discontinuance.[49] Similar problems are associated with operations that are dangerous but are commonly performed nonetheless.[50]

Such problems are intensified in medical operations involving expensive technologies, such as "intensive care" units, which in many respects are the vanguard of high-tech medicine.[51] Although many of these devices have considerable technological sophistication, they are very expensive, and the degree to which they relieve suffering and postpone death is not always clear.[52] Statistical studies of the welfare of patients suggest that improvements made by these procedures are often modest at best, and in some cases the procedures represent no improvement at all over having the patient cared for at home.[53] Ivan Illich, a social philosopher, has commented on a number of instances in which the benefits produced by high technologies are questionable, with medical "improvements" a major case in point. His book *Medical Nemesis*[54] argues that increasing "medicalization" of American society does not necessarily lead to greater healthiness.

In the United States "new and improved" is a phrase commonly used to describe consumer products. But are the products we use today better than those of previous generations? Consider books. Today books are generally made to last for a much shorter time than they once were. I have several books from the 1700s that are in very good shape. Few books printed today will last 200 years! The paper they are printed on, the inks used for printing, and the way they are bound all result in early obsolescence. What is new is not always better, but it may be cheaper to produce. Today books printed on 100 percent rag paper (as were most eighteenth-century books) would be much more expensive than current ones, which usually have only a fractional rag content. Thus many current "modern" products are in fact cheap substitutes for older, higher-quality ones. The matter of whether or not "new" is "better" involves several questions.

What Does the Invention Do? In most cases the answer to this question is relatively straightforward, but sometimes it is not. What the invention does in terms of its transformations of matter and energy is obvious; in what social role it should perform these transformations is not so obvious. The potential uses of the digital computer, for instance, constituted a large set of possibilities that were only slowly realized. The invention's "mission" in society is often obscure even to the inventor, whose idea of the technology's real function in society may be wildly off base. The telephone was originally expected to be an instrument of mass communication; the television was envisioned to be primarily an instrument for point-to-point communication. Their actual roles have turned out to be reversed.

By What Criteria of Performance Should the Invention Be Judged? Technological systems of any complexity have a number of performance features. Let's consider the steam engine, for instance. An engine develops a certain power, requires a certain amount of fuel, weighs a certain amount, takes up a certain amount of space, has characteristic noise and vibration features, and so on. What importance should be given to each of these characteristics in determining whether any particular engine is better than some alternative?

Just such an issue arose during the American Civil War over the steam-engine designs of Benjamin Franklin Isherwood, who, as engineer-in-chief of the U.S. Navy, was in charge of steam-engine design at the beginning of the conflict. He realized that during the war a large number of poorly trained engineers would serve in the Navy because it would be impossible to gather enough skilled ones. So he purposely designed for the Union fleets engines that compensated for this lack of talent:

With durability and reliability as his guiding principles, Isherwood designed engines which could withstand the manhandling of the clumsiest of mechanics. His engines were immensely strong, and consequently, immensely heavy. Fuel economy and power, though desirable qualities, took second place to simple mechanical dependability. In a period of rudimentary steam technology, when low steam pressures were necessary, Isherwood refused to employ more advanced theories which utilized high pressures and great degrees of steam expansion. The success of engines built on more sophisticated principles was too questionable, considering the low caliber of most engineers and the tremendous demands placed continuously on naval steam engines. His engine designs, therefore, became an easy target for those who criticised Navy steamers for lack of power or economy and for the excessive weight of their machinery.[55]

Thus, whereas reliability might seem to be the most important criterion to some engineers, other qualities might seem more important to other engineers. Unless one invention is superior to another on *all* criteria, then the superiority of one system over another is usually debatable. A consensus that progress has occurred can exist only if all parties judge using the same criteria.

Sometimes the criteria that are used to judge an innovation "superior" may seem quite arbitrary. During the Civil War many in the Union Navy debated about the form its ironclad ships should take: Should they have a low freeboard and turrets like John Ericsson's famous *Monitor,* or should they be higher ships with casemated sides like the *New Ironsides?* Each type of ship functioned best in a particular environment— the monitors as coastal defense vessels and the casemated ships as cruisers on the open sea. In the end the enthusiasts of the "monitor craze" won, a victory that some naval historians would subsequently view as misguided.[56]

Similarly, the initial adoption of firearms by the English army would seem to have taken place for superficial reasons. At a time when bows and arrows were still superior to guns in a number of characteristics, the British nonetheless adopted firearms largely because guns were considered the weapon of professionals. In addition, the increasing popularity of certain sports accompanied the decline of archery practice, which was an essential component of the military use of bows.[57] Prestige was also a strong motive for the adoption of clocks by many towns in Europe and Great Britain at a time when these instruments were far from reliable.[58]

How Is the Technology to Be Tested? Even if one can agree on the criteria to use in evaluating a given system, there is still the matter of actually testing it to see how it performs. Few who have had experience with research and development think that testing is a straightforward business. Consider, for instance, the following questions:

Is the technology to be tested under laboratory or field conditions?

What are the specific test procedures (protocols) to be used in the testing?[59]

Have those who are testing the technology really figured out what it is good for?

How many "bugs" remain at the time of testing and how serious are they (that is, how long will it take to fix them)?

How much training is necessary for the testing personnel? Have they really learned to use the device properly?

The issues raised by these questions assume particular importance when we consider a number of classic instances of inconclusive testing: the failure of the Bessemer process with high-phosphorous iron ores; Rudolf Diesel's initial failure to produce a consistently workable engine; the short life of Norman Shumway's first heart transplant patient. The great problem with testing is that different interpretations of the results are possible. Was the test a failure? Can the "bugs" that were responsible for problems be ironed out? The designers of the Fermi I reactor, which suffered a serious meltdown, thought that the problems could be corrected. Whereas some critics saw the meltdown as potentially catastrophic, the reactor's sponsors claimed it was simply a glitch that interrupted an otherwise worthwhile program.[60] Did the repaired reactor meet all required standards? Many would argue that the test conditions were too favorable and that the system would not work outside this narrow envelope.[61] Another example of controversy in testing is the rejection by a Navy board of Isherwood's battleship *Wampanoag,* the "fastest ship in the world," after outstanding sea trials.[62]

The issue of experimentation can be very complicated. We have seen how Isherwood favored rugged steam engines for the Union fleet. He also believed that steam engines generally ought to have a long cutoff—that steam should be forced into the cylinder well into the stroke, rather than cutting it off early, so that the steam's natural expansion would do most of the work. This preference contradicted engineering principles that had wide support at the time. Isherwood favored the long cutoff as a result of a lengthy course of experiments he had conducted on engines in various naval applications. Others, however, were violently opposed to the long cutoff and impugned both Isherwood's ideas and his basic competence. One rival, Edward Dickerson, constantly attacked the Navy in print for wasting its money on Isherwood's ideas. The press, including the *New York Times,* aided these attacks. Only after repeated trials in which Dickerson's engines failed to perform as well as Isherwood's was Dickerson silenced.[63]

"Better," then, is relative and is basically a matter of perception. Among the many sources of perception is tradition. In any long-standing group or collectivity, some kinds of truths will come to seem self-evident. These "obvious" truths often appear to be certainties simply because no one has seriously questioned them.[64] In earlier times a good deal of technological knowledge was of this character. In relatively stable societies it is understood that certain methods are the desirable ways to do things, and these ways are unquestioned. In these cases the justification for believing in the superiority of a given technology is the belief of others who support the tradition.

The same principle can also be applied to innovations. Very often what is attractive about innovations is that others we respect think they are superior. We have seen how "prestige" considerations influenced the adoption of firearms and clocks. These innovations were adopted because important groups in the broader social context considered them to be improvements. And for the most part, such socially established norms about what is superior must prevail because most individuals and groups have neither the time nor the expertise to determine experimentally what is best.[65] It is interesting to note that the consideration of alternatives is often in and of itself a radical act. To inquire—to suspend judgment—is a reflex of creative people and highly creative organizations.[66] Inquiry is often the prelude not only to innovation, but also to primary invention stimulated by discontent with current arrangements.

The Diffusion of Innovations

One of the most interesting topics in the sociology of technology concerns the mechanisms by which technology is transferred from one part of the society to another. This transfer can take place within a society, between societies, or even within a single firm. These "technology transfers" are important, but here we will given them only the most summary treatment because the subject is too large for the space that is available. Literally thousands of studies in at least three distinct research traditions have explored innovation diffusion.[67] A single work summarizing this field would be enormous, although Everett Rogers has written several outstanding reviews.[68]

Quite a variety of studies have been done on the nature of the adopters of technology—why they adopt, the kinds of adoption decisions they make, what kinds of things they adopt, the pattern of adoptions over space and time, and so forth. These studies cover a great deal more than technology, including virtually every kind of social and administrative practice from family planning to Dial-A-Ride. Here we will note a few of the major conclusions of these studies.

As we might expect, adopters of innovations tend to be socially "central" in a variety of ways: They tend to be better educated, to have more contact with the mass media and with change agents, and to be opinion leaders.[69] Larger firms and managers of larger units tend to find out about innovations sooner, and they tend to adopt innovations more rapidly.[70] Readily adopted innovations are those that offer some kind of comparative advantage to users and that tend to be easier to observe, understand, and try out.[71]

A variety of innovation patterns also exist. For instance, slow diffusion may occur as a result of geographical or social proximity, a process often studied by geographers; alternatively, relatively more rapid diffusion can be stimulated by sponsors.[72] Then there is a very interesting form of "generative" diffusion, described by Donald Schon, in which a diffusion center interacts with peripheral elements.[73] Over time, the study of technological diffusion has involved increasingly larger subject areas, and it seems at times to extend to the entire sociology of technology.

As we shall see in the following chapters, a number of issues relevant to diffusion, especially sponsorship, are important to our discussion, and we will touch on many of them. There is a large literature devoted specifically to technological diffusion, and interested readers are encouraged to consult the more thorough works cited in this chapter's notes.

Conclusion

To introduce new technology is to threaten a society with institutional change. Even the suggestion of new technology can be threatening to a society. One of the key issues, then, concerns what people expect the new technology to do. Those threatened by the proposed changes will resist the new technology intellectually and socially. On the other hand, those who think that the new technology will allow them to do things they currently cannot do, will feel and act very differently toward it. Often innovation becomes a "tug of war" between the innovative and conservative forces.

Not all innovations are useful, and there may be reason to resist some of them in whole or in part. Thus it would be wrong to view conservatism as a purely negative force; some things ought to be resisted. But innovation and conservatism work on different logics, which we have called generative rationality and calculative rationality, respectively. It may be very difficult, given this difference in outlook, for the two sides to communicate. For this and many other reasons, the process of innovation is often a struggle, albeit a very interesting and important one. We will consider other parts of this struggle when we discuss technology assessment and related issues in the final chapters of the book.

Notes

1. See the important essay by Nathan Rosenberg, "Problems in the Economist's Conceptualization of Technological Innovation," in his *Perspectives in Technology* (Cambridge: Cambridge University Press, 1977).

2. George B. Baldwin, "The Invention of the Modern Safety Razor: A Case Study of Industrial Innovation," *Explorations in Entrepreneurial History,* Vol. IV (Dec. 15, 1951), pp. 73–102. A splendid essay.

3. See Lynn White, "The Act of Invention: Causes, Contexts, Continuities, and Consequences," *Technology and Culture,* Vol. 3, No. 4, pp. 486–500.

4. Ronald E. Rice and Everett M. Rogers, "Reinvention in the Innovation Process," *Knowledge: Creation, Diffusion, Utilization,* Vol. 1, No. 4 (June 1980), pp. 499–514.

5. Boelie Elzen, "Two Ultracentrifuges: A Comparative Study of the Social Construction of Artifacts," *Social Studies of Science,* Vol. 16, No. 4 (Nov. 1986), pp. 621–662. On the con-

cept of framing, see Wiebe Bijker, "The Social Construction of Bakelite: Toward a Theory of Invention," in W. E. Bijker, T. P. Hughes, and T. Pinch (Eds.), *The Social Construction of Technological Systems* (Cambridge, Mass.: MIT Press, 1987), pp. 159–187.

6. Bernhard J. Stern, "Resistances to the Adoption of Technological Innovations," in National Resources Committee, *Technological Trends and National Policy* (Washington, D.C.: U.S. Government Printing Office, 1937), pp. 39–66. This is a basic essay that anyone interested in resistance to new technologies should read.

7. A recent collection is Christopher Cerf and Victor Navasky, *The Experts Speak: The Definitive Compendium of Authoritative Misinformation* (New York: Pantheon Books, 1984). The reader should treat this material with some circumspection, for the authors have not always been careful to trace quotations to their original sources.

8. Simon Newcomb, "Is the Airship Coming?," *McClure's Magazine,* Vol. 17, No. 5 (1901), pp. 432–435; see also Eugene Garfield, "Negative Science and 'The Outlook for the Flying Machine,'" *Current Contents,* No. 26 (June 27, 1977), pp. 5–27.

9. Arthur C. Clarke, *Profiles of the Future* (New York: Holt, Rinehart & Winston, 1984), pp. 15–36. I highly recommend Clarke's discussion because he writes so well and he lived through some of the failures in question. Interestingly enough, before I read Clarke's book I had decided on exactly the same two categories, which makes me think they may be the correct ones.

10. M. D. Sturge, "The Impossible in Technology: When the Expert Says No," in Philip J. Davis and David Park (Eds.) *No Way: The Nature of the Impossible* (New York: Freeman, 1987); pp. 111–138.

11. David A. Hounshell, "Elisha Gray and the Telephone: On the Disadvantages of Being an Expert," *Technology and Culture,* Vol. 16, No. 2 (Apr. 1975), pp. 133–161.

12. Francis Jehl, *Menlo Park Reminiscences* (Dearborn, Mich.: Edison Institute, 1936), p. 231.

13. See Harold A. Linstone, "Eight Basic Pitfalls: A Checklist," in H. A. Linstone and M. Turoff (Eds.), *The Delphi Method: Techniques and Approaches* (Reading, Mass.: Addison-Wesley, 1975), pp. 573–586.

14. George Iles, *Leading American Inventors* (New York: Henry Holt, 1912), pp. 60–61. Copyright 1912 by Henry Holt. Reprinted with permission of the publisher.

15. Peter Marris, *Loss and Change* (London: Routledge and Kegan Paul, 1974).

16. See, for instance, Wolfgang Schivelbusch, *The Railway Journey: The Industrialization of Time and Space in the 19th Century* (New York: Urizen Books, 1979); Schivelbusch catalogues many of the imaginary dangers of railway travel. See also Stern, "Resistances," p. 42.

17. William Markowitz, "The Physics and Metaphysics of Unidentified Flying Objects," *Science,* Vol. 157 (Sept. 15, 1967), pp. 1274–1279. This article is a gold mine for students of rhetoric and psychology.

18. Actually, the best guarantee that UFOs could not be real would be if they were reported to look like these same chemical rockets, which certainly could not get from one star system to another.

19. Markowitz, "Physics," p. 1276.

20. Stanton Friedman, a well-known lecturer on UFOs, refers to UFOs as "earth excursion modules." But Friedman was part of the aerospace community and had actually worked on interstellar rockets. I suspect that Markowitz, then a professor of physics at Marquette University, had not had this highly relevant experience.

21. Elting Morison, "Gunfire at Sea," in his *Men, Machines, and Modern Times* (Cambridge, Mass.: MIT Press, 1966); pp. 17–44.

22. John Walsh, *One Day at Kitty Hawk* (New York: Crowell, 1975), p. 247.

23. Elting Morison, *Admiral Sims and the Modern American Navy* (New York: Russell and Russell, 1968), p. 505.

24. Rodrigo Garcia y Robertson, "The Failure of the Heavy Gun at Sea, 1898–1922," *Technology and Culture*, Vol. 28, No. 3, (July 1987), pp. 539–557.

25. Gregg Easterbrook, "What Caspar Weinberger Could Learn from the Falklands," *Washington Monthly*, Sept. 1984, pp. 43–48. Nor was this the first time that a ship had been sunk by a naval missile: In 1967 the Israeli destroyer *Eilat* was sunk by a Russian Styx ship-to-ship missile. But learning is slow.

26. William McLean, inventor of the Sidewinder missile (see Chapter 6), was so convinced of the vulnerability of surface ships to missiles that after 1954 he turned his inventive attentions solely to underwater craft. Interview with Howard Wilcox, February 1987.

27. John H. Cushman, Jr., "Exocet Attack in Gulf Prompts a Hard Look at Fleet Strategy," *New York Times*, May 31, 1987, Section 4, p. 3. Copyright 1987 by the New York Times Company. Reprinted by permission.

28. I have been told that in the case of the two British ships sunk by Exocet missiles during the Falklands conflict, neither missile actually exploded. These ships were thus sunk by "dud" missiles. If a relatively small "dud" missile can sink ships so easily, one wonders what a large, exploding missile might do.

29. John Lehman, Secretary of the Navy from 1981 to 1987, did not encourage free examination of such sacred naval doctrines as the superiority of aircraft carriers; Daniel Charles, "The Navy After Lehman: Rough Sailing Ahead?" *Science*, Vol. 236 (Apr. 3, 1987), pp. 22–25.

30. Frank Greve, "Navy Duping Itself on Strengths, Experts Say," *Detroit Free Press*, May 31, 1982, Section A, pp. 1, 2.

31. Steven Strasser, David Martin, James Doyle, Mary Lord, and John Lindsay, "Are Big Warships Doomed?" *Newsweek*, May 17, 1982, pp. 32–44.

32. "USS *Theodore Roosevelt* Joins Active Service as Fifteenth Carrier," *Washington Post*, Oct. 16, 1987, Section A, p. 21.

33. Evan Thomas, "Carrier Power: America's Floating Forward Bases," *Time*, May 5, 1986, pp. 14–19.

34. Such a use of rhetoric is discussed in Tim Clark and Ron Westrum, "Paradigms and Ferrets," *Social Studies of Science*, Vol. 17, No. 1 (Feb. 1987), pp. 3–33.

35. An example of such tactics is the use of "go-fast" boats by drug runners at sea and cellular phones by their counterparts on land; see Elaine Shannon and Aric Press, "Running Silent, Running Fast," *Newsweek*, Oct. 27, 1986, pp. 94–95; and Terry Johnson, Hugh Brooks, Karen Springen, and Todd Barrett, "Crime: Dialing for Dollars," *Newsweek*, Sept. 14, 1987, p. 42.

36. William Marbach, Carl Robinson, Peter McAlevey, and Karen Springen, "High Tech on the High Seas," *Newsweek*, Oct. 20, 1986, pp. 66–67.

37. Vertical integration occurs when a firm includes many phases of its production process under its own roof. For instance, an automobile company that is vertically integrated is more likely to own the firms that supply parts for its cars, instead of ordering them from

independent firms; see William J. Abernathy, Kim B. Clark, and Alan M. Kantrow, *Industrial Renaissance: Producing a Competitive Future for America* (New York: Basic Books, 1983).

38. Robert England, "As Ford Constellation Rises, Taurus Is the Brightest Star," *Insight,* July 13, 1987, pp. 15–17.

39. John Holusha, "Advice for a Humbled Giant," *New York Times,* Dec. 7, 1986, Business Section, pp. 1, 32.

40. This issue of keeping other groups "on board" is lovingly examined in Bruno Latour, *Science in Action* (Cambridge, Mass.: Harvard University Press, 1987), Chapter 3.

41. See Thomas Kuhn, *The Structure of Scientific Revolutions,* 2nd ed. (Chicago: University of Chicago Press, 1970), pp. 111–135.

42. A leading contemporary believer in following one's hunches is Tom Peters, whose book *Thriving on Chaos* (New York: Knopf, 1988) provides aid and comfort to corporate experimenters. He recommends a large number of "fast failures" (pp. 258–266), a concept utterly foreign to calculative rationality because its basic orientation is to avoid any failures in the first place.

43. That inventors are often extremely persistent is shown by examples in Chapter 5. Other examples are readily apparent. The discoverer of meteorites, Ernst Chladni, said this about his investigative style:

> People have often asked, *by what stroke of luck* I came to make certain discoveries. But luck never favored me; to obtain success, I always had to employ an opinionated persistance.
>
> E. F. F. Chladni, *Traité d'Acoustique* (Paris: Courcier, 1809), p. iv.

44. In industry this strategy is known as "betting on the come," that is, betting that the right ideas will come along when one needs them. Thomas Edison used this strategy in developing the electric light; he took a tremendous gamble by promising the invention before he had it perfected. See Robert Friedel and Paul Israel, with Bernard S. Finn, *Edison's Electric Light: Biography of an Invention* (New Brunswick, N.J.: Rutgers University Press, 1986), p. 13. Fortunately, this strategy paid off for him. But it can also backfire. In Livermore Laboratory's unsuccessful attempt to build a nuclear fusion reactor, the right ideas failed to materialize. John Clarke and Edwin Kintner, officials at the U.S. Department of Energy, noted that many people believed strongly in the project.

> Explains Clarke: "We were always betting that somebody would come up with a clever idea down the road." Ever the optimist, Kintner was also fond of quoting the father of the atomic submarine, Vice Admiral Hyman G. Rickover: "Where there is no vision, there is no future."

See William Booth, "Fusion's $372 Million Mothball," *Science,* Vol. 238 (Oct. 9, 1987), p. 154. Unhappily, in this particular case, vision was not sufficient to guarantee a future. In industry, however, "betting on the come" is a common and often useful strategy.

45. Howard Wilcox, personal communication. It is notable that in developing the steam railroad, some engineers believed that an engine would not "pull" on a smooth track, so they invented a toothed track. Actual experiment, however, rapidly demonstrated that trains would pull on a smooth track. So in this instance the engineers had constructed an imaginary difficulty for themselves. See Samuel Smiles, *The Life of George Stephenson* (Boston: Ticknor and Fields, 1858), pp. 80–85. It was this kind of problem that McLean was trying to avoid.

46. Another of McLean's coworkers, Douglas Wilcox, told me that "McLean's only supervisor was the laws of nature."

47. Early science also suffered from such *a priori* doubts. The philosopher Sextus Empiricus (who lived circa A.D. 200) brought together a number of writings of the Greek Skeptics, a collection that asserts the difficulty of establishing anything with certainty in science. If one took these arguments seriously—and many intellectuals did—science would seem to have little hope. Fortunately, experience proved that many of these doubts, even if well founded in theory, could be ignored in practice. Science has developed in spite of philosophical difficulties of this kind. See Richard H. Popkin, *The History of Scepticism from Erasmus to Descartes* (Assen, Netherlands: Van Gorcum, 1960).

48. Science cannot always provide easy answers to such questions. For an idea of the complexities introduced by new products, see "AD-X2: The Difficulty of Proving a Negative," in U.S. House Subcommittee on Science, Research, and Technology, Franklin P. Huddle (Ed.), *Technical Information for Congress,* 3rd ed. (Washington, D.C.: U.S. Government Printing Office, 1979), pp. 23–72.

49. Tamar Lewin, "Despite Criticism, Use of Fetal Monitors Persists," *New York Times,* Mar. 27, 1988, Section I.

50. For instance, see Associated Press, "Study Critical of Antistroke Operation: Some Patients Harmed by Carotid Endartectomy, Rand Says," *Washington Post,* Mar. 24, 1988. More generally, see Frederick Mosteller, "Innovation and Evaluation," *Science,* Vol. 211 (Feb. 27, 1981), pp. 881–886.

51. Louise B. Russell, *Technology in Hospitals: Medical Advances and Their Diffusion* (Washington, D.C.: Brookings Institution, 1979).

52. Bryan Jennett, *High-Technology Medicine: Benefits and Burdens* (Oxford: Oxford University Press, 1986).

53. Russell, *Technology in Hospitals,* pp. 65–70.

54. Ivan Illich, *Medical Nemesis* (New York: Bantam Books, 1977).

55. Edward William Sloan III, *Benjamin Franklin Isherwood, Naval Engineer* (Annapolis, Md.: United States Naval Institute, 1965), pp. 33–34. Copyright 1967 United States Naval Institute. Reprinted with permission.

56. Sloan, *Isherwood,* pp. 49–77.

57. Thomas Esper, "The Replacement of the Longbow by Firearms in the English Army," *Technology and Culture,* Vol. 6, No. 3, (Summer 1965), pp. 382–393.

58. Carlo Cipolla, *Clocks and Culture* (London: Collins, 1967), pp. 41–43.

59. For many medical treatments, randomized clinical trials may yield very different results than do unsystematic observations; see John Tukey, "Some Thoughts on Clinical Trials, Especially Problems of Multiplicity," *Science,* Vol. 198 (Nov. 18, 1977), pp. 679–684; and other articles in this special issue on "Biostatistics in Medicine," pp. 675–705.

60. For differing views on the accident, see John Fuller, *We Almost Lost Detroit* (New York: Ballantine Books, 1975); and E. Pauline Alexandersson (Ed.), *Fermi: New Age for Nuclear Power* (Lagrange Park, Ill.: American Nuclear Society, 1979).

61. Many military systems are in fact given "middle of the envelope" tests that are easy for them to pass; see Timothy Noah, "It's Hard to Flunk a Weapons Test," *Newsweek,* Sept. 7, 1987, p. 18; and Wayne Biddle, "How Much Bang for the Buck?" *Discover,* Sept. 1986, pp. 50–63. On the other side, systems that do not have sufficiently influential sponsors can be subjected to whatever testing is necessary to get them to fail, as happened with some of the testing on the M-16 rifle.

62. Elting E. Morison, *Men, Machines, and Modern Times* (Cambridge, Mass.: MIT Press,

1966), pp. 98–122. The context of the events Morison describes is more fully spelled out in an essay by Lance C. Buhl, "Mariners and Machines: Resistance to Technological Change in the American Navy 1865–1869," *Journal of American History,* Vol. 61, No. 3 (1974), pp. 703–727. Dean Allard was helpful in bringing this reference to my attention.

63. Sloan, *Isherwood,* pp. 105–157.

64. John W. Meyer and Richard Rowan, "Institutionalized Organizations: Formal Organizations as Myth and Ceremony," *American Journal of Sociology,* Vol. 83, No. 2 (1977), pp. 340–363.

65. Wiebe Bijker and Trevor Pinch refer to traditional use as the "stabilization of artifacts" in their essay "The Social Construction of Facts and Artifacts," in W. E. Bijker, T. P. Hughes, and T. Pinch (Eds.), *The Social Construction of Technological Systems* (Cambridge, Mass.: MIT Press, 1987). Comparable observations have been made in organization research; see John Meyer and Brian Rowan, "Institutionalized Organizations: Formal Structure as Myth and Ceremony," *American Journal of Sociology,* Vol. 83, No. 2, pp. 340–363.

66. Some organizations test routinely, but they are the exceptions. The outstanding example of such an experimentally oriented organization is the Honda automobile company.

67. The three traditions come from geography, economics, and sociology, respectively; see "The Diffusion of Innovations," in Patrick Kelly and Melvin Kranzberg (Eds.), *Technological Innovation: A Critical Review of Current Knowledge* (San Francisco: San Francisco Press, 1978), pp. 119–150. This is a thorough, and somewhat overwhelming, review.

68. See Everett Rogers, *The Diffusion of Innovations,* 3rd ed. (New York: Free Press, 1983). Rogers is the major intellectual synthesizer in this field, and his book is both well written and interesting.

69. Rogers, *Diffusion of Innovations,* pp. 271–311; see also James S. Coleman, Elihu Katz, and Herbert Menzel, *Medical Innovation: A Diffusion Study* (Indianapolis: Bobbs-Merrill, 1966).

70. At least this is the conclusion of some studies by Edwin Mansfield, cited in Kelly and Kranzberg, *Technological Innovation,* p. 126. Others, however, have not found this relationship; see, for instance, L. Nabseth and G. F. Ray, *The Diffusion of New Industrial Processes* (Cambridge: Cambridge University Press, 1974), pp. 307–308.

71. Rogers, *Diffusion of Innovations,* pp. 210–240.

72. For more on sponsored diffusion, see Lawrence A. Brown, *Innovation Diffusion: A New Perspective* (London: Methuen, 1981), pp. 100–151.

73. Donald A. Schon, *Beyond the Stable State* (New York: Norton, 1973), pp. 80–115; also see Rogers's comments on this topic in *Diffusion of Innovations,* pp. 333–346. An example of such a system, a clearinghouse for farm innovations, is described in the anonymous article "Program Helps the Disabled Farmer," *New York Times,* July 18, 1988.

CHAPTER 9

The Sponsorship
Of Technology

In this chapter we confront one of the key
issues of this book: how technology is socially shaped and supported. Many people
view technology as autonomous, as a self-
directed force.[1] From this viewpoint the evolution of technology is something that drives
and directs itself, rather like a chain reaction,
which, once begun, goes inevitably in a certain direction. This view seems difficult to sustain, however, once we've looked in detail at
what actually happens. Far from proceeding
automatically, technologies are pushed along
or retarded by those who take responsibility for their management and benefit from
their growth or stagnation; technologies are
shaped by the institutions that manage them.

Thus far we have largely discussed how
technologies originate and spread. But technologies require support from the groups or
organizations that invest resources in developing them, diffusing them, and managing
them. These sponsoring groups will be the
focus of this chapter.

Each technology, in the process of its development, is likely to develop an accompanying technological community. These technological "support groups" have considerable power in influencing specific developments. Not only can such groups sustain particular technologies; they can also hamper the growth of others. The institutions that assume (or usurp) the control of particular technologies then shape the destiny of a society by deciding which technologies will operate in it and what their configurations will be. We call an institution that is the major force in this process the technology's *sponsor*. If we are to understand the dynamics of technologies in society, then we must understand the dynamics of these sponsors, which may be makers, users, regulators, and so on. This line of inquiry will lead us back to the basic issue of the adequacy and appropriateness of our society's institutions for creating, managing, and coping with technology.

How Sponsors Shape Technology

We take as a basic principle the view that things reflect their makers. We expect a Steinway piano to possess certain qualities because the maker is known to operate according to certain high standards and principles.[2] Fishing rods made from Tonkin cane by craftsman John Kusse are considered to be among the best in world because of the loving care with which he makes them by hand.[3] But devices also reflect forces other than their makers, including the availability of materials, previous owners, the nature of the client for whom the technology was designed, social traditions, and legal constraints.[4] From initial conception to final use, every technology is shaped—pushed and pulled—by a host of influences.[5] Because designers and innovators respond to these outside influences, the technology reflects those who exercise this influence, which may come from market demand, from legal authority, or from access to other key resources. The physical act of making a device, then, is influenced by a series of institutions and forces that affect what will be made and how it will be used (see Figure 9-1). In the following sections we will make several observations about these institutions and forces.

The Nature of Sponsors. *The evolution and structure of a technological system are shaped by the social institution that sponsors it.* An everyday example of this principle is a person's home, especially when the homeowner had a hand in designing the house.[6] When someone buys an already-built house, its design reflects the actions of the builder and the house's previous owners. The longer a person has the house, the more its contours, colors, furnishings, and general physical condition reflect the owner's characteristics. And then there is the basic fact that the owner *chose* the house in the first place. We could go a

Figure 9-1

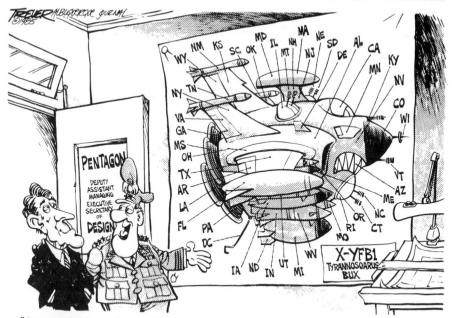

"AT LAST! A WEAPONS SYSTEM ABSOLUTELY IMPERVIOUS TO ATTACK: IT HAS COMPONENTS MANUFACTURED IN ALL 435 CONGRESSIONAL DISTRICTS!"

SOURCE: *Reprinted by permission of John Trever,* Albuquerque Journal.

step further by suggesting that the nature of the eventual homeowner shaped the technology in the first place; put another way, the technology was designed for its intended users.

The same principles apply on a much larger scale for public systems. The development of nuclear power in the United States took place under heavy sponsorship by the federal government. The government's early interest and support encouraged utilities to develop nuclear power in the face of serious financial problems that otherwise would have severely retarded its growth. In particular, federal help with fissionable materials and insurance problems made a critical difference.[7] The Price-Anderson Act of 1957 allowed private firms to ignore the most serious concern, a "catastrophic accident," the costs of which would exceed private insurance companies' ability to pay. The dangers, both known and suspected, were downplayed to the public.[8]

Similar government influence has taken place in programs to control insect pests using chemical secretions called pheromones. Insects communicate using pheromones as chemical signals, and their breeding behavior can be altered by placement of pheromones into the environment. The result is less breeding and thus fewer insects. But federal regulation of pheromones has been indecisive, sending confusing signals to the sponsors of pheromone technology.[9] Although the Environmental Protection Agency takes the position that pheromones are less dangerous than the usual chemical poisons used on insects, pheromones are classified as a pesticide, and pesticides are carefully regulated because of past experience. Sponsors of pesticide chemicals are uncertain about how to use biological substances like pheromones. "We all know how to replace an old chemical with a new chemical but not how to replace a chemical with a pheromone," a chemical industry representative is quoted as having said.[10] Furthermore, because of the absence of dead bugs that chemical poisons usually produce, growers distrust the effectiveness of pheromones. Thus, in spite of recent impressive tests of pheromones, their use has been limited.

Sponsors' Agendas. *Technologies are designed according to the wishes of their sponsors.* We should note here that sponsors and users are not always the same. Whereas houses and other consumer products are maintained by the people who use them, quite a number of large, important technologies are owned, managed, or otherwise controlled by people other than the users. Factory equipment, public buildings, military weapons systems, transport systems, and the like have keepers who are distinct from those who use them. Consequently, some users have relatively little influence on technology with which they come in contact, with alienation and stress as predictable results.[11] Furthermore, we often find that technology is controlled not by individuals but by organizations and groups.

The Internal Dynamics of Sponsors. *The fate of a technology is often shaped by the internal dynamics of the institution that sponsors it.* The classic example here is the ill-fated history of the battleship *Wampanoag,* built by the U.S. Navy shortly after the Civil War. Designed by Benjamin Franklin Isherwood, then Engineer-in-Chief of the Navy, the *Wampanoag* was, in its day, the fastest ship in the world. Even though it was an exemplary weapons system, it did not fit well into the worldview of the more traditional naval officers, who resented its novel design, its dependence on steam, and its designer. With these attitudes on the part of its keepers, it is not surprising that this beautiful and effective ship languished from misuse and was eventually discarded by the Navy.[12]

In some cases the directors of technological systems are genuine "hands-on" managers who have the sense and skill to manage the system well. Arthur Squires points to the atomic bomb and the fluid catalytic cracking process as examples of systems developed by what he calls technological "maestros."[13] When such people are not in charge, however, system development is often badly handled, and serious problems and failures are likely. Squires cites the problems with the M-16 and the *Challenger* explosion as examples of the consequences of badly managed systems.[14]

Relationships Among Sponsors. It is not only what goes on inside the institution that is important, but also the sponsoring institution's relationship to others in the social structure. Thus *the fate of a technology is often determined by the sponsoring institution's pattern of relationships with sur-*

rounding organizations. Solar power is an excellent example of the importance of a sponsor's relationships.[15] This technology, which shows great promise, is largely undeveloped because the advocates and developers of solar power are groups with relatively little social influence. Although these groups have often been able to influence educated members of the public, solar power as a technology remains in a primitive stage because far more powerful groups do not want it developed.[16] The gas, oil, and nuclear power industries consider solar power a threat and have blocked its commercial development. Had these large, powerful industries developed solar power in the first place, its fate would be very different.

Clearly, a technology's development and success are intimately connected with the social institutions that sponsor it. It is tempting to say that its success is bound up with that of its beneficiaries, but the benefits and beneficiaries of a technology are not always clear. Technologies of great value may be stubbornly resisted even by those who stand to benefit by them.[17] *Expected* costs and benefits, however, are quite important considerations in technological decisions.

Sponsors' Strategies. *The fate of a technology may rest with the sponsor's strategy for using it.* Institutions sponsor certain technologies and resist others.[18] Because inventions become innovations only with the aid of sponsors, a technology's sponsoring institution may determine its fate. The crucial matter, then, is "the soil on which the new seed falls." If the innovation finds fertile soil—a powerful and resolute sponsor—it may take root, grow, and blossom. In many cases, however, the innovation finds barren ground—an unsupportive sponsor; this may occur most frequently when the invention requires considerable further development, which many do. Such was the case, for instance, with the diesel engine.[19]

A sponsor's strategy for the use of a technology may include what has been called "heterogeneous engineering"—designing the community of users at the same time the hardware is designed.[20] Michel Callon has written of the social engineering used by a technology's sponsors to provide an environment in which the technology can operate.[21] For instance, sponsors sometimes limit the applications of their devices through poor product introduction or by aiming it at the wrong market. In a series of articles on the telephone, Claude Fischer has shown how A.T.&T.'s strategy shaped the way in which the telephone developed.[22] Rural customers tended to be neglected, even though they were very interested in telephone service; they were left to be dealt with by independent telephone companies in many cases. Henry Ford, on the other hand, consciously marketed his automobile to rural Americans because he shared their values.[23] The railroads systematically neglected passenger trains, because they made more money on freight. The result has been that more intercity passenger traffic moves by airplane and by automobile.[24] In a more humble example, Glennis Yeager, in trying to encourage the adoption of condoms as a birth control measure in Pakistan, discovered that the men did not like the white condoms distributed by population control agencies. White is the color of purity, and thus to use a white condom for sex was embarrassing. Yeager subsequently ordered them "in every color of the rainbow," which very much improved their acceptance. The orange ones were liked best.[25]

Sponsors' Economic Powers. *In some cases sponsors may hold so much economic power that they may be able to influence social choice, not just among devices, but among entire technologies.* In the United States this has been true of the automotive industry in relation to public transportation. Companies that manufactured buses bought city trolley car companies and had the tracks ripped up to prevent future competition. In California, National City Lines (a holding company formed by General Motors, Standard Oil of California, and Firestone Tire and Rubber) bought Pacific Electric, the world's largest interurban railway. It then proceeded to rip up the 1,100 miles of its tracks.[26] This move has been credited with helping to create Los Angeles's freeway problems.

The creation of the nuclear power industry meant developing not only a technology, but a powerful sponsor as well.[27] Now any new energy technology must face not only the tremendous sunk costs of its own plants and equipment, but the powerful and well-staffed lobby of the nuclear power industry, including an extensive network of scientific advisors. Electrical utilities have lobbied successfully against the allocation of public funds to develop solar power.[28] It was once common practice for central electricity-generating stations to insist that local customers rip out their own generators as a condition of service.

The Fate of Sponsorless Technologies.
The absence of a sponsor may doom certain technologically excellent products. In the military it is customary to refer to weapons systems that lack powerful sponsors as "orphans." An example was the F-20 "Tigershark" fighter plane, which was developed independently by a contractor without Air Force sponsorship. The Tigershark was largely disregarded, and finally killed, by the Air Force,[29] in spite of its apparent abilities and its low cost as compared to the more prestigious F-15 and F-16 fighters.[30] Something similar would have happened with the Sidewinder missile, again developed without Air Force sponsorship (it was a Navy missile), except that brilliant salesmanship was used to win its adoption.[31] Nonetheless, it is regarded by some in the Air Force as an inferior missile, in spite of combat experience to the contrary.[32] The AR-15 (later M-16) rifle was a victim of similar pressures.[33]

Just as there are "orphan airplanes" without powerful sponsors, there are also "orphan drugs" that either have a clientele that is too small to make mass production profitable or present hazards that would expose their makers to potentially crippling lawsuits. Many people for whom certain drugs are literally life-and-death matters thus cannot get these drugs because they have no sponsor. Developing countries need some vaccines that can be produced only in developed countries, but the companies there have no economic incentive to do so.[34] Some large, powerful companies have occasionally produced such drugs as a public service, but not all such drugs can find sponsors. Some pharmaceutical companies also have been aided by the Orphan Drug Act of 1983, which gives monopoly power to firms in return for producing drugs for which there is a small demand. Unfortunately, some of the companies so aided have used their monopolistic position to charge high prices.[35]

Because of the threat of lawsuits, some firms have withdrawn from the production of otherwise profitable drugs, thereby stranding those who need them without protection.[36] Drugs are not the only product orphans stranded by the expansion of liability suits. Such medical products as intrauterine devices (IUDs) and vaccines have been strongly affected. Claims filed against the A. H. Robins company in the wake of Dalkon Shield IUD problems total $2.48 billion,[37] an amount larger than the gross receipts of many drug companies. Merck & Company, for instance, is the only manufacturer of a combined measles, mumps, and rubella vaccine; all of its competitors have left the field. Because of legal liability more than six firms have stopped making football helmets in the last ten years, leaving only two manufacturers.[38]

Some inventions do not find a sponsor because they originate with individuals. Because most organizations in our society are set up to interface with other organizations, an individual with an original idea often cannot get a sponsor to develop it. A variety of social institutions are being developed to cope with this problem, although the solutions so far must be considered inadequate. One example of such a solution is a program shared between the Office of Energy Related Inventions of the National Bureau of Standards and the U.S. Department of Energy. The approximately 200 proposals sent in to the Office of Energy Related Inventions each month pass through two sets of evaluators, and only about 1 percent of them eventually get funding. This is not a very high rate, but it is typical of this kind of operation.[39] The Department of Energy also has another special office to handle "orphan" projects, the Division of Advanced Energy Projects.[40] The annual budget for this agency is only about $10 million, which is a tiny fraction of the money spent on the projects of more powerful sponsors of conventional alternatives; it has only one full-time professional on its staff.

The problem of technology "orphans" is thus a long way from being solved. Technologies that lack sponsors are underpromoted, but technologies that do have sponsors may well suffer from the opposite problem—an excess of zeal on the part of their sponsors.

This discussion of technology shaping by sponsors suggests the existence of certain common configurations of sponsorship. Each of these configurations exists due to unusual power in the hands of clients, regulators, or makers. It is to these topics that we turn next.

The Configurations of Sponsorship

Dominant Clients. Obviously, firms try to produce products that will sell, and they use a variety of techniques to determine what customers want. When there is a single or very powerful client, its preferences constitute the major influence in shaping the product. This kind of relationship is evident between manufacturers of military products and their clients, whose demands are often met only at great cost.[41] The products of American firms that design for the military are often inferior to those that design for the civilian market.[42] Commercial products must compete with the products of other firms; military purchases reflect the often indulgent whims of a single buyer. Thus commercial products are often far less expensive, finicky, and unreliable than their military counterparts.[43]

When General Dynamics, which had been designing aircraft for the military, tried to operate in the civilian market, it soon experienced a corporate crisis. Because it was not familiar with operating in a competitive environment that required high efficiency, it suffered severe losses when it began to compete against other civilian-sector firms.[44] In another example, Lockheed designed both the L-1011 jumbo jet and the C-5A military transport during the same period; the L-1011 was designed for the civilian market, the C-5A for the Pentagon.[45] The former is a high-performing "star" and the latter is unquestionably a "dog." The reasons for the success of the one and the failure of the other reflect the relative efficiency of market forces in developing a "quality" technology compared with the pathologies that accompany American military procurement. If commercial products are not delivered on time, the producer suffers; if military products are not delivered on time, the client suffers and pays more.[46]

The influence of powerful clients is also evident in the railroad industry. Before the coming of the diesels and the accompanying standardization, steam locomotives were produced in distinctive styles for specific railroads:

> The steam locomotive was unique in that it was first and foremost an individual. No two designs were exactly alike. Each railroad exercised its own individual desires upon the locomotive foundries, and the resultant products were the symbols of the railroad on which it would operate. . . . The roster of 7667 steam locomotives which the Pennsylvania [system] carried in 1920 consisted completely of machines designed and built by the railroad either in its own shops or in the huge Eddystone foundries of Baldwin. The locomotives carried the unmistakable trademarks of the railroad—a stubby, fat boiler with a huge square firebox, and a long square tender atop which a doghouse sat, reserved for the head-end brakeman. All locomotives were built for power—for power was what was needed to enable the plush tuscan-red passenger limiteds and long trains of coal and merchandise to cross the rugged ridges of the Allegheny Mountains.
>
> The New York Central had no rugged mountains to cross. It was advertised proudly as the "Water Level Route." Stretching from the broad Midwestern plains to the shores of the Great Lakes and down through the valleys of the Mohawk and Hudson Rivers, there were no grades of any significance. As a result, every New York Central locomotive had the graceful lines of a whippet, designed by that railroad over three generations. Speed was their forte, and whether it be behind a 75-car train of merchandise freight or the plush *Twentieth Century Limited,* the Mohawk freight locomotives or the Hudson passenger locomotives swept along the lake shores and down the Hudson Valley night and day, the living personification of a great railroad.[47]

Clearly, clients are often influential at the design stage of a technology. Quite a few industrial production systems originated first with a major client, who then convinced a manufacturer to make the product.[48]

Dominant Regulators. A sponsor may be a dominant regulatory agency, standard-setting institution, or pressure group. In Chapter 1 we saw how the Soviet government's attitude toward technology is shaped by potential impacts on Soviet institutions. Governments' awareness that technologies can both preserve and threaten institutions is widespread. But regulatory agencies may themselves set policies that shape a technological field.

One of the most striking examples of regulatory shaping of technology is the political wrangling concerning air collision-avoidance systems in airliners. According to Knight-Ridder Newspapers, by 1975 the Honeywell Corporation had developed a simple, inexpensive system designed to prevent airplane collisions. The system would have been carried on each airliner and would have been independent of ground controllers. However, the Federal Aviation Administration (FAA) allegedly stifled this system for air-crash avoidance because it might result in layoffs of ground personnel and because the FAA itself was developing a different system. The agency repeatedly stressed the problems with Honeywell's system while promoting the virtues of its own. Much of this propaganda was dishonest and distorted facts known to the FAA.[49]

It is common for interest groups to lobby regulatory agencies because legislation can be a means by which companies restrict competitors. Because the Food and Drug Administration serves as a gate-keeper for medicines, drugs, and medical devices, very considerable pressures are brought to bear to influence this agency's opinions and policies. Similarly, the tobacco industry regularly exerted pressure on the Office of the U.S. Surgeon General to prevent the office from issuing reports that would link cancer and smoking,[50] and the National Rifle Association lobbies regulatory agencies to permit easy access to handguns, rifles, and automatic weapons in the United States.

In recent years, American roads have been used by more and more trucks with double trailers, some of them weighing around 80 tons.[51] These trucks pose a more serious hazard than single-trailer trucks because they are bigger and less stable. By 1982, 14 states had passed laws barring double-trailer trucks from the roads because their instability seemed to be related to increased traffic hazards. But in that year Congress passed a law allowing doubles on all interstate highways, thus overturning the states' efforts to restrict double trailers. The federal law was passed after strong lobbying by the trucking industry.[52]

Monopolistic Makers. The nature of the market in which a device is produced is critical in determining its qualities. Monopoly control of markets allowed the enormous Roche Pharmaceutical Company to set industry standards and prices.[53] The influence of A.T.&T. on the telephone system is another classic example of such control.[54] I.B.M.'s role in the computer industry in earlier years had this kind of impact, although I.B.M. was not, strictly speaking, a monopoly. I.B.M.'s role changed of necessity as smaller computers were introduced by a host of competitors.[55] Development of the telegraph in the United States was aided by the ability of its sponsors to attract private financial support for its use.[56] This support and the pooling of patents allowed use of the telegraph to grow rapidly in a secure economic environment. A similar situation existed for the development of Bessemer steel-production.[57]

In some ways the telephone provides a useful study both of the value and the limitations of powerful sponsorship. The developing monopoly of the telephone business by A.T.&T. initially provided a secure environment for growth. In the long run, however, the monopoly tended to slow equipment innovations at the same time that it helped to provide uniform service. Expiration of the telephone patent was temporarily instrumental in bringing about a rapid growth of independent telephone networks.[58] Eventually, however, A.T.&T. managed to co-opt the independent networks in a web of regulation and coordination.[59] Only with the Carterphone decisions, which allowed users to hook up to the network a non-A.T.&T.-manufactured phone, was this process to be dramatically reversed, again bringing a generative burst of invention and innovation.

More elements of this process can be seen in the history of Bell Laboratories, one of the great research institutions of the world.[60] Because Bell Labs is owned by A.T.&T., its research is stimulated by the needs of the telephone system, and this helps ensure that its inventions have a good chance of being used. Bell Laboratories has been an important center for invention, but its importance can be exaggerated, for many of its contributions might have come, if somewhat later, from other sources.[61]

For instance, the transistor was invented at Bell Laboratories by William Shockley, John Bardeen, and Walter Brattain in 1947. Because Bell Labs is a quasi-public institution, it was unable to exploit this invention exclusively, but was forced instead to present it to the electronics industry as a whole.[62] Still, Bell's research resources allowed it to devote intense time and effort early on toward exploitation of the transistor, and because Bell Labs was a powerful sponsor, the transistor was assured ample development and publicity. The discovery of the transistor was both a victory of "pure" research and a victory of the sponsorship of such research by an organization that expected it to pay off. And it did. Even the public announcement of the transistor served Bell, for it stimulated others to invent devices that the telephone network could use.

Two graduate students at Purdue, Seymour Benzer and Ralph Bray, came close to inventing the transistor themselves.[63] Had Purdue provided a higher level of support for their research, Purdue might now be given the credit for inventing the transistor. And, given time, undoubtedly others would have also discovered how to make transistors. Similarly, creation of the integrated circuit did not depend on a single inventor. The "chip" was a multiple invention, although the devices that Jack Kilby and Robert Noyce invented differed in detail.[64]

But let us follow the transistor story a bit further. When the transistor was invented, a flourishing electronics business based on vacuum tubes already existed in the United States. The first years of the transistor business were fairly rocky, for it proved difficult to manufacture the finicky semiconductors, and manufacturers frequently had to sort their output into transistors that worked and those that didn't.[65] Vacuum tube firms did take some interest in semiconductors and did much to develop transistors, but new firms were destined to take over most of the market because they had no conflict of interest between old and new technologies. By 1957 64 percent of the U.S. semiconductor market was held by firms new to the electronics business.[66] The founding of new firms, then, was an important element in the rapid growth and development of semiconductors. Had the new technology remained in the hands of radio tube companies, its growth would have been retarded. Clearly, the market conditions in the United States help to generate new technologies and speed their development.

A similar flowering of integrated circuit technology took place in California's "Silicon Valley" of Santa Clara County. Small firms that spun off from larger ones became the carriers of new innovations in a continual process of competition and renewal. Rapid movement of people between firms discouraged excessive consolidation and conservatism, for whenever a valuable new idea was being blocked by a company's management, the affected managers could start their own company to exploit the innovation. Large amounts of "venture capital" made financing relatively easy.[67] Silicon Valley thus provided an ideal atmosphere for the generation of creative projects. It was there that the Apple II, the first commercial personal computer, was developed. Both Silicon Valley and Boston's Route 128 development have been seen as models for technology hothouses elsewhere.[68]

How New Technologies May Spawn New Sponsor Institutions

New technologies frequently require new institutions to support them. Inserting a novel technology into a traditional setting is likely to cause disruptions. An interesting study by Ulf Berggren contrasts the diffusion into medicine of two imaging technologies: CAT (computerized axial tomography) scans and ultrasonography.[69] The development of CAT scanning required a high financial investment but few organizational changes. This may have much to do with its later acceptance by the medical community. Berggren's study shows that CAT scanners, which use crystal receivers instead of film for X-ray imaging of the body, integrated easily into standard radiological departments and were rapidly adopted and developed. Little institutional change was required, although CAT scanners were quite expensive (around $1 million apiece) and represented a major investment for hospitals. Ultrasound, by contrast, did not have the huge institutional support and was considered a rather "low-status" technology, for even though it represented an appreciable technological advance, it did not have the "high-tech" ambience or impressive presence that CAT-scan machines did. Ultrasonography required a much higher degree of personal skill and judgment, and thus its interpretation seemed more subjective. The use of acoustic imaging was less familiar to physicians than were X rays; its effective development required new departments, not just new machines; and its applications were also less obvious. Accordingly, its diffusion was much slower.

Even with CAT scanners, however, the way the technology should fit into the radiography department is far from clear. A study of two hospitals by Stephen Barley showed that very different organizational arrangements could follow the introduction of a new CAT-scan machine.[70] In one hospital, which Barley called "Suburban," physicians were forced to allow technicians greater freedom of action because the technicians had already gained greater experience with the technology. Thus the technicians actually took responsibility for most routine scanning, although it was still the physician's responsibility to interpret the results. At the other hospital ("Urban"), however, technicians were allowed to gain mastery over the technology much more slowly because the radiologists insisted on much closer supervision. Interestingly, in neither hospital did a genuine "team spirit," which would have allowed better coping with technical uncertainties, develop.

The introduction of new technologies may provide an occasion for important kinds of social change. For instance, the introduction of typewriters into offices also brought women into them.[71] Previously, women simply had not been employed in offices.[72] (Note that the clerks in Charles Dickens's tale "A Christmas Carol" are males.) The telephone also led to another new occupation for women: the telephone operator. Initially, telephone operators in the United States, were young boys. But they proved to be too offensive to customers (they used bad language), so they were replaced by women, who developed a new standard of telephone etiquette. Female operators forced male businessmen, often over their objections, to observe this etiquette as well, for businessmen who were not capable of "telephone courtesy" found themselves disconnected.[73] The "voice with a smile" became a social institution.

When anesthesia was introduced into the operating room, a new medical speciality, anesthesiology, developed. Surgeons, formerly the undisputed lords of the operating room, now had to tolerate another physician who would operate increasingly complicated anesthesia devices.[74] Similarly, the development of radiology led to an influx of radiologists, who are also, we should note, physicians.

Because new technologies often cause such changes, they may be resisted by individuals and groups that favor the status quo. We have already noted that the introduction of new technologies may leave whole communities isolated and entire occupations idled. Economic hardships for certain groups may follow the introduction of new technologies, as was the case with diesel locomotives. In another example the effect of permanently pressed shirts and trousers was to remove much of the trade from steam laundries. The development of the safety razor vastly reduced the number of barbershops. Use of the linotype machine harmed many veteran printers who could not adjust to the new machines. Compositors by the thousands were thrown out of work, and many took their lives in despair.[75] These negative impacts took place in spite of strenuous resistance to and later partial control of linotypes by the International Typographical Union.[76]

Thus new institutions may be more efficient at developing and exploiting new technologies. Although many traditional radio-tube companies became producers of transistors, they did not, as we noted earlier, become the leading transistor firms.[77] Rather, the new firms that grew up around transistors became dominant. Although airplanes might have been used by railroads in some new, expanded transportation industry, their use instead led to the development of a separate airline industry. Thus new technologies can spawn the development of new organizations and new social forms. These new social forms, in turn, become the masters of the technology and shape its further evolution, until they are mastered by yet newer technologies and *their* sponsors.[78]

How Sponsors Shape Beliefs About Their Devices

One of the effects of sponsorship is that information about and evaluation of the device or technology may be manipulated and controlled by the sponsor. My first recognition of this fact took place in the late 1960s, when I was reading an article on solar power in *Newsweek,* an influential magazine with a huge national audience. One chart in the article projected a relatively bleak outlook for the contribution of solar power for the nation's future energy needs. It was only by chance that I glanced at the minuscule attribution under the chart, which said "Source: American Petroleum Institute." Although they might not be obvious to many people, the implications of my observation are very significant: Because the American Petroleum Institute is controlled by the petroleum industry, it could hardly be expected to promote the prospects for solar power.

We expect commercial makers of products to exaggerate their worth.[79] It has been estimated that the U.S. pharmaceutical industry, for instance, spends about $5,000 per physician per year to promote its products,[80] and similar behavior is common to all kinds of sponsors. In promoting atomic power, the Atomic Energy Commission has shown films stressing the safety of atomic power plants to some 40 million Americans.[81] The U.S. military is also prone to exaggerate the performance and ignore the failings of favored weapons systems. This kind of campaigning not infrequently extends to the rigging of tests for "shoot-outs"—head-to-head trials between two rival technologies—a custom that goes back at least to Eli Whitney and his "interchangeable" rifle parts.[82] In the cases of both the Sidewinder missile and the M-16 rifle, the military establishment attempted to rig tests in favor of the officially approved systems. In the case of the Sidewinder this attempt proved unsuccessful.[83] These tactics were successful, however, with the M-16 rifle: In spite of the protests of the inventor, his rifle was altered, under bureaucratic direction, from a highly efficient weapon to a scandalously bad one.[84]

The tobacco industry has labored long and hard to prevent publication of information suggesting that tobacco might be harmful to the health of smokers. Apparently their efforts have included the use of "disinformation" in an attempt to confuse the public. During a lawsuit against three tobacco companies in the death of a smoker, evidence was introduced to the effect that "a brilliantly conceived and executed strategy" had been used by the industry for 20 years to counter assertions that smoking causes cancer and other illnesses. This phrase was used in a memo written by Frederick R. Panzer, a vice-president of the Tobacco Institute, to the Institute's president, Horace R. Kornegay. According to the memorandum, two tactics—creating doubt about studies that showed adverse effects and demanding further studies—were used to shape public opinion and delay legislative action. In another instance marketing of a notably safer cigarette was rejected so that the public would not get the idea that currently marketed forms were dangerous.[85]

Marxists often contend that this sort of behavior is largely an effect of capitalism. I strongly suspect instead that such manipulation of information is typical of sponsorship, regardless of the prevailing economic system. Thus "socialist" bureaucracies may promote technologies they sponsor, whether they do so because the technologies fit in with agency policy or because the technologies support national economic plans.[86] And the tendency toward dishonesty (whether promotion or "disinformation") is even stronger in situations that link the sponsor's fate to its chosen technology, as is so often the case with U.S. military systems.[87]

We might even go a step further: An agency's involvement with a certain technology may lead to bias that is very detrimental to the agency's mission. People with such a bias may find change not only undesirable but even unimaginable, as we have seen with Navy sponsors of battleships and aircraft carriers. Those who object to the technology may find themselves hounded out of jobs and even blacklisted, as were physicists John Gofman and Arthur Tamplin after they portrayed too vividly the dangers of radiation from nuclear power plants.[88] No doubt something similar would have happened to Rachel Carson, whose *Silent Spring,* a book on the dangers of pesticides, infuriated the agricultural chemical industry. But because she didn't have a government job, she couldn't be fired.[89] The most celebrated case of this kind was the dismissal of A. Ernest Fitzgerald from the Department of the Air Force because of his willingness to hold up bureaucratic blunders to public scrutiny.[90]

The health risks associated with asbestos have been widely publicized in the United States, and in January 1986 the Environmental Protection Agency proposed that it be banned. But because asbestos is important to the economy of Canada, the asbestos industry and the government of Canada have launched an expensive public-relations effort to assure the public that the dangers of asbestos have been exaggerated.[91]

Sponsors have learned that the images associated with a product are important for its success. Films that feature charismatic heroes and heroines have often accidently helped to sell particular products. An interesting example of this phenomenon in the United States is the film *Dirty Harry,* in which Clint Eastwood portrayed an overzealous detective who carried a .44 magnum revolver. This weapon, useful largely as a backup gun for those hunting something large—say, a polar bear—is so powerful that in firing it, even with ear protection, hearing loss can result from the conduction of sound through bone. Nonetheless, apparently the movie caused a sudden brisk demand for this awesome weapon. Operators of an Ann Arbor, Michigan, firearms store indicated that they sold the only two .44 magnums they had ever sold shortly after this film played locally. One .44 magnum was returned after its new owner fired it only once; that was enough for him.

Other products have been unintentionally promoted by films. After the appearance of *Shaft,* starring Richard Roundtree, leather coats became a sudden fad among blacks.[92] After *Blowup* there was a sudden demand for cameras, and so on. According to a popular tradition, the film *It Happened One Night* almost destroyed the American undershirt business. During one scene in the film, Clark Gable removed his shirt and was seen not to be wearing an undershirt. This apparently made many American males feel that if Gable did not need an undershirt, they did not need one either. The impact on the manufacturers—the sponsors—was said to have been devastating.

Many manufacturers now try to produce this kind of effect by giving out free cars or watches (to name but two items) to television and film stars, thus creating "non-commercials" that are nonetheless influential in selling products.[93] Few TV watchers are aware that many products that appear on television or in films reflect sponsors' conscious attempts at promotion, rather than any actor's choice or a studio's need for realism. Sociologists would do well to study the impact on social norms of such practices. Although some studies have suggested that advertising largely produces effects through influencing opinion leaders, who then influence others, there may be certain direct influences that are worth considering as well.[94]

Sponsorship
and Responsibility

Sponsors may shape the product, manufacture it, and distribute it, but if a product does not work or if it has serious defects, how will the manufacturer know? How well does this web of contacts between sponsor and user act to transmit bad news? These are very important questions in a technologically developed society, for producers and consumers are often far away from one another.

In the worst case the producer's firm is a transient operation that produces something dangerous and then packs up and leaves the scene. Unhappily, this sometimes takes place with firms that dump toxic chemicals, whose effects may not be felt until the remote future. On the other end of the spectrum are firms whose long existence and reputation are incentives to having a strong interest in making sound products and replacing those found to be defective.[95]

The kind of diligence sponsors must show was exhibited by the biochemist Paul Ehrlich in the promotion of the drug salvarsan as a cure for syphilis. Ehrlich discovered what he called a "magic bullet" after laborious experimentation, and the drug was released for public use in 1910. For many researchers this would have been enough. But after its release, Ehrlich continued to monitor the effects of the medicine:

> It was for this purpose that Ehrlich and Farbwerke Hoechst supplied some 60,000 samples to doctors, free of charge. Approximately 10,000 patients had been treated with Salvarsan, and Ehrlich continued to concern himself tirelessly with matters pertaining to dosage, injection technique, and the ideal composition of the injection solution.[96]

We might well wish that such diligence was universal, for interest in the evaluation and fine-tuning of medical innovations is very important. Many promising surgical procedures turn out upon careful examination to have either no real gain over standard methods or to actually represent a worse alternative.[97] Drugs can also present unexpected problems. The sedative thalidomide, often given to pregnant mothers, had not been tested sufficiently to detect its potential for causing birth defects. In Germany some 10,000 malformed babies were born to mothers who had used the drug before its sale was halted.[98]

We have already mentioned the issue of product liability suits. But the use of lawsuits is a very primitive and clumsy method of ensuring product safety and performance. Not only is the use of liability likely to drive fearful companies from the market, but it is also likely to penalize the large, responsible firms to the benefit of fly-by-night operators who run many of the dirtiest operations. Clearly, we need a better method of ensuring safety. The best guarantee, of course, is a genuine concern on the part of producers for designing a safe product and making it properly. We do not know how many firms in fact imbue their employees with such an attitude, or how many are willing to suffer whatever costs are necessary to make a product safe. Unfortunately, Americans tend to depend too much on the court system and not enough on a sense of moral correctness,[99] which often leads to the attitude that whatever is not actually illegal or does not lead to prosecution is legitimate. This is all the more damaging in a society that exerts strong pressures on people to make money and to be financially successful. We need to emphasize other values, including a willingness on the part of manufacturers to make sure that everything they ship is safe.

I remember a situation uncovered by a quality control study I did some years ago. A firm making automobile springs had discovered it could sneak through substandard springs on close-out orders. These springs would not last as long as properly tested springs, and the company was taking a big risk in shipping them. Even more serious, however, was that many of these springs had metal "burrs" on them that could cause a serious gash if workers assembling a car ran their hands over the spring. The workers in this firm were quite pleased that they had found a way of shipping these defective parts and getting the customer to accept them. This kind of attitude is particularly dangerous in a high-tech society in which often very complicated systems depend on the functioning of small parts. And we can never be sure about where a computer chip will end up—it could be inside an airplane or a medical machine used to monitor anesthesia.

Faulty parts for hospital machines are particularly frightening because those who depend on them are often helpless. In the early 1980s the Puritan-Bennett Corporation, which manufactures medical equipment, began to discover that two of its respirator machines, the Foregger 705 and 710, had a problem with failure during use. One of the turret valves would fail, and the patient would sometimes get straight anesthetic unmixed with oxygen. Between 1980 and 1983, 22 instances of valve failure were reported, yet the corporation waited until the summer of 1983 to notify the health community that there was a problem with the respirators. Meanwhile, three women had died of anesthetic overdoses, apparently caused by the machines.[100]

Because the parts of our society are so interdependent, we require a high standard of concern about the technologies we use. The products we manufacture, regulate, and promote must have integrity. They should work and should do something worthwhile, and they should not present unnecessary hazards to their users or to those who come in contact with them. This sense of responsibility is essential for our safety and our survival.

When nuclear reactors are built with substandard parts, they imperil everyone. When industries fail to take obvious measures to make their products safer, such as the automobile industry's refusal to use polarized glass in headlights, we all suffer. In the chapter on technological accidents we will consider some of the more dramatic examples of such failures. But important also are the millions of day-to-day decisions about the products and processes that provide the basic things on which our existence depends. We must learn that things are highly connected and that whatever we do will have an impact somewhere. This awareness—this sense of being a member of an immense community—is needed by more of the organizations in our society.

The issue of responsibility is especially important for public bodies that regulate technology. Industries in capitalist countries have done a great deal of damage to the environment, but at the same time the freedom of information in these countries has contributed to important measures designed to halt this degradation. These countries have had a large measure of public debate and they have developed powerful private organizations to protect the environment.[101] In communist countries degradation of the environment has not had this kind of scrutiny because private organizations have been systematically discouraged. This critical blindness has led to policies that are so irresponsible it is difficult to believe they actually exist.[102] Now, such organizations are developing apace with increasing freedom of speech in these countries.

These changes bring home a signal point: Responsibility may grow with the development of feedback channels, but it does not necessarily follow from mere awareness. Still, it is difficult to exercise responsibility without such awareness, and we are likely to get sponsors to increase their sense of responsibility only if we can get them (and others) to monitor the effects of their actions.

Conclusion

To understand how technology operates in society, we must examine its supporting institutions. No technology stands alone; around it are a host of organizations and networks that design it, manufacture it, distribute it, promote it, and regulate it. We must understand the internal and external dynamics of these institutions, then, if we are to understand the fate of the technology.

The phenomenon of sponsorship is obvious is capitalist societies, but it is no less important in socialist societies. Bureaucracies, like businesses, often make decisions about which technologies they will support and promote. Such decisions can be even more fateful in socialist societies because bureaucracies in them are often monopolies unopposed by other organizations. Even voluntary associations may become sponsors of technologies. So we must learn more about how sponsors develop, evolve, and disappear, a subject we will explore in the next chapter.

Notes

1. Langdon Winner, *Autonomous Technology: Technics-out-of-Control as a Theme in Political Thought* (Cambridge, Mass.: MIT Press, 1985).

2. Larry McShane, "Steinway Holds the Key to Success," *Ann Arbor News,* May 17, 1988, Section D, pp. 1–2.

3. C. P. Crow, "Profiles: The Only Way," *New Yorker,* June 22, 1987, pp. 34–44.

4. Much of this chapter was stimulated by one seminal paper: Trevor Pinch and Wiebe Bijker, "The Social Construction of Facts and Artifacts: Or How the Sociology of Science and the Sociology of Technology Might Benefit Each Other," *Social Studies of Science,* Vol. 14, No. 3 (Aug. 1984), pp. 399–441.

5. A very different formulation than the one offered here can be found in John Staudenmaier, "The Politics of Successful Technologies," in S. H. Cutcliffe and R. C. Post (Eds.), *In Context: History and the History of Technology; Essays in Honor of Melvin Kranzberg* (Bethlehem, Pa.: Lehigh University Press, 1989), pp. 150–171. Staudenmaier shows how the "constituency" of a technology influences its design, maintenance, and impact stages.

6. Tracy Kidder, *House* (Boston: Houghton Mifflin, 1985).

7. John F. Hogerton, "The Arrival of Nuclear Power," *Scientific American,* Feb. 1968.

8. Daniel Ford, *The Cult of the Atom: The Secret Papers of the Atomic Energy Commission* (New York: Simon & Schuster, 1982), pp. 17–82.

9. William Booth, "Revenge of the Nozzleheads," *Science,* Vol. 239 (Jan. 8, 1988), pp. 135–137.

10. Booth, "Nozzleheads," p. 137.

11. See also Charles Perrow, "The Organizational Context of Human Factors Engineering," *Administrative Science Quarterly,* Vol. 28, No. 4 (Dec. 1983), pp. 521–541.

12. Elting Morison, "Men and Machinery," in his *Men, Machines, and Modern Times* (Cambridge, Mass.: MIT Press, 1967), pp. 98–122.

13. Arthur Squires, *The Tender Ship: Governmental Management of Technological Change* (Boston: Birkhauser, 1986), pp. 13–42; see also Arthur Squires, "Maestros and Duffers: The Good and the Bad in the Management of Technology," talk delivered at Virginia Tech, Blacksburg, Va., Apr. 2, 1987 (available from Squires).

14. Concerning the *Challenger,* see also Joseph J. Trento, *Prescription for Disaster* (New York: Crown, 1987).

15. Ray Reece, *The Sun Betrayed: A Report on the Corporate Seizure of U.S. Solar Energy Development* (Boston: South End Press, 1979).

16. Barry Commoner, *The Politics of Energy* (New York: Knopf, 1979), pp. 32–48.

17. I remember my own reaction to the pocket calculator. I was one of the minority of students to get really proficient with a slide rule. A pocket calculator seemed ridiculously easy to use, almost like cheating; there was no glory in it. So I was one of the last people to learn how to use one.

18. For an outstanding and very comprehensive treatment of this issue, see Ronald G. Havelock, *Planning for Innovation Through Dissemination and Utilization of Knowledge* (Ann Arbor, Mass.: Institute for Social Research, 1971).

19. Donald E. Thomas, *Diesel: Technology and Society in Industrial Germany* (Tuscaloosa: University of Alabama Press, 1987), pp. 119–151.

20. Computer buffs once distinguished among "hardware, software, and meatware," the latter referring to people. I find this offensive and prefer the more innocuous "peopleware."

21. Michel Callon, "Society in the Making: The Use of Technology as a Tool for Sociological Analysis," in W. Bijker, T. P. Hughes, and T. Pinch (Eds.), *The Social Construction of Technological Systems* (Cambridge, Mass.: MIT Press, 1987), pp. 83–103.

22. Claude Fischer, "Touch Someone: The Telephone Industry Discovers Sociability," *Technology and Culture,* Vol. 29, No. 1 (Jan. 1988), pp. 32–61; "The Revolution in Rural Telephony," *Journal of Social History,* in press; "Technology's Retreat: The Decline of Rural Telephony, 1920–1940," paper presented to the American Sociological Association, 1985; "The Diffusion of the Telephone and Automobile, 1902 to 1937," unpublished paper, University of California, Berkeley, 1986.

23. Raymond M. Wik, *Henry Ford and Grass-Roots America* (Ann Arbor: University of Michigan Press, 1973); pp. 14–33.

24. Robert Fellmeth, *The Interstate Commerce Omission: Public Interest and the ICC* (New York: Grossman, 1970), pp. 285–310.

25. Chuck Yeager and Leo Janos, *Yeager: An Autobiography* (New York: Bantam Books, 1985), pp. 394–395.

26. James J. Flink, *The Car Culture* (Cambridge, Mass.: MIT Press, 1975), p. 220.

27. See H. Peter Metzger, *The Atomic Establishment* (New York: Simon & Schuster, 1972); and Mark Hertsgaard, *Nuclear Inc.: The Men and Money Behind Nuclear Energy* (New York: Pantheon Books, 1983).

28. Barry Commoner, *The Politics of Energy* (New York: Knopf, 1979), pp. 32–48; see also

Engelbert Broda, "Creative Innovation in the Exploitation of Solar Energy," *Impact of Science on Society,* No. 134/135.

29. Gregg Easterbrook, "The Airplane That Doesn't Cost Enough," *Atlantic Monthly,* Aug. 1984, pp. 46–56; see also Mark Thompson, "Northrup Scraps Fighter Jet Program," *Detroit Free Press,* Nov. 18, 1986.

30. Chuck Yeager expressed a high opinion of the F-20 in the caption to a picture in his autobiography, *Yeager.* Curiously, there is no text to elaborate the point.

31. John Fialka, "Weapon of Choice," *Wall Street Journal,* Feb. 15, 1985, pp. 1, 30.

32. Easterbrook, "The Airplane," pp. 49–51.

33. James Fallows, *National Defense* (New York: Random House, 1981), pp. 76–95; and Edward Ezell, *The Great Rifle Controversy* (Harrisburg, Pa.: Stackpole Books, 1984), pp. 162–228.

34. Anthony Robbins and Phyllis Freeman, "Obstacles to Developing Vaccines for the Third World," *Scientific American,* Nov. 1988, pp. 126–133.

35. Andrew Pollack, "High Cost of High-Tech Drugs Is Protested," *New York Times,* Feb. 9, 1988, pp. 1, 47.

36. N. R. Kleinfield, "Orphan Drugs: Caught in Limbo," *New York Times,* July 20, 1986, Section F, pp. 1, 27.

37. Barnaby J. Feder, "What A. H. Robins Has Wrought," *New York Times,* Dec. 13, 1987, pp. 1, 17.

38. Michael Brody, "When Products Turn into Liabilities," *Fortune,* Mar. 3, 1986, pp. 20–24.

39. *Small Firms, High Technology, Innovation, and Inventors,* hearings before the Investigations and Oversight Subcommittee of the Committee on Science and Technology, U.S. House of Representatives (Washington, D.C.: U.S. Government Printing Office, 1981).

40. Eric J. Lerner, "Nurturing Energy Infants and Orphans," *IEEE Spectrum,* July 1984, pp. 55–58.

41. Merton J. Peck and Frederic M. Sherer, *The Weapons Acquisition Process: An Economic Analysis* (Boston: Harvard Graduate School of Business Administration, 1962).

42. Fallows, *National Defense,* pp. 76–95.

43. "The Pentagon's Designer Defense," *Ann Arbor News,* Apr. 26, 1985 (originally published in *Washington Post*).

44. Richard Austin Smith, *Corporations in Crisis* (New York: Doubleday, 1963), pp. 63–96.

45. Easterbrook, "The Airplane," p. 52.

46. For an example of the point, see A. Ernest Fitzgerald, *The High Priests of Waste* (New York: Norton, 1973), p. 174.

47. From *Focus: The Railroad in Transition,* by Robert S. Carper, A. S. Barnes & Co., San Diego, CA. 1968, at p. 15. Reprinted by permission of the publisher.

48. Eric Von Hippel, *The Sources of Innovation* (New York: Oxford University Press, 1989), pp. 11–27.

49. James R. Carroll, "FAA Reportedly Stifled Air-Crash Avoidance Plan," *Detroit Free Press,* Oct. 27, 1986, Section A, pp. 1, 14.

50. For echoes of this struggle, see Eliot Marshall, "Tobacco Science Wars," *Science,* Vol. 236 (Apr. 17, 1987), pp. 250–251.

51. William E. Schmidt, "Sharp Rise in Truck Crashes Prompts Action Across U.S.," *New York Times,* Dec. 7, 1986, pp. 1, 15.

52. Paul Judge, "Double Trouble," *Ann Arbor News,* Sept. 22, 1986, Section D, pp. 1, 2.

53. Stanley Adams, *Roche Versus Adams* (Glasgow, Scotland: Fontana/Collins, 1985), pp. 14–15.

54. Joseph C. Goulden, *Monopoly* (New York: Pocket Books, 1970).

55. Katherine Fishman Davis, *The Computer Establishment* (New York: McGraw-Hill, 1981); and Stephen T. McClellan, *The Coming Computer Industry Shakeout: Winners, Losers, and Survivors* (New York: Wiley, 1984).

56. Robert L. Thompson, *Wiring a Continent* (Princeton, N.J.: Princeton University Press, 1947).

57. Elting Morison, "Almost the Greatest Invention," in his *Men, Machines, and Modern Times* (Cambridge, Mass.: MIT Press, 1967), pp. 123–205.

58. Claude Fischer and Glenn Carroll, "The Diffusion of the Telephone and Automobile, 1902 to 1937," unpublished paper, University of California, Berkeley.

59. Joseph C. Goulden, *Monopoly* (New York: Pocket Books, 1970).

60. Jeremy Bernstein, *Three Degrees Above Zero: Bell Labs in the Information Age* (New York: New American Library, 1984).

61. Nor was Bell Labs completely forthright about what it thought of potential innovations in the telephone field. Its scientists were perfectly capable of stating publicly that some of the technologies on which they were furiously working had little promise. See examples in Richard Dunford, "The Suppression of Technology as a Strategy for Controlling Resource Dependence," *Administrative Science Quarterly,* Vol. 32, No. 4 (Dec. 1987), pp. 521–525.

62. Ernest Braun and Stewart MacDonald, *Revolution in Miniature* (Cambridge: Cambridge University Press, 1978), pp. 54–55.

63. Braun and MacDonald, *Revolution,* pp. 44–45.

64. T. R. Reid, *The Chip: How Two Americans Invented the Microchip and Launched a Revolution* (New York: Simon & Schuster, 1984).

65. Braun and MacDonald, *Revolution,* pp. 61–82.

66. Braun and MacDonald, *Revolution,* p. 68.

67. Everett M. Rogers and Judith K. Larsen, *Silicon Valley Fever: The Growth of High-Technology Culture* (New York: Basic Books, 1984).

68. Annalee Saxenian, "Silicon Valley and Route 128: Regional Prototypes or Historic Exceptions?" in M. Castells (Ed.), *High Technology, Space, and Society* (Beverly Hills, Calif.: Sage, 1985), pp. 81–105.

69. Ulf Berggren, "CT Scanning and Ultrasonography: A Comparison of Two Lines of Development and Dissemination," *Research Policy,* Vol. 14 (1985), pp. 213–223.

70. Stephen Barley, "Technology as an Occasion for Structuring: Evidence from Observations of CT Scanners and the Social Order of Radiology Departments," *Administrative Science Quarterly,* Vol. 31, No. 1 (Mar. 1986), pp. 78–108.

71. Alan Delgado, *The Enormous File: A Social History of the Office* (London: John Murray, 1979), pp. 37–56.

72. Alan Delgado notes that
    earlier, in 1872, the [British] Post Office experimented in employing male and female staff *and putting them in the same room.* "It was considered to be a hazardous experiment," wrote a senior official at the time, "but we never had reason to

regret having tried it. . . . It raises the tone of the male staff by confining them during many hours of the day to a decency of conversation and demeanour which is not always to be found where men alone are employed."
See Delgado, *Enormous File,* p. 39.

73. Hebert N. Casson, *The History of the Telephone* (Chicago: McClurg, 1910), pp. 154–159. Businessmen who could not give up their crude language were "cast out as being unfit for a telephone-using community."

74. For example, see John P. Bunker, *The Anesthesiologist and the Surgeon: Partners in the Operating Room* (Boston: Little, Brown, 1972).

75. George Iles, *Leading American Inventors* (New York: Henry Holt, 1912), pp. 425–426.

76. George E. Barnett, *Chapters on Machinery and Labor* (Carbondale: Southern Illinois University Press, 1976), pp. 3–29.

77. Braun and MacDonald, *Revolution,* p. 68.

78. Richard Foster, *Innovation: The Attacker's Advantage* (New York: Summit Books, 1986).

79. See, for example, Gerald Carson, *Cornflake Crusade* (New York: Rinehart, 1957).

80. Alison Bass, "Changing Doctor's Habits," *Technology Review,* Nov./Dec. 1983.

81. Ford, *Cult of the Atom,* p. 48.

82. Robert S. Woodbury, "The Legend of Eli Whitney and the Interchangeable Parts," *Technology and Culture,* Vol. 1, No. 3, pp. 235–254.

83. Fialka, "Weapon of Choice," p. 1.

84. Ezell, *Rifle Controversy,* pp. 200–228.

85. Donald Janson, "Lawyer Says Secret Tobacco Data Show Industry Hid Dangers of Smoking," *New York Times,* Mar. 13, 1988, Section 1, p. 18.

86. See the review by David Joravsky of Douglas Weiner's *Models of Nature: Ecology, Conservation, and Cultural Revolution in Soviet Russia,* in *Science,* Yol. 245, (Aug. 4, 1989), pp. 541–543.

87. A. Ernest Fitzgerald, *The High Priests of Waste* (New York: Norton, 1972), pp. 21–58.

88. John Gofman and Arthur Tamplin, "Dr. John W. Gofman and Arthur R. Tamplin," in R. Nader, P. J. Petkas, and K. Blackwell (Eds.), *Whistle-Blowing* (New York: Grossman, 1972), pp. 55–74.

89. Frank Graham, *Since Silent Spring* (Boston: Houghton Mifflin, 1970), pp. 59–76.

90. Fitzgerald, *The High Priests of Waste,* pp. 224–282.

91. Bill Powell, Jerry Buckley, and Mary Hager, "The Case for Asbestos," *Newsweek,* Sept. 29, 1986, pp. 40–41.

92. Charles Michener, "Black Movies," *Newsweek,* Oct. 23, 1972, pp. 74–82.

93. See Peter Funt, "How TV Producers Sneak In a Few Extra Commercials," *New York Times,* 1974 (exact date not recorded); and David Friendly, "Selling It at the Movies," *Newsweek,* July 4, 1983, p. 46.

94. A famous study by Elihu Katz and Paul F. Lazarsfeld, *Personal Influence: The Part Played by People in the Flow of Mass Communications* (New York: Free Press, 1955), emphasized the role of "opinion leaders" in shaping political opinions, choices of films, and food preferences. It argued that the mass media tended to influence the opinion leaders, who then influenced others. It would be interesting to know to what extent such forces are important in regard to the effects discussed here.

95. An excellent example is the Canadian aircraft producer De Havilland. Their Comet jetliners originally developed serious structural problems that caused a tragic series of crashes. When the problem was fixed and a new line of jetliners was brought out, there was a temptation to call them something completely new. But Geoffrey De Havilland decided to retain the name Comet because he wanted his customers to know that he was not ashamed of his product. See Henry Petroski, *To Engineer Is Human: The Role of Failure in Engineering Design* (New York: St. Martin's Press), pp. 176–180.

96. Ernst Baümler, *Paul Ehrlich: Scientist for Life* (New York: Holmes and Meier, 1984), p. 180.

97. J. P. Bunker, D. Hinkley, and W. V. McDermott, "Surgical Innovation and Its Evaluation," *Science,* Vol. 200 (1978), pp. 937–941.

98. Edward W. Lawless, *Technology and Social Shock* (New Brunswick, N.J.: Rutgers University Press, 1977), pp. 140–148.

99. Samuel Florman disagrees with this point of view, arguing instead that in a highly competitive economy, law needs to be the primary mechanism for ensuring safety. I find what he has to say very cogent, but I am not entirely convinced. See his remarks in Vivian Weil (Ed.), *Beyond Whistle-Blowing: Defining Engineers' Responsibilities; Proceedings of the Second National Conference on Ethics in Engineering* (Chicago: Illinois Institute of Technology, 1983), pp. 83–89.

100. J. A. Farrell, "Firm Knew of Anesthesia Machine Failures," *Denver Post,* Aug. 12, 1984, pp. 1, 16, 17.

101. David Sills, "The Environmental Movement and Its Critics," *Human Ecology,* Vol. 3, No. 1 (Jan. 1975), pp. 1–42.

102. Hilary French, "Industrial Wasteland," *World-Watch,* Vol. 1, No. 6 (Nov.–Dec. 1988), pp. 21–30.

How Sponsors Evolve

In the previous chapter we saw how important sponsors are to the development and fate of technologies. It is important, then, to examine how these institutions grow and develop and to understand why they can become so difficult to change. To provide a focus, we will consider ways in which the size and market dominance of sponsors allow them to advance their chosen technologies and subdue potential rivals. Although we will primarily choose examples from the United States, we could just as easily analyze other capitalist economies and socialist economies as well.

As we noted in the previous chapter, sponsorship of technologies is a generic social phenomenon. The difference in sponsorship in the two major economic systems is that monopoly is often *planned* into certain centrally directed economies (such as that of the U.S.S.R.), whereas in the United States monopolies are generally illegal and have a precarious existence. It should be obvious, then, that the real problem with respect to technological sponsorship is not capitalism *per se,* but rather powerful economic establishments that operate without sufficient external controls. These large establishments naturally seek (using a calculative rationality) to administer their environments in a way that will benefit them.

Softsoap: A Case Study
of Market Forces at Work

One day in 1978 Robert Taylor, the head of the Minnetonka Corporation, was driving to work in Minnesota. Because Minnetonka specializes in consumer products, Taylor happened to be thinking about bar soap: How ugly it was, how it messes up bathrooms, and so forth. And suddenly he thought: "Why not [develop] a high-quality liquid soap that comes in an attractive bottle?"[1] So he decided to do some laboratory tests to develop the product, and when these were successful, he decided to put the product on the market. This was the beginning of Softsoap. Minnetonka was still a small company at this time, "geared to hippies and people with luxury bathrooms"; its major asset was the entrepreneurial spirit of Robert Taylor.

Taylor knew when to put his money on a winner. When Softsoap was ready to go on the market, Minnetonka ran an $8-million campaign to advertize the product, a risky decision because the company's net worth was only $8 million. But the risk paid off, and the company's annual revenues went from $25 million to $96 million within two years. The company projected a $400-million market and expected that they would get half of it.

The liquid-soap market turned out to be worth only $120 million, and nearly 100 other companies decided to compete against Minnetonka. The powerful competition included such consumer-product giants as American Home Products and Proctor and Gamble. For two years Minnetonka lost money; its market share dropped to under 30 percent, and the payroll had to be cut from 550 to 315. Taylor decided to fight back. He redesigned the bottle to a smaller size (so that it could sell at a lower price) and in new colors and floral designs. Only by these astute packaging tactics did Minnetonka finally manage to achieve a 38-percent market share by the end of 1986.

Clearly, then, economic power strongly influences the ability to innovate in our society, for without the very risky commitment that Robert Taylor was willing and able to make to the marketing of Softsoap, his product might never have survived in the marketplace. Examples like this one show that in certain fields, a technology may be rapidly developed, sponsored, and thrust onto the market, and may then suddenly experience a rapid reshuffling of sponsors. In this reshuffling, the strong firms are more likely to survive. Whereas on one hand this shows a system that is very responsive to consumer wants, it also illustrates the dominance of large, powerful sponsors. With a less able company chairman, Minnetonka might well have disappeared; it may yet be absorbed by some larger organization.

Similarly, in 1976 the Apple II computer was developed in a garage by two young entrepreneurs, Stephen Wozniak and Steven Jobs. In 1977 the Apple II was introduced to the market with considerable fanfare. It was the first personal computer for which the user did not need to be a programmer. It rapidly became synonymous with "personal computer."[2] Other firms, however, rapidly joined the market. In 1981 I.B.M., the market giant in larger computers, introduced its own personal computer, the I.B.M. PC. In spite of the technical mediocrity of the I.B.M. machine, it rapidly garnered a big chunk of the personal-computer market, especially the business market. From 1981 to 1983, I.B.M.'s share of worldwide sales went from 3 percent to 28 percent.[3] Apple's influence on personal computers has continued to be strong, but it is no longer the market leader.[4]

A society's ability to satisfy its citizens' wants depends on the structure of such institutions as markets. In the case of Softsoap, there was a strong incentive for Minnetonka to develop a product that would allow it to expand its product line. And indeed Softsoap was very profitable. But in the end, market forces decided Minnetonka's fate. Whether Softsoap is superior to other forms of liquid soap is unclear; what is certain is that its fate is intimately linked with the fate of its sponsor, Minnetonka. Similarly, the struggle between I.B.M. and Apple for the microcomputer market has also influenced the products made available to consumers; in this case it seems that competition has much improved the products of both.

The powerlessness of the individual in the face of these powerful market forces deserves some thought. Although competition does improve products in some ways, the structure of the market does not always foster competition that is helpful to the consumer. Let's proceed by considering market evolution in a more systematic way.

How Markets Evolve

Technology is frequently produced by private corporations, and when it is, the nature of the corporations and their positions in a market frequently affect the technology. We call these positions "niches," a term borrowed from biology; a corporation's niche is its role in the ecology of the market. We can consider markets to be in one of four phases of evolution:

1. *Nascent phase.* In this phase the product has usually just been invented, and a company or division has been created to bring it to market. Product variety is low, and the technology frequently has a large number of defects or "bugs." Nonetheless, demand is frequently brisk and prospects are seemingly unlimited. This situation encourages the entry of other firms into the market.

2. *Generative burst.* During this phase a large number of firms enter the market because they can either develop a comparable product internally or can acquire another firm that is already manufacturing the product. All players hope to cash in on the continuing brisk demand. The technologies generated in this phase are accordingly diverse, incorporating divergent and often contradictory subsystems. A wide variety of products is offered, of which only a few are likely to survive in the long run.

3. *Market shakeout.* During this phase a number of the firms that originated during Phase 2 leave the market because they were unable either to acquire a sufficient market share or to generate sufficient profits to continue the struggle. These firms have lost their niches in the market through bankruptcy or acquisition by more successful firms, as happened in the electrical industry in the nineteenth century (see Figure 10-1). The technologies now tend to narrow to the few used by the successful firms, although these firms are likely to borrow useful ideas from the products that disappeared.

4. *Maturity.* In this phase only a few firms are high-volume producers or well-known brand names. Economists call this condition "oligopoly." The firms of this dominant oligopoly tend to control the mainstream versions of the product. Other firms remaining in the market have niches in which they cater to specialty interests that cannot support high-volume production and thus may escape the interest of the major firms. Products are now standardized, and subsequent changes tend to be only marginal improvements.

The evolution of the market for mainframe computers illustrates these phases.[5] The digital computer was invented during World War II in England, Germany, and the United States. Although initial perception of demand was unpromising—at one time I.B.M. thought total U.S. demand would be five machines—commercial development rapidly led to smaller and more efficient machines, thus providing far greater commercial possibilities. The first commercial computers were developed by the Eckert-Mauchly Computer Corporation, whose founders had worked on the government ENIAC (Electronic Numerical Integrator and Computer) project. At first their product, known as UNIVAC, was synonymous with computers. But soon other firms that had related technologies, such as N.C.R. and I.B.M., began to offer their own machines as well. Although computer firms proliferated, I.B.M. established a market dominance that has continued to the present day. In the meantime many of the firms that tried to carve out a niche in the computer market failed to make a profit in it.[6] In the mid-1960s the computer market in the United States was known to many as "I.B.M. and the Seven Dwarfs."

Similarly, as videotape cassette recorders grew in popularity, so did stores offering rental of the cassettes. But as stores proliferated, competition forced prices down. As this was being written, a massive shakeout and consolidation of the cassette-rental market was taking place. We can expect to end up with a few large chains supplying most of the cassettes that consumers rent.[7]

From the point of view of the technology user, each of these market situations represents different benefits and hazards. For instance, if we want maximum choice,

Figure 10-1 The Development of the General Electric and Westinghouse Companies, 1872–1896: How Market Shakeout Can Shape Technological Firms

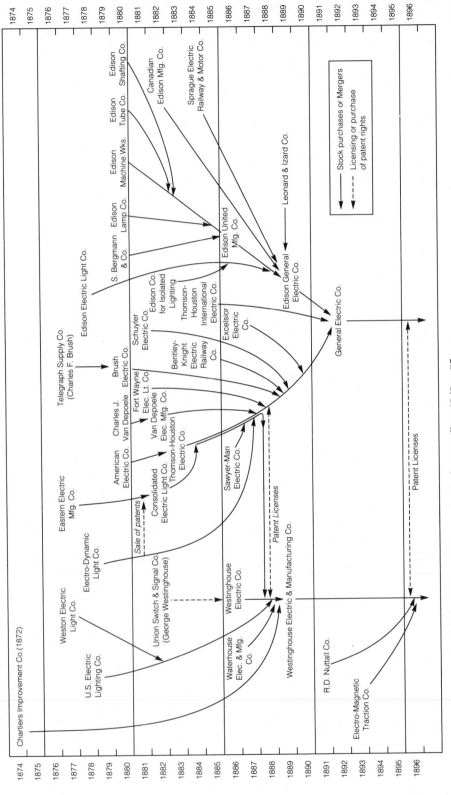

SOURCE: *From Arthur A. Bright, The Electric Lamp Industry, New York: Macmillan, 1949, p. 85.*

then Phase 2 is likely to offer the best alternatives. Phase 4 is likely to offer the best situation if we want to have a device repaired or if we need to connect the device in question to some other device. The best prices are likely to occur during Phase 3. The major virtue of Phase 1 is that acquiring a device during this phase may give us or our organization an advantage over those who don't have it. But being "first on the block" means also that we will have to live with system "bugs" that later models will lack.

Firms may also concentrate their efforts on one of these market phases. Some "prospector" firms may specialize in bringing new products to market because they are unable to be highly efficient in producing them compared to others. Other "defenders" may not be the first producers but may depend instead on high volume at low production cost to maintain their market niches.[8]

It is also possible for the market to revert from the "mature" phase, as apparently happened in the automobile industry.[9] As a result of the oil embargo of 1973 and the Iranian crisis of 1979, American automobile firms discovered they were producing vehicles that were too large for much of the consumer market. Government regulation also forced them to begin producing smaller cars, contrary to their desire to maximize profits by building big ones.[10] As Americans began to buy smaller cars, they began to observe differences in equality between American and Japanese cars, a difference resulting from intensive Japanese efforts to improve their products.[11] For years American firms had done little to improve the overall quality of the models they offered. In general, it would be fair to say that the American firms had become preoccupied with the business aspect of their firms to the neglect of their products.[12]

But consumers are less interested in the firms' "bottom line" than they are in their products, and when consumers have a choice, they tend to choose products that suit their needs, not the firms' needs. Thus in automobiles, as in electronics and cameras, Americans increasingly buy Japanese products. Only in the last few years have some American automobile firms become committed to radical change.[13]

In some markets, such as food or clothing, Phase 4 may never be reached, and in many markets there are only a few survivors in the long run. This means that initially the user will have a choice based on a selection process. The problem is that this "natural selection" is a selection of firms, not devices. The most profitable firms, not always the best devices, survive; the firm that wins may not have the superior product, but rather the most efficient production system. Said one author of a shakeout of breakfast cereal companies: "So of all who felt the urge to agitate 'the food question' at a profit, the greater number proved to be inadequate in the areas of finance, of production, or of merchandising."[14] All too often, this is the fate of good products with inefficient sponsors. Almost everyone can cite some beloved product that is no longer made because its sponsor went out of business. We have electric refrigerators in the United States, rather than gas refrigerators, because the firms financing the former were better capitalized or better managed than firms responsible for gas refrigerators.[15]

Conglomerates and
Limited Choices

Even as new firms are constantly being born, markets strongly tend to become oligopolies.

A generative burst of activity brings many new firms into the market, but in the end only a few are left, many of which may well be owned by other firms. If they are, they may be part of a conglomerate—a giant company owning subsidiaries in a variety of different markets.[16] Because the companies in these different markets have products with different market dynamics, the managers of a conglomerate cannot be expert in all of them. Thus they are tempted not to be product-oriented, but instead to manage the firms according to financial criteria; they are concerned less with what they are producing than with whether those products are making money.

This kind of problem is vividly brought home by examples in a variety of industries. In book publishing, for instance, certain publishers will not even consider books unless they can be mass marketed through large chain bookstores such as B. Dalton or Waldenbooks.[17] Film studios are often owned by conglomerates, whose decisions about what to produce are made more often according to abstract financial criteria than on any deep knowledge of the product.[18] Inside large firms managers seem to be more reluctant to become involved with actual products than to concern themselves with the financial "bottom line" of profit and loss.[19]

The kind of problem involved was made apparent by Jacob Rabinow, a prolific inventor. Rabinow invented a feedback-controlled tone arm that moved straight across a phonograph record rather than pivoting from a single point. After trying unsuccessfully to make a profit on manufacturing the turntable through a company of his own, Rabinow sold his patents to the Harman-Kardon Company. Even though Rabinow has lived in America most of his life, the nine companies that now make his phonograph are all outside the United States. He explains:

> And the reason is this. Harman-Kardon had a license from me. Harman-Kardon was eventually owned by Beatrice Foods. Beatrice Foods owns 500 companies, and they manage them with 200 people. I asked Harman-Kardon, "Did you ever meet with the top management?" And he said "No."
>
> So the situation is that we have tremendous conglomerates where a single U.S. company can have 200 subsidiaries. RCA owns Hertz, a publishing house, a food company, a bank, and incidentally a radio company that makes radio sets, or used to. You have a situation where, because of the size of the business, they cannot bother with individual little companies. So Harman-Kardon did not make good record players, so it was sold to the Japanese and it is now a Japanese company. Only the name remains. Everything is designed in Japan and sold under the Harman-Kardon name.[20]

The availability of consumer goods, then, is limited by the nature of the institutions that produce these goods. Should sponsors decide, as did Henry Ford, that we can only have black cars, then (for a while anyway) we will get only black cars. Even when foreign competition arises, it may be temporary. American automobile firms, for instance, are beginning to merge with Japanese automobile firms; if this trend continues, the market may return to a "mature" phase. We must face the realistic prospect that fewer and fewer conglomerates will control more and more products, thereby reducing the channels by which new ideas are considered. And as ideas become fewer, the variety of things we have will decrease as well.

The Evolution of Other Kinds of Sponsors

Private firms are, of course, only one kind of sponsor. As we have noted, many kinds of organizations may become sponsors, including groups of users and governmental organizations. We have seen how the armed services, for instance, exert their preferences for specific types of weapons systems. In the development of numerical control of machine tools, for instance, the Air Force exerted a decisive influence on the nature of the developing technology. The Air Force preferred a sophisticated form of numerical control, which was difficult for most manufacturers to use. The push toward this more sophisticated version, as opposed to the simpler record/playback system, probably would not have taken place without heavy Defense Department sponsorship.[21] Regulatory agencies exert similar influences. As they change their rules, technologies are likely to change also, as they did as a result of the Pure Food and Drug Acts in the early part of this century.

Government initiatives in technology can also be defeated by other sponsors of current technology, as occurred when the government attempted to boost research and development in the building industry. The Civilian Industrial Technology (CIT) program was an attempt to help "backward" industries by developing and encouraging the uses of new materials and techniques. But the building industry, one of the "backward industries" the program was supposed to help, strongly resisted it. The industry succeeded in making the program only a shadow of its original intention. Donald Schön explains why the CIT program met such fierce resistance:

> [N]ot only suppliers and unions, but contractors, building inspectors, purchasers, the building press (through advertisers) and government agencies that set standards for the approval of mortgages—all have an investment in the technology of bricks. Their livelihoods, the ways of working with which

they are familiar, and the standards they know how to write and enforce, depend on the prevailing technology, which is based on the brick.[22]

So when the CIT program threatened to introduce new and unfamiliar technologies, the entire web of sponsors of brick-building technology fought back—successfully.[23]

Nonprofit organizations may also be important in shaping technological policies and practices. Clearly, the National Rifle Association, once a hobbyists' association, has lobbied strongly to maintain private ownership of guns, for instance.[24] Planned Parenthood has encouraged the use of condoms and birth-control pills. The development of space travel was strongly aided by the rocket societies of several countries.[25] The use of computers has been aided by computerization movements.[26] And nonprofit organizations may influence important technological decisions in such fields as energy and the environment through lobbying and public education and by managing the distribution of the devices themselves. Because such voluntary associations are sometimes quite large, they may be very powerful. The story of a Minnesota powerline, to which we turn next, is an object lesson in how the nature of the sponsor may shape the form taken by the technology.

The Minnesota Powerline Wars: How a Sponsor Shapes Its Technology

In the late nineteenth century rural cooperatives proliferated in the United States.[27] These cooperative societies provided lower prices for their members by buying in bulk, and they often offered important advice and assistance to the small farmer. Many people who grew up in small towns are familiar with the Farm Bureau or the local "co-op." By the late twentieth century many of these cooperatives, which originally were democratic organizations, had become large, powerful bureaucracies. One of the fields in which cooperatives had developed was the provision of electric power, thanks to loans and technical expertise available from the Rural Electrification Administration (REA).[28] Originally an example of "democracy on the march," the character of the electrical co-ops changed. Because of their size, their increasing bureaucratization, their competition with other utilities, and their access to large amounts of federal funds through the REA, these cooperatives were no longer as responsive to their members. They now resembled typical large business firms and reacted to their members and their environment much as other business firms would react.

In 1972 two rural electric power cooperative associations, composed of some 34 individual cooperatives, decided to build a powerline from South Dakota to Minnesota.[29] This line, which was to provide power generated near coal mines in South Dakota to the Minneapolis metropolitan area, would be made up of towers, some over 150 feet tall, that would support high-voltage direct-current wires. The laws for the construction of such monumental public works require that the organizations contemplating them invite public feedback, and public hearings were scheduled in communities likely to be affected. Because of public pressures to keep the powerline out of wilderness and "scenic" areas, the line was often routed across agricultural land. Thus many of those who chose to participate in public hearings were farmers on whose land the powerline towers would be located.

The farmers rapidly discovered that participation in the hearings did not necessarily result in the ability to influence ultimate decisions. Farmers came to public hearings and protested the line's existence and its proposed location. These protests, though often emotional and sometimes eloquent, had little effect on the final decisions regarding siting. Facing the unpleasant aesthetic, medical, and economic effects on them, the farmers found themselves powerless to affect important decisions that were being made about their lives and their farms. It was even more ironic that many of the farmers were co-op members: The very organization dedicated to assisting farmers was bringing them distress.

The farmers tried organizing into larger and larger associations over a considerable territory, but this proved to be of no avail in the face of the superior organizational, legal, and political abilities of the powerline interests.[30] In desperation the farmers resorted to obstruction. They interfered with the crews building the line and delayed arrival of materials. When the line was built in spite of these tactics, they resorted to even more drastic measures—toppling the towers. "Bolt weevils" removed the bolts that held the tower sections in place. Over a dozen of the 150-foot towers came crashing down. Farmers shot out the glass insulators and even the line itself with high-powered rifles. Local law-enforcement agencies were unable to cope with the problem; state and federal agents were called in, but the sabotage continued.[31] In the end, however, the cooperatives prevailed; the line was built, and it remains.

The farmers had faced a problem typical of modern societies: the reduced responsiveness of large-scale organizations. Even though many of them were co-op members, the farmers struggled against the influence of a large, powerful sponsor, and they were beaten because as an organization increases in size, each person working in it or belonging to it has much less power over it. As Karl Marx and sociologists Georg Simmel recognized, this factor of size makes the individual feel that the organization is an objective fact, distinct from him or her. This new "social reality" must then be confronted on its own terms.

The powerline is but one of a number of impositions a technological society is likely to make on its members. Because many technological processes have economies of scale,[32] there are many operations in which a large installation—a power plant, a powerline, an incinerator, or an airport, for examples—is difficult to avoid. But where is the installation to be located? Increasingly, no one wants the installation located near them: "Not in my back yard!" (NIMBY) has become a rallying cry for many groups and communities opposed to a nuclear waste dump, a mental hospital, or a prison.[33] Often this means that sites for such facilities are chosen less for technological reasons than for political acceptability; the less-resistant community is more likely to get the undesirable installation.

The growth of giant, soulless bureaucratic organizations is one of the major unresolved problems of our time.[34] We must cope not only with the technology, but also with the organizational juggernauts that have developed along with it. A major challenge we face is getting these organizations to become more responsive and more responsible.

How Technological Niches Evolve

In previous sections we have seen how markets evolve; a similar evolution in the distribution of the devices themselves also occurs. Just as firms find niches in a market, technologies find niches in the ecology of human activities. Natural selection of technologies, however, does not mean that the best technologies survive; rather, it means that *those technologies will survive whose firms make a profit producing and selling them.*

Ralph Waldo Emerson said, "Build a better mousetrap, and the world will beat a path to your door." We have seen how false this idea is in relation to inventions, and it can be equally false with respect to commercial products. A firm may make a fine product and may even sell the product widely, but the continued existence of the firm depends on commercial, not technological, success; the quality of the product does not guarantee the survival of the firm.

Technologies and firms have complex relationships. A firm can seize the market niche of an inefficient producer, even though the former's product is inferior, or a technology may be sustained by a new business firm. Conversely, a technology may lose its niche while its sponsor remains; even though carriages disappeared, carriage builders survived, often by switching to the production of automobile bodies. And not only are there choices among products, but among the technologies themselves. One technology may be challenged by others that perform similar functions.

Exactly such a challenge took place recently in the recorded music field, with the invention of the compact disc. The ability of the new technology to sample audio waveforms digitally at 44,100 times a second accounts for what some regard as its spectacular performance. Its convenience and durability are undoubtedly important as well, but these features had also been strong selling points for cassettes; one cannot play a vinyl record in a car. The disc players were designed to be compatible with existing amplifiers and turntables, which lowered the cost of conversion, and the appeal of the new technology was further enhanced when manufacturers agreed on a single standard format.

In 1983 compact digital recording discs appeared on the shelves of U.S. music stores. Within two years, 15 million compact discs a year were being sold.[35] In 1986, this amount more than doubled.[36] Sales of vinyl long-playing records suffered accordingly, as they had earlier as a result of competition from cassette tapes. Although music listeners owned some 80 million turntables in the United States in 1986, vinyl was rapidly becoming an "obsolete" medium, and some industry specialists predicted that

> it could disappear as early as 1990, though it's more likely to linger some-
> what longer, the province of small, independent labels, collectors and die-
> hards who refuse to make the move to a new technology.[37]

Meanwhile, some audiophiles have decried the artificial and "unsatisfactory" quality of digitally recorded music. One critic, James Boyk of Performance Recordings in Los Angeles, which produces high-quality analog (as opposed to digital) recordings on vinyl record discs, has strongly criticized the sound performance of digital recordings.[38] Boyk contends that analog recording better represents the music as heard live by the ear. His own recording studio uses only tubes in the recording devices that transform auditory signals into grooves on records; solid-state devices cannot perform as well as tubes, according to Boyk.[39]

Boyk cited an experiment in which Sheffield Records used ribbon microphones to record a concert on each of five different storage media. Whereas only the direct-to-disc recording was comparable to the "direct feed" from the microphones, only the three analog tape recordings were considered acceptable by a sample of listeners. The digital recording, however, in spite of using a specially selected and modified digital recorder, was not acceptable to the listeners, for it tended to blend timbres and textures, "homogenizing" the sound of the different instruments. The digital recording, Boyk said during a radio interview, was "not only inferior, but musically unacceptable. In fact, I'll go so far as to say that I found it antimusical."

So at least some listeners were unlikely to switch to compact discs on principle, whereas others were unwilling or financially unable to give up their turntables. Thus in 1986 recording stores carried LP's, cassettes, and compact discs, while expecting that the bulk of their business would soon consist of compact disc sales, just as it currently consisted of cassette sales.[40]

At the time the compact disc boom was developing, however, a threat to this recording medium appeared in the form of digital audio tape (DAT). The sponsors of the other recording media in the United States considered DAT a threat because it would allow personal recording of digitized compact discs. The recording industry accordingly began lobbying to block the sale of DAT. The issue was still unresolved at the time this textbook was written.[41]

Our discussion of the recording industry has now brought us to an important observation: Technologies that ostensibly do the same thing, such as playing recorded music, may have very different *functional niches* among users. This layering of technologies, in which "old" technologies exist side by side with new ones, is a natural result of financial and emotional investments in systems that can be superseded (but usually only in some respects) by others.

Consider the case of the piano. Although new electronic versions promise greater mobility and versatility, there is still considerable interest in mechanical pianos, especially those of premium quality. After Steinway Pianos was sold to CBS, an independent consortium of investors bought the company from CBS because they were afraid the traditional quality of the brand would be affected.[42] Still, overall sales of pianos have fallen off. Although some 167,000 pianos were sold in 1986, 2.5 million portable electronic keyboards, 350,000 synthesizers, and 84,000 other keyboards were sold in the same year.[43] Americans are not losing interest in making music, but apparently they are losing interest in making music with mechanical pianos.

These struggles over pianos mirror other debates about the way music should be performed: Should music be performed on instruments of the time at which the music was written, or on current instruments?[44] Should the orchestra's pitch be set electronically by a black box or by the "A" note of the oboe?[45] And so forth.

One of the most salient struggles for technological niches has been that between AM and FM radio.[46] FM radio began as an underdog, but its superior sound quality and more sophisticated programming led to a gradual changeover of the listening audience. Whereas FM had 40 percent of the radio market in 1976, it had 70 percent in 1987. Meanwhile, AM has not disappeared, for it has certain advantages for rural listeners: It has more range than FM, especially at night, and is not blocked by hills and valleys the way FM is. AM is also trying to modernize with better programming, improved sound quality, and stereo. One of the difficulties with AM stereo was broadcasters' inability to agree on a common system; at one time five incompatible systems were being used. This also made it difficult for manufacturers to produce radios that would pick up AM stereo. The number of systems is now down to two, and a common, voluntary standard may soon emerge. Another problem is that as AM stations have proliferated, crowding of the airwaves has degraded sound quality. The Federal Communications Commission is now reconsidering its policies for allocating band space, which were based on older technology and, many in the industry feel, now need rethinking.

A major factor in determining the shape of a new technology is of course the current technology with which it must "interface." Thus, for instance, compact disc players had to interface with conventional amplifiers and speakers, so that someone switching to CD recordings need not change the whole system. Similarly, with the advent of the computer terminal came an opportunity to change to a keyboard arrangement that would allow faster typing than is possible on the QWERTY keyboard with which we are all familiar. But the familiarity was just the problem: No firm in the business was willing to face the enormous customer resistance to a computer keyboard that would not be compatible with typewriters already in use.[47] Perhaps even more important, no one was ready to face the resistance of tens of millions of secretaries, typists, and others who had learned to type on a QWERTY keyboard. So the changeover to computers retained a notoriously inefficient keyboard.

Figure 10-2 The Use Over Time of Optical Devices in 68 Sea-Serpent Sightings

Time Period	Sightings with Telescope	Sightings with Binoculars
Before 1826	••••• ••••• ••••	
1826–1850	••••• •	
1851–1875	•••••	••
1876–1900	••••• ••	••••• ••••• •••
1901–1925		••••• ••••• •
After 1925	•	••••• ••••

SOURCE: *Based on studies by Ron Westrum and Teresa Crabtree.*

Some technological niches and forms of technology may thus remain stable over substantial periods of time, and we may see technologies of very different vintages existing side by side. An example is the use of small optical devices for long-distance observation. The telescope was invented some four centuries ago and has had a variety of applications ever since. Binocular telescopes, or "binoculars," although invented at roughly the same time, were brought into common use by improvements that began about 1825. Unlike telescopes, binoculars tend to have terrestrial and generally very practical applications—war, navigation, sporting events, and entertainment ("opera glasses"). In the middle of the nineteenth century, as binoculars became more common, they began to exist side by side with telescopes. Unlike most types of technology today, telescopes could be maintained and used for all practical purposes almost indefinitely.

In marine communities and on shipboard, both telescopes and binoculars have their uses for observing distant objects. Sometimes the objects in question are rather odd—sea serpents, for instance. Because the marine animals called sea serpents generally are rather large and travel rapidly (for instance, at 10 or more knots per hour), they might be visible over a considerable distance.[48] In such a situation the observer would naturally reach for the appropriate device—the trusty telescope. In the course of processing some data on sea-serpent sightings, I discovered a rather distinct historical trend: Over time there was a slow but steady trend toward the use of binoculars. The data in Figure 10-2 were assembled from 68 sea-serpent sightings in which an optical device was mentioned.[49] These data show a very gradual replacement of telescopes by binoculars, rather than a sudden transformation.

Other data might indicate a similar replacement of waterwheels by steam engines, carriages by automobiles, and mechanical clocks by electrical ones.[50] A theoretical example of this kind regarding illumination devices is shown in Figure 10-3.

It would be interesting to see whether in our own time such niches develop more rapidly overall, or just in such high-tech fields as electronics. Living room couches, for instance, do not seem to have such a rapid obsolescence, but perhaps this is just because we are unaware of the technology of couches. The changes over time of the niches of many everyday objects would certainly be a fruitful subject to explore in future research.

Figure 10-3 Theoretical Popularity of Types of Artificial Illumination in Pennsylvania from 1850 to 1950

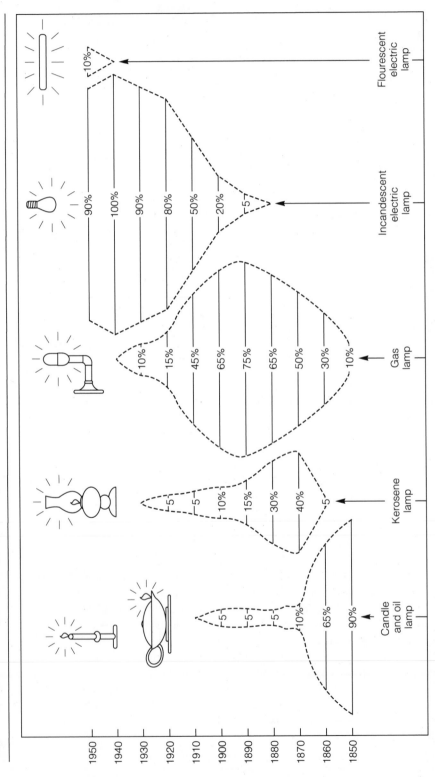

SOURCE: *William J. Mayer-Oakes, Prehistory of the Upper Ohio Valley, Anthropological Series No. 2 (Pittsburgh: Annals of the Carnegie Museum, 1955, Vol. 34), p. 179. Reprinted by permission.*

Conclusion

Just as technologies themselves evolve, the institutions that develop, support, and evaluate them also evolve. If we are to understand why certain technologies, whether handguns or nuclear power, seem persistent, we must inquire into the institutions that keep these technologies in place. We must inquire also into why the institutions that might support other, vanished technologies no longer exist, and why technologies that might be valuable to some groups don't get developed at all. Technologies and their supportive institutions are strongly intertwined; a technology is supported by its sponsor, and its sponsor's existence is supported by the continuance of the technology. When these institutions change, technologies may change also, and to change the technologies, we may have to change the institutions.

Notes

1. Steven Greenhouse, "Minnetonka's Struggle to Stay One Step Ahead," *New York Times*, Dec. 28, 1986, Business Section, p. 8. The account in this chapter is based entirely on Greenhouse's article.

2. Doug Garr, *Woz: Prodigal Son of Silicon Valley* (New York: Avon, 1984), pp. 77–88.

3. Michael Moritz, *The Little Kingdom: The Private Story of Apple Computer* (New York: Morrow, 1984), p. 321.

4. One article described the Apple Macintosh II as "the best designed microcomputer yet"; see Cary Lu and Ellen W. Chu, "The New Microcomputers: Strategies for Macintosh Owners," *High Technology*, Aug. 1987, pp. 48–49.

5. This sketch is loosely comparable with that in David Dale Martin, "The Computer Industry," in Walter Adams (Ed.), *The Structure of American Industry* (New York: Macmillan, 1977), pp. 285–311.

6. Katharine Davis Fishman, *The Computer Establishment* (New York: McGraw-Hill, 1982), pp. 155–228.

7. N. R. Kleinfeld, "A Tight Squeeze at Video Stores," *New York Times*, May 1, 1988, Section F, p. 4.

8. On market strategies, see Raymond E. Miles and Charles C. Snow, *Organizational Strategy, Structure, and Process* (New York: McGraw-Hill, 1978), pp. 31–80.

9. William J. Abernathy, Kim B. Clark, and Alan M. Kantrow, *Industrial Renaissance: Producing a Competitive Future for America* (New York: Basic Books, 1983), pp. 44–56.

10. Generally, the cost of building a car changes relatively little with the size of the car, as compared to the difference in selling price. Thus, a car company makes more money in building bigger cars.

11. David Halberstam, *The Reckoning* (New York: Morrow, 1986), pp. 301–318.

12. Robert M. Hayes and William J. Abernathy, "Managing Our Way to Economic Decline," *Harvard Business Review,* July-Aug. 1980, pp. 67–77.

13. See, for instance, Eric Gelman, "Ford's Idea Machine," *Newsweek,* Nov. 24, 1986, pp. 64–66.

14. Gerald Carson, *Cornflake Crusade: From the Pulpit to the Breakfast Table* (New York: Rinehart, 1957), p. 191.

15. Ruth Schwartz Cowan, *More Work for Mother: The Ironies of Household Technology from the Open Hearth to the Microwave* (New York: Basic Books, 1983), pp. 127–145.

16. Jon Didrichsen, "The Development of Conglomerate and Diversified Firms in the United States, 1920–1970," *Business History Review,* Vol. 46, No. 2 (Summer 1972), pp. 202–219; see also "The Conglomerates' War to Reshape Industry," *Time,* Mar. 7, 1969, pp. 75–80.

17. Thomas Whiteside, *The Blockbuster Complex: Conglomerates, Show Business, and Book Publishing* (Middletown, Conn.: Wesleyan University Press, 1981), pp. 39–48.

18. Pauline Kael, "The Numbers: Or Why Are Movies So Bad?" *New Yorker,* June 23, 1980, pp. 82–93.

19. Halberstam, *The Reckoning,* p. 479.

20. Testimony of Jacob Rabinow, in *Small, High Technology Firms, Inventors, and Innovation,* hearings before the Subcommittee on Investigations and Oversight of the Committee on Science and Technology, U.S. House of Representatives (Washington, D.C.: U.S. Government Printing Office, 1981), p. 398. In a development germane to this topic, RCA was acquired in February 1986 by General Electric, which then proceeded to sell the entire RCA consumer electronics business. This business was making a modest profit; its "crime" was that it could not "dominate" the world market. Consumer electronics at RCA had been a $3.5 billion business; had it been the major business of a single firm, might not something different have happened? One wonders.

21. David Noble, *Forces of Production: A Social History of Industrial Automation* (New York: Knopf, 1984).

22. Donald A. Schon, *Beyond the Stable State* (New York: Norton, 1971), p. 42.

23. See also Dorothy Nelkin, *The Politics of Housing Innovation: The Fate of the Civilian Industrial Technology Program* (Ithaca, N.Y.: Cornell University Press, 1971).

24. William E. Schmidt, "Pressure for Gun Control Rises and Falls, but Ardor for Arms Seems Constant," *New York Times,* Oct. 15, 1987, Section IV, p. 5.

25. Frank H. Winter, *Prelude to the Space Age: The Rocket Societies 1924–1940* (Washington, D.C.: Smithsonian Institution Press, 1983).

26. Rob Kling, "The Mobilization of Support for Computerization: The Role of Computerization Movements," *Social Problems,* Vol. 35, No. 3 (June 1988), pp. 226–243.

27. John Daniels, *Co-Operation: An American Way* (New York: Covici, Friede, 1938), pp. 109–151.

28. Richard A. Pence (Ed.), *The Next Greatest Thing* (Washington, D.C.: National Rural Electric Cooperative Association, 1984).

29. See Barry M. Casper and Paul David Wellstone, *Powerline: The First Battle of America's Energy War* (Amherst: University of Massachusetts Press, 1981); most of the information reported on this case comes from this classic study of the clash of technological sponsors and democratic institutions.

30. Luther P. Gerlach, "Dueling the Devil in the Energy Wars," in Paul P. Craig and Mark D. Levine (Eds.), *Decentralized Energy* (Boulder, Colo.: Westview Press, 1981), pp. 79–105. There is a good deal of wisdom in this article.

31. Sheldon Mains, "The Minnesota Power-Line Wars," *IEEE Spectrum,* July 1983, pp. 56–62.

32. Strictly speaking, an "economy of scale" exists when the unit-cost of producing something goes down when one produces more of it. Obviously, this is the rationale for mass production, for instance.

33. William Glaberson, "Coping in the Age of Nimby," *New York Times,* June 19, 1988, Business Section, pp. 1, 25.

34. See, for instance, Morris L. Ernst, *Too Big* (New York: Little, Brown, 1940).

35. David Pauly, Penelope Wang, and Patricia King, "A Compact Sonic Boom," *Newsweek,* Dec. 16, 1985, pp. 48–50.

36. Richard Harrington, "The Vinyl Days," *Washington Post,* Oct. 26, 1986, Section F, pp. 1, 5.

37. Harrington, "The Vinyl Days," Section F, p. 1.

38. James Boyk, in an interview with Allen Young on the University of Michigan's WUOM program "Background," recorded sometime during 1985. Even though it is easy to suspect Boyk of having a vested interest in preferring the analog medium, his credentials and expertise are compelling; nor is he alone in his criticisms. See Gerald Brennan, "You Can Sample a Waveform, but You Can't Sample a Feeling," *Ann Arbor News,* Apr. 26, 1986, Section D, pp. 1, 2.

39. Wayne Biddle, "Seduced by the Pure Music of Virgin Commies," *Discover,* May 1987, pp. 68–75.

40. Harrington, "The Vinyl Days," Section F, p. 2.

41. Jeff Copeland, Eleanor Clift, and Monroe Anderson, "The Sound of Money," *Newsweek,* Oct. 5, 1987, pp. 72–73. Although I'm not an expert in this subject, it appears to me from an overview of recording technologies that manipulation of the market is more common in music than elsewhere; see Oliver Read and Walter L. Welch, *From Tin Foil to Stereo: Evolution of the Phonograph* (Indianapolis: Howard Sams, 1977).

42. Edward Rothstein, "For the Piano, Chords of Change," *New York Times,* Sept. 27, 1987, Arts and Leisure Section, pp. 1, 28.

43. Rothstein, "Chords of Change," pp. 1, 28.

44. John Rockwell, "The Symphony Looks Back to the Future," *New York Times,* Feb. 22, 1987, Arts and Leisure Section, pp. 1, 26.

45. Donal Henahan, "Finding the Perfect 'A' Sparks a Pitched Battle," *New York Times,* Aug. 30, 1987, Arts and Leisure Section, pp. 19, 28.

46. John Burgess, "A New Day Dawns for AM Radio Stations," *Washington Post,* Feb. 28, 1988, Section H, p. 2.

47. Stephen Jay Gould, "The Panda's Thumb of Technology," *Natural History,* Jan. 1987, pp. 14–23.

48. See Bernard Heuvelmans, *In the Wake of the Sea-Serpents* (New York: Hill and Wang, 1968).

49. For data sources, see Ron Westrum, "Knowledge About Sea-Serpents," *Sociological Review Monographs,* Vol. 27 (1979), pp. 293–314. I am also indebted to Gary Mangiacopra, who collected many of the reports by combing old newspapers. Although the data here are offered in a humorous spirit, they are nonetheless real data, carefully compiled by my student Teresa Crabtree and me.

50. Nebojsa Nakicenovic, "The Automobile Road to Technological Change: Diffusion of the Automobile as a Process of Technological Substitution," *Technological Forecasting and Social Change,* Vol. 29 (1986), pp. 309–340.

PART FOUR

Users of Technology

The Technology/User Interface

One of the important tasks of a social science of technology is to study the ways technology is actually used. Researchers in psychology who do this describe their work as "human factors engineering," but there seems to be no general term for this kind of study in sociology.[1] Such work, nonetheless, is very important.[2] If we are to understand how to improve people's ability to use technology, and how to improve the technology itself, we must first find out how people use it. Not only does the study of what people actually do with devices provide many clues about what needs improvement, it is also interesting in and of itself. How people learn to use a technology, how people's degrees of skill are distinguished, what virtuosos do, and how different people cope with the same device—all this is useful knowledge that enriches our understanding of the human condition.[3]

In most of the previous chapters we have examined technology in its material form.

Here we must consider technology not only as a *thing* with which to perform operations, but also as an *ability* to carry out operations. Thus technology here has two aspects, as device and as technique. An example may make this clear. If we gave a pocket calculator to one of the few remaining "primitive" peoples on the earth, it would be absolutely useless to them, except perhaps as a decoration. Without the knowledge of how to use and maintain the device, it would possess little value. Thus a technology is not simply a collection of implements; it is also human understanding of how to use them.

We have learned how a technology is developed, nurtured, and deployed by a series of social groups, and that the dynamics of these groups is crucial to technology's impact. But what happens to technologies once they are in use? In this chapter and the next, we will explore what people do with devices and techniques once they have them. To aid our discussion we introduce the idea of a "technical community"—the totality of people who are involved one way or another with a given technology, including designers, manufacturers, repair people, and users.[4] Now let's explore the nature and identity of the users of technology.

Types of Users

We might consider a "user" to be anyone who is in direct contact with the technology in question. However, this definition is too broad, because both the dentist and the dental patient would be users of dental technology in this sense. Consider some possible categories of users:

1. Users can be people who have chosen to use a technology. In the United States most automobile drivers fall into this category. In this case the automobile represents their solution to problems of personal transportation, movement of materials, and recreational driving.

2. Users can be those for whom the technology has been chosen, as is typically the case with secretaries and factory workers. Such users are still responsible for operating the technology, and although they may make small modifications, they are forced to carry out tasks with tools others have chosen for them.

3. Users can be a client population, such as children playing in a park, hospital patients, or college dormitory students. For these users, technology is something to be reacted to, not something to operate. When most people who live in cities confront the technical features of their environment, they interact with technology chosen and operated by someone else. We will not pay much attention to this category of users here.[5]

We need to distinguish among the kinds of users through a more articulated concept. We ordinarily use a variety of words to describe different kinds of users:

Figure 11-1 Level of User Control Over a Technology

passenger	operator	jockey
victim	tender	driver	virtuoso

operator, driver, passenger, victim, jockey, tender, virtuoso. These words connote different degrees of *agency,* or degrees of control that an individual possesses in relation to the device. Some users exercise a very high degree of skill and virtuosity in controlling the device. Others are at best passive victims in relation to the technology. A passenger and a race car driver both use automobiles, but their level of control is very different. Figure 11-1 shows a scale that reflects the user's degree of control over a technology.

In thinking about driving, flying, cooking, operating machinery, or making music, this kind of gradation of abilities is easy to envision. In what follows, it will be helpful to keep this sort of user spectrum in mind. In this chapter we will be more concerned with users toward the right end of the scale, but we will touch upon other users as well.

The Nature of Users' Skills

Learning to Use a Technology. As people use a technology, they become increasingly skillful. This acquisition of skill has been very carefully studied by psychologists.[6] But there are also important social aspects of this acquisition of skill. These may include learning from others who are more skilled in certain habits of thought, jargon, and values associated with carrying out the task.

Socialization to a technological community usually includes learning the verbal and nonverbal communication practices that characterize skilled practitioners. In the process of becoming a railroader, Luis Kemnitzer discovered that a variety of verbal and nonverbal signals were used to communicate messages and orders. Few if any of these signals were explained to him; instead, he had to learn them from the context of use, an important socialization ritual.[7] Interestingly enough, he also found that railroaders tended to use the same terms to describe much of the nonrailroad world. Many of us readily recognize such a use of jargon by computer buffs, who talk about something that "doesn't compute" or about "interfacing" with someone else. In medicine, jargon is used not only to identify distinct conditions, but also to mystify and withhold information from patients who otherwise would have more control over their treatment. Perhaps the key point about jargon is that we learn it at the same time we learn to use the technology, and so our conception of the devices is shaped by the words and metaphors we use to describe them.

Table 11-1   Proportion of Electricians Owning Certain Tools by Year of Apprenticeship

Year of Apprenticeship	"Square D" Voltage Tester	"Lufkin" Rule	"Channel Lock" Pliers	"Klein" Sidecutters	Total
1	38%	56%	81%	75%	16
2	72%	58%	83%	69%	36
3	79%	68%	79%	89%	28
4	81%	83%	87%	87%	47
5	87%	84%	88%	96%	68
Total	77%	74%	85%	86%	195

SOURCE: © 1979 Sage Publications. Rerinted with permission.

Learning appropriate tool habits is also important. In an astute study of the socialization of construction electricians, Jeffrey Riemer described the visual cues by which veteran electricians recognize other veterans.[8] Journeymen are distinguished from apprentices by three main features: their possession of certain informally approved "brand" tools; the presentation of tools and wearing of appropriate clothing; and knowledge of electrical trade jargon. The journeymen were more likely to wear bib overalls and to have their names on their hard hats, and to carry certain tools of the "approved" brands. As apprentices worked on the job, they came to share a journeyman's opinions about what kinds of tools are best. Riemer was able to show that, as apprentices gained experience, the equipment they used increasingly approached that used by the journeymen (see Table 11-1). Also, the more experienced electricians tended to carry fewer, better-chosen tools with them than did the inexperienced apprentices.

It is not necessarily true, however, that "a few, well-chosen tools" mark the veteran in all fields. The array of tools possessed depends on the requirements of the job. In his study of ambulance personnel, Donald Metz found that veteran Emergency Medical Technicians (EMTs) tended to have a greater array of tools, often as a result of the variety of situations they expected to encounter. "Nearly all the experienced EMTs have equipment holsters loaded with a variety of lights, forceps, scissors, bite sticks, knives, wrenches and keys. Some have their own stethoscopes."[9] Many of these tools had been "liberated" from hospitals to which the ambulance crews had delivered patients. Some EMTs even carried some specialized tools in their personal "trauma boxes" in their cars, just in case they needed to provide assistance to the sick or injured when they were off duty. Even so, excessive and ostentatious tool display by "hot dogs" was frowned on.

In another study Chandra Mukerji found that student filmmakers tended to share with faculty a perception that certain kinds of movie cameras were superior to others.[10] Furthermore, both faculty and students tended to limit access to the more prestigious equipment. The equipment was evaluated by how closely it approximated professionals' equipment, which used larger film and was correspondingly more expensive. Thus, although some very interesting films could be made with 8mm film and cameras—and of course, made much more cheaply—students valued access to 16mm equipment much more highly. This value judgment, Mukerji notes, produced an "artificial scarcity" of equipment suitable for film-making.

Indicators of Skill. The facility with which the individual can use a device is ordinarily indicated by such words as competence, skill, and virtuosity. Professionals may recognize indicators of skill not evident to outsiders. In his outstanding book *The Racing Driver,* Dennis Jenkinson describes a number of characteristics of first-class drivers that would be difficult for an outsider to assess, including the willingness to drive at "ten-tenths" of the racer's ability.[11] Similar issues regarding higher degrees of skill impassion the aficionados of most competitive sports, as well as music and art critics and even professional criminals.[12]

Interestingly enough, the virtuosos may not always be aware of how their own outstanding performances are produced. A careful study of expert trumpet players, for instance, showed that the experts were not good at assessing how much force they or others used to produce the sounds they made.[13] This meant the experts were less able to understand how much force a pupil needed to exert to get the trumpet to sound right. Thus expertise in use does not always mean expertise in training others. We might wonder in how many other areas highly skilled experts may possess erroneous ideas about how the task "has to be" performed.

Certain kinds of devices may require high skill levels to get them to operate at all. Harry Collins found that scientists trying to build certain kinds of lasers from written instructions were often unable to do so without "hints" from those who had already built them.[14] Similarly, one of my students mentioned a grandmother's cake-making technique that seemed impossible to describe in the form of a written recipe; if one had not actually *seen* how the cake was baked, one couldn't do it. Office and factory folklore is filled with stories of devices that will only perform for certain people. The reverse may also be true: Certain people have a kind of "black thumb"—they will cause otherwise resilient equipment to "crash" for no apparent reason.[15]

An individual's use of a device may be distinctive enough so that people can distinguish the user, in the way the device is used, from other practitioners. Telegraphers, for instance, can recognize each other's "fist," or "stuff"—the way the telegraph key is hit.[16] I am reliably informed that computer software often bears distinctive marks of its creators. When computer programmers who have worked with very restrictive computer architectures are transferred to systems with more options, they may still continue to program as if the same limitations were in force.

Flow States. Skill may be shown in the user's ability to get the device to perform in ways that others cannot. This is sometimes done with considerable and evident effort, but in other cases what is conspicuous is the smoothness of the performance and the evident control over the system by the virtuoso. For instance, in the study of trumpeters, researchers discovered that expert players could often generate massive force with little sign of effort.[17] Studs Terkel describes a car parker who, although one-armed, was an outstanding driver on the lot.[18] Test pilots often display an almost unbelievable coolness of attitude in situations that are potentially lethal and require immediate decisions.[19] In an interview with the journalist Walter Shapiro, Major Russ Stromberg described a near-fatal experience while taking off from the aircraft carrier *Tarawa* in a Harrier jump jet:

> I was very surprised by the evolution of the thing. Everything went into slow motion. . . . After about one second, about 75 feet after I started rolling. . . . I knew I was in deep trouble. As the plane approached the bow, I turned off the "limiter" switch and took off all the power limitations on the engine. Still, there was no response. There was no overt decision to eject. It was just the next thing to do. . . . I couldn't tell by feel alone that I got a good grip on the handle of the ejection seat. I had to take time—even if there wasn't a lot of it—to look. I finally pulled the handle when I was about 30 feet above the water.[20]

Expert race drivers develop a level of oneness with the car so that driver and car seem a single system. They enter into what some psychologists call "flow states," in which the driver concentrates only on the task to the exclusion of irrelevant surrounding stimuli.[21] The most conspicuous mark of a flow state is the smoothness of performance. Driver and machine develop a natural rhythm of action; the driver seems almost magically to anticipate approaching obstacles and acts effectively to avoid them in highly coordinated motions that take place at very high speeds. Flow states are common occurrences in other kinds of skilled sports and gaming behavior—for instance, among mountain climbers and chess players.[22]

General Chuck Yeager, a renowned combat and test pilot, provides a very good description of a flow state:

> It's almost impossible to explain the feeling. It's as if you were one with that Mustang, an extension of the damned throttle. You flew that thing on a fine, feathered edge, knowing that the pilot who won had the better feel for his airplane and the skill to get the most out of it. You were so wired into that airplane that you flew it to the limit of its specs, where firing your guns could cause a stall. You felt that engine in your bones, felt it nibbling toward a stall, throttle wide open, getting maximum maneuvering performance. And you knew how tight to turn before the Mustang snapped on you, a punishment if you blundered. Maximum power, lift, and maneuverability were achieved by instinctive flying: you knew your horse. Concentration was total: you remained focussed, ignoring fatigue or fear, not allowing static into your mind. Up there, dogfighting, you connected with yourself. That small, cramped cockpit was exactly where you belonged.[23]

User Compatibility: Technotonic and Technostressing Technologies

An important concern for users is the fit between their needs and the technology. This matter has two important aspects: the primary goal served by the technology and the users' pleasure in its use. Both of these aspects are likely to affect the individual's emotional reaction to the technology. In the chapter on design we discussed the relation of technology to primary needs. Here we discuss users' reactions to the technology itself.

As a case in point, let's consider the automobile, a technology that for most Americans serves a multitude of purposes and carries a multitude of cultural and personal meanings.[24] Given the structure of American cities and services, for most Americans the automobile is considered a necessity for personal travel and for transporting such things as groceries. The automobile each of us uses, furthermore, is not purely a matter of free choice; rather, it is usually a device chosen from a limited set of possibilities. Our financial resources limit the choices we can make, as do the constraints of family needs, the distance to our job, and the pressures of the environment within which the cars are used. Yet even when all these constraints are taken into consideration, we still have some choice. At this point some other features are likely to enter in: How much pleasure is the technology likely to give? What does the car signify about its user? The choice of a car is thus influenced in part by how much pleasure and symbolic benefits its prospective buyer anticipates from its use.

This intrinsic pleasure in the use of the technology itself is called technotonicity. A technology is *technotonic* to the extent that is pleasurable and reinforcing to the user. Technotonicity is the opposite of *technostressing,* a situation in which using the technology causes stress. Such stress might exist when the technology is not challenging enough to use, thus causing boredom, or when it is too challenging, causing anxiety.[25] When the technology's demands are relatively equal to the users' ability, however, then a sense of "flow" is likely. It is also possible that what causes the technology to be tonic or stressing is intrinsic not to the device itself, but only to its meaning in the user's social context. Let's consider some of the elements that go into making a device technotonic:

1. *A device is technotonic to the extent that it gives the user a sense of control, of mastery over the environment.* This is the feeling that a computer hacker gets sitting at the keyboard of a favorite machine or that a "fighter jock" gets when sitting in the cockpit of a combat jet. Air traffic controllers often get a strong "high" from managing the complexities of a sky filled with planes, stereotypes to the contrary.[26]

2. *A device is technotonic to the extent that its use reflects a high degree of skill.*
    When a skilled violinist picks up the instrument that he or she uses, there is an
    immediate feeling of pleasure from the sense of mastery over the device. Studs
    Terkel interviewed a grocery store checkout clerk whose use of the cash regis-
    ter was as good as a speed typist's use of a typewriter.[27] In another of his cases,
    a stonemason describes his satisfaction in having a skill unique among his
    friends and neighbors.[28] Even in the kitchen, some degree of skill is important.
    Some 30 years ago, Pillsbury marketed a cake mix that produced a very good-
    tasting cake, and all the cook had to do was add milk. The product sold very
    badly, though, because cooks like to feel they are really *doing* something.
    When the product was reformulated to require the cook to add both milk and
    an egg, sales rocketed.[29]

3. *A device is technotonic when its appearance or use evokes aesthetic pleasure.*
    Technologies that are pleasing to the eye may be used simply for this reason,
    even when their performance may be inferior to others from a purely technical
    point of view. Automobiles that are large, attractive, and comfortable may be
    considered desirable by certain users, even though they may handle badly at
    speed. Such cars give users a feeling of "luxury," and people can often be seen
    to relax when getting into them.

4. *A device may be technotonic because of the associations it evokes.* It is obvious
    that family heirlooms often have a value to their users considerably greater
    than their appearance or performance would merit. Often a technology will be
    preferred because those who use it have a high social status or are otherwise
    admired.[30] Films often create fads for particular devices, as was the case for the
    Smith and Wesson .44 magnum revolver after the film *Dirty Harry* was re-
    leased. Weight lifters can be extremely choosy about the kinds of weights they
    lift. Although concrete-filled vinyl barbells probably are just as effective in creat-
    ing muscle, professionals, and those who wish to identify with them, choose
    barbells and dumbbells made of iron.[31] The use of cellular phones has become
    such a high-status activity that some 40,000 fake phones—they look like the real
    thing but don't work—have been sold to customers eager to improve their
    images.[32] The possession of a packed rolodex file is a more visible indication of
    status than is a computer file.[33]

    The Physicist Helmut Kirchner claims that in atomic particle physics the qual-
    ity of the experiment is rated not only by how important a question is being
    investigated, but also by how large a device is used to do the measurement.[34]
    Kirchner claims that because of the nature of experimental research with par-
    ticle accelerators, those who do it are likely to have more prestige because they
    alone have access to the large machines. There has been some question in
    recent years about the value of using accelerators versus the value of other,
    smaller-scale, forms of physics research. But the dominance of the researchers
    who use accelerators is firmly in place.[35]

    Use of a device may also recall situations which are pleasurable for the
    user. A rock singer told me that she always liked the plug that went from her

guitar into the amplifier, because she associated inserting the plug with the excitement of beginning a concert.

We can consider the inverse of these features to constitute a set of characteristics that make a device technostressing:

1. *A device is technostressing if it makes its users feel helpless.* Videocassette record-ers (VCRs) often prove very difficult to program, and thus may be a source of serious stress for their users.[36] Once they break down, they are also very tricky to repair, another source of stress.[37] Similar hassles await car owners, whose vehicles, according to an experienced mechanic, require preemptive servicing three times a year.[38] Automobiles that chronically break down are very stressful to use, especially if the breakdown causes serious inconvenience.

2. *A device is technostressing when using it demonstrates the user's lack of skill and knowledge of the device.* The introduction of personal computers into the workplace in the 1980s caused a great deal of stress. Workers were suddenly confronted with a novel technology they did not understand and were afraid they would be unable to master. They regularly experienced problems that they could not solve, and thus were made to feel stupid.[39] When a device be-haves in a nonintuitive manner—a sign of poor design—its bewildered user may become nervous and concerned.[40]

3. *A device may be technostressing if it is ugly or evokes bad sensations.* In the workplace noisy machines or foul odors may cause workers stress. Home-makers or housepersons may hate an appliance whose appearance bothers them but is too expensive to be replaced. Male friends who are single and who drive conspicuously degenerating cars have told me how much these cars are a "turn-off" for prospective female companions.

4. *A device may be technostressing if it achieves disrepute through association.* People who become upwardly mobile may develop a strong aversion to de-vices that remind them of their more humble roots. The market for Volkswagen automobiles was damaged by its association with the "flower children" of the 1960s. Similarly, the "zoot suit" (a flashy garment of extreme cut) was often associated with people of low status and was avoided by middle-class Ameri-cans because of these associations.

The technotonic and technostressing aspects of devices need a great deal more exploration. While there are numerous studies of stress in the workplace, much more needs to be learned about technology's role at home and in public places. Since this is a book on the sociology rather than the psychology of technology, our brief discussion here will have to suffice.[41]

Informal Knowledge and
Skill in the Workplace

Much of the knowledge used in the workplace may involve skills gained through personal experience with the technology. Such skills may be apparent in workers' unusual ability to understand and interpret the behavior of devices and processes. In an interesting article by the French researchers Terssac and Christol, the authors described how ostensibly unskilled workers in fact learn to manage complex technologies. Terssac and Christol showed that workers in the factories studied often were able, from long experience, to interpret the behavior of the machinery and materials in ways completely unanticipated by their managers or by the designers of the automatic systems. The ability to interpret signals that meant nothing to the uninitiated was very valuable to the proper functioning of the system.[42]

> Faced with a logic of conception which is often strange to them, the workers develop task procedures which are necessary to get the job done. These informal procedures, unforeseen by the designers, are often felt unacceptable by them because, for them, such initiatives, such modifications in operating procedures threaten to perturb the proper functioning of the system. That there are several ways of accomplishing the task intended seems to be excluded from the logical system of many designers: everything rests, for them, with a standardization and homogenization of operating procedures. This standardization often corresponds to a model of human behavior which postulates the stability of the "functioning of the human operators" and their interchangeability.[43]

A fascinating incident of this kind was related to me by my colleague John A. Gordon, a paint chemist.[44] In 1938 Gordon was working at California Ink in Berkeley as a research chemist. Near Gordon's laboratory was a room full of 250-gallon reactor kettles in which varnishes were made. The temperature of these kettles was monitored by thermometers enclosed in copper sleeves. One day the supervisor of these kettles, Vern "Pappy" Maynard, came into Gordon's laboratory. "John A., the thermometer in the kettle ain't reading right. It should say 600 degrees, but it only says 575," Maynard informed him. Although Maynard, a man in his 60s, had little formal education, Gordon respected his ability, so he said "Let's go see" and grabbed an accurate glass thermometer to take with him. The two went to the reactor room, where Gordon proceeded to check the temperature in the kettle with the glass thermometer.

"What do you read now?" Gordon asked Maynard, who could see only the kettle thermometer. Maynard said the kettle thermometer now read 585 degrees, but the real temperature was 610 degrees. Gordon's test thermometer read 610 degrees.

"How did you know?" the amazed Gordon asked Pappy. "I spit in the kettle, didn't you see me?" was his prompt reply. "If I had chicken feathers, I'd have been more accurate." Evidently, Maynard had learned to interpret varnish's temperature by its reaction to the spittle; specifically, he used the noise made as an indicator of temperature. He claimed this method gave him accuracy to within plus or minus 5 degrees. With chicken feathers, it was the appearance that did it; if the temperature was correct, the feathers turned brown.

These important technical skills were not universally respected at California Ink. A later supervisor of the varnish room didn't think much of his workers' ability to act on their own, and he told them explicitly to "Do what I tell you to do; you're not paid to think." The workers obeyed his orders to the letter, resulting in a serious decline in quality. When the supervisor was fired and Gordon took over the varnish room, every tank had off-grade material in it. Because he respected the workers' skills, he asked them what to do. "Leave it to us," they told him, "we know exactly what to do. What do you want first?" He told them what he wanted, and within two weeks the quality was again within acceptable limits.

Implications for Designers. It is important to note that these kinds of skills have their limits. When reactor kettles with closed lids were used, workers' abilities to interpret the reactions within them dropped. In fact, "the workers nearly burned the place down twice." This probably suggests a need to give workers increased feedback for the processes under their control. The desire for such feedback explains why the Mercury astronauts wanted a window to enable them to see out of the capsule. Although the trajectory of the capsule was entirely determined from the ground, the window made the astronauts feel that they had some control over what was going on.[45]

These examples suggest that we need to pay attention to providing users a "feel" for what is going on with a technology. Such an ability to grasp what the technology is doing may be essential for getting it to operate safely and effectively. There is a lesson for designers here: instead of allowing these signals to be emitted by chance or improvised by the workers, the signals could be intentionally amplified to make them easier for workers to interpret.

The Location of Knowledge. One dilemma that faces designers is where to locate knowledge necessary to get the job done correctly. A variety of forces, including higher labor costs, managers' desires for control, and rapid reaction requirements may lead to highly automated systems. In these systems the knowledge resides primarily in the hardware or software with which workers interact. While such automation may be desirable—consider word processing—there may also be negative consequences.

For instance, the psychologist Nehemiah Jordan noted very different reactions from radar crews when the new SAGE (Semi-Automated Ground Equipment) system went into operation in the 1950s. Even though the old equipment was often temperamental and required frequent repairs, its crews were often able to use their considerable skill to keep it working. The SAGE crews, by contrast, had equipment that was largely a "black box" and required only infrequent monitoring, but they could seldom effect the repairs themselves. Jordan observed that the crews of the manual sets often were quite proud of their skills, whereas the SAGE crews often seemed bored and uninspired by their much more sophisticated equipment.[46] In automating the job, designers sometimes neglect the roles played by motivation and skill; a bored team is not likely to be a good team.

Much technology is designed to avoid the need for workers' skills. For instance, during World War I, the United States wanted to produce aircraft engines in quantity. In trying to determine the best way to mass-produce engines, a survey was made of European practices. The Americans noted that even though European engines were often well made, much of the skill for building them was kept in the technicians' heads, a situation very different from that in the United States, where most of the knowledge had been committed to drawings and high-tolerance manufacturing processes. In the design of the Liberty engine, which the United States manufactured by the thousands, the American approach of clearly spelling out requirements enabled the engines to be manufactured by several different firms.[47]

The Role of Intuition. Making tacit knowledge explicit, however, may be difficult for some forms of expertise. A great deal of the judgment used by experts relies upon intuition, which is to say thought processes that are largely unconscious and that lead thinkers to conclusions by ill-defined steps.[48] Thomas Kelly, a dam engineer for Southern California Edison (SCE), was an expert at diagnosing the problems of Vermilion dam high in the Sierra Nevada mountain range.[49] Extensive experience with similar dams had given Kelly, age 55, useful insights that he used to intuit problems. But Vermilion was a particularly large and difficult dam to keep in order. As Kelly's remarkable skills in understanding it sharpened, SCE executives began to worry about what they would do if he ever left. So they decided to develop a computer program that would simulate Kelly's thought processes, a so-called expert system. Similar expert systems had already been used in medicine and in a variety of fields. Importing two "knowledge engineers" from Texas Instruments, the company carried out a $300,000 project to build a mechanical clone of Thomas Kelly's diagnosis system. The project has had reasonable success, but how good a substitute for Kelly the computer program proves to be, only time will tell.

The High-Performance Team

Skills may be the possession of a team rather than an individual. What factors, we wonder, make some teams perform well and others less so? One obvious possibility is learning time; the longer a team has learned to function as a unit, the better the team's performance is likely to be.[50] Another possibility is that high-performance teams result from certain structural features, such as a free flow of information.[51] In discussing the characteristics of high-performance teams, Peter Vaill notes that they achieve a kind of oneness with the technology, which makes men and machine seem part of a single system.[52]

This level of skill is particularly evident in the performances of aerial demonstration teams, such as the U.S. Navy's Blue Angels or the U.S. Air Force's Thunderbirds.[53] Flying in tight formation requires outstanding coordination and leadership skills; mistakes can be fatal. And of course as teams practice together, they get better with experience.

Certain jet fighters, such as the F-4 Phantom, are designed with two seats, because there is too much for a single pilot to handle. But working together requires mutual respect and teamwork, without which trouble is likely to ensue. One way to provide this teamwork is to keep fighter teams together, so they learn to become a unit. For instance, there is some evidence that the Navy policy of keeping Phantom jet fighter teams together produces better coordination between pilots than does Air Force policy, which is to shuffle team members freely.[54]

Fruitful team interaction may also enhance creativity. A pianist once described to me how it felt to play with a musical partner (a guitarist) with whom she "connected" musically:

> Some of the things we did musically are simply unbelievable. It's hard to put them in words. They just *happened.* In many ways our communication was almost telepathic; we just knew what the other person was going to do. Things happened between us that were incredibly beautiful. In many respects a musical relationship may be more intimate even than marriage.

Here there was more than simply skill, there was joint creativity. No doubt similar observations would be made by members of jazz ensembles. Playing music with others demands high coordination; the results are evident in good performances, particularly of ensembles and orchestras that have worked together for some time.

Such coordination is equally important in high-tech contexts such as the flight decks of airliners, where the effective coordination of a team that includes the pilot, copilot, and sometimes a flight engineer or radio officer is crucial. Coordination cannot be taken for granted. Many aircraft captains (the captain is the senior pilot, who sits in the left-hand seat) adopt a "macho" attitude and expect the copilot and engineer simply to follow orders.[55] This style of leadership is ill-suited to a complex technology, which not only requires good coordination—especially in emergencies—but also full use of the cognitive skills of the entire team.[56] Simulation studies by NASA have shown that aircrews that communicate more make fewer errors, and that the effective aircrews also tend more to acknowledge each others' communications more, rather than simply listening passively.[57] One of the most remarkable features of the television series "Star Trek" was the use of *coordinate leadership,* in which leadership shifts among leaders according to whose skills are most appropriate at the moment.[58] How organizations and groups can best use their total cognitive resources is an important challenge as technology becomes more complex.

In hospitals serious problems can arise when physicians and nurses communicate poorly, whether about patient medications or during surgery. Many physicians

adopt an attitude of omniscience, regarding nurses as inferiors who need enlightenment from the more highly trained physicians. This attitude strongly interferes with
effective transfer of information among nurses, paramedical personnel, and physicians.[59] Decisions concerning medication are often faulty when medical personnel
fail to communicate effectively, and patients can suffer as a result.

A conflict arises here between organizational traditions and technical needs dictated by the situation. The tradition in medicine is for sharp differences in status
between doctors and nurses. But often the traditional, hierarchical, calculative approach proves inadequate in the face of complex technologies. These new technologies require flexibility, higher communication and decision-making skills, and a willingness to adopt generative strategies to cope with uncertainty. Yet often there has
been too little effort to train professionals to cope with the situation. Thus the complete cognitive resources of the team are not used to attack the problem.

The High-Performance Organization

Skill may also be considered a property, not
just of small groups, but of whole medical organizations. In the 1960s a national study in-
vestigated the effectiveness and safety of the anesthetic halothane. In the process of
conducting this study, two of the researchers, Lincoln Moses and Frederick
Mosteller, made a very interesting discovery. The hospitals in the study had very different mortality rates for the same operation, even after differences in the kind of
patients had been controlled for.[60] One obvious explanation might be that the personnel of some hospitals were more skillful at performing the operations than
others. In 1979 a second group of researchers showed that the volume of operations
performed was one of the factors in success; hospitals that performed more operations, especially heart operations, tended to have lower mortality rates.[61] Subsequent
studies also supported this finding. In 1986 another national study showed that hospitals that performed operations more frequently were more likely to be successful.[62] Because success seems to be associated with how often the operation is performed, it appears that in surgery, as in many other matters, practice improves skill.[63]
This finding does not simply involve getting better by doing something more often.
Rather, those hospitals that have a high volume of operations of a certain type are
likely also to develop techniques and practices that increase effectiveness of the
operation. In turn, more patients are referred to these hospitals by physicians who
perceive the hospital as particularly adept.[64]

For instance, in a 1988 study of success rates for coronary bypass surgery in California, hospitals that performed more operations tended to have much higher success rates (see Table 11-2). At the hospital with the lowest death rate, Kaiser Founda-

Table 11-2   Coronary Bypass Mortality Rate Versus Number of Operations Performed in California Hospitals

Number of Operations	1985	1986
Less than 100/year	6.94%	7.4%
100–200/year	6.05%	6.8%
Over 200/year (for 1985)	3.94%	—
200–350/year (for 1986)	—	5.2%
Over 350/year (for 1986)	—	4.3%

SOURCE: *Adapted from Robert Steinbrook, "Hospital Death Rates for Five Surgeries Vary,"* Los Angeles Times, *March 27, 1988, Part I, pp. 3, 35; Robert Steinbrook, "Heart Surgery Death Rates Found High in One in Six Hospitals,"* Los Angeles Times, *July 24, 1988, Part I, pp. 3, 33.*

tion Hospital in Los Angeles, a comprehensive system of postoperative care was in effect. Three cardiologists were on 24-hour shifts to prevent the postoperative complications that are often responsible for deaths. Other hospitals, by contrast, often left postoperative care in the hands of interns and younger residents. Kaiser Hospital also carefully analyzed its previous problems—the only hospital in the study that did so—to maintain its high efficiency.[65] This kind of "reflective practice" is typical of high-performance teams in general. Interestingly, another study found that hospitals where physician performance was closely scrutinized tended to have lower mortality rates than those where it was not. In a short-term study of 58 hospitals in Massachusetts, Stephen Shortell, Selwyn Becker, and Duncan Neuhauser found that death rates were lower in hospitals where more control was exercised over how physicians practiced.[66]

In the future, accreditation of hospitals may depend on performance. Rather than considering only whether a hospital has the *potential* to provide high-quality care, it may soon be judged on how well its patients actually fare. Hospitals that do not perform well may have their accreditation removed by the national Joint Commission on Accreditation of Hospitals.[67] Furthermore, because success rates of operations improve with practice, some experts have argued that certain kinds of operations should be concentrated in a few regional hospitals. Because performing such operations carries considerable prestige, there is competition among hospitals that wish to perform them. In Michigan, a hospital must now get a Certificate of Need before carrying out an organ transplant; these are used by the state to concentrate certain high-risk operations in a few hospitals. The University of Michigan Hospital—one of perhaps ten hospitals in the country performing transplants of all the major organs—is working hard to develop a "national profile" in transplant operations.[68]

The hospital hopes that by repeated experience in this area, it will develop skills that make it outstanding.

Of course, skill is important in a variety of organizational settings in which the technology is complicated and faultless performance by humans is important. One fascinating study of an organization's technological skills was conducted by researchers at the University of California at Berkeley. The authors, Gene Rochlin, Todd La Porte, and Karlene Roberts, studied the handling of aircraft carrier landings.[69] The authors spent considerable time on aircraft carriers and were astonished at the proficiency of crews who would land hundreds of planes in a single day under conditions that were extremely unfavorable. This level of skill is even more amazing when the high turnover of naval crews is considered. The authors described a number of reasons for this situation, including a powerful system of informal training, a high level of oral communication, and a high redundancy of checks. One of the more interesting aspects of the study was the apparent suppression of rank during dangerous operations. Because skill and knowledge are what count during the risky business of carrier landings, the organization's hierarchy is ignored in favor of an emphasis on accuracy and safety in the task.

The same dominance of task requirements over the organization's status structure is mentioned in an important paper by Wernher von Braun, an expert in rockets and a technological maestro.[70] Von Braun felt that the organization's skill in developing a missile required using the team's intellectual talents without respect to hierarchy or scientific specialty. He noted that "a 'bug' in a guided missile does not care how the development agency has been organized, and who was responsible for what." Thus it is important that a rocket team put the development task first and its own internal structure and status concerns second. Similarly, because open lines of communication are very important, von Braun emphasized that if errors are made, one should reward rather than punish those who admit to them. In one case an engineer accidently disabled a Redstone missile without realizing what he had done. After the missile crashed, the organization prepared to redesign the rocket to avoid the problems suspected to have caused the crash. At this point the engineer approached von Braun to report that he thought he might be responsible for the failure. When this proved to be the case, von Braun sent the man a bottle of champagne to emphasize the importance of honesty. This response reflects von Braun's gifts for organization; most organizations would have punished the engineer severely. Note, by the way, the effect that this act was likely to have on future information flow.

It is worth noting that organizations can represent an important skill bank and that keeping them intact may be valuable for maintaining skills at a high level. One of the problems with continuously dismantling organizations in the private sector of American society is that these skill banks may be lost. Consider the sale of RCA to General Electric and the subsequent sale of its consumer electronics business to Thomson, S.A., a French organization. Although after the RCA takeover GE had 23 percent of the American market for television sets, this was not considered a stake worth saving, and the entire $3.5 *billion* consumer electronics business was disposed of. Many questions arise: Should RCA have been sold to GE in the first place? Would the consumer electronics business, which seemed unimportant to GE, have seemed worth saving to another management group? What happened to all the RCA consumer electronics engineers in the process?

One of the larger concerns about American industrial management is the level of its involvement with the technologies it uses. People tend to maintain and develop the things they cherish. But increasingly, it seems that American industrial managers are more oriented to making money than to making things.[71] In a situation like this, the products are bound to suffer.[72] One of the striking differences noted by many observers of Japanese management practices is that personal investment in technology runs from top to bottom in Japanese companies.[73] Willingness of American managers to acquire this kind of knowledge and the associated skills seems strikingly absent. A return to immersion in particular technologies has been suggested by more than one team of experts.[74] The transition from a "hands-on" style of management to a more abstract one contributed to the destruction of the space shuttle *Challenger,* which is discussed in the chapter on technological accidents.[75] Arthur Squires has suggested that effective management of government technology programs requires top managers who are technological maestros, but the truth is that commitment to the technology is probably required from bottom to top to work well.[76]

Conclusion

In this chapter we have shown the importance of the user's skill and attitudes vis-à-vis technologies. Just as it is important that technologies be designed for their users, it is also important to realize that users may have very different reactions to them. Some may acquire almost professional skills with the technologies; others may find them difficult or impossible to handle, a situation that should concern us. If we create a world in which technologies are not user-friendly, and a society of users who are unskilled, very high levels of stress may result, leading to damaged people and social revolts.

Linking the skill of the user to the technology is the task of social institutions. Two approaches are possible: adjust the people or adjust the technology. Society recognizes the need to educate users in some areas, but not all. Although we have drivers' education for automobile users, many other complex technologies make demands equal to or greater than automobiles. How is the education for these technologies to be provided? In other cases the technology may need to be made more user-friendly, so that it causes users less stress and harm. Liability suits have made some manufacturers (such as those who make toys) more aware of "human factors," but much more needs to be done.[77]

Ultimately, both of these approaches need to be used. We must have better-educated users and technology that is better suited to human needs. Perhaps our schools ought to create courses in "systems education." In the meantime, we will experience a good deal more "computer phobia," "technostress," "hacking," and tinkering with the technologies deployed. We can expect increasing demands to be made on designs with respect to safety, but we can also expect some of the responses to these demands to be inadequate, leading to more accidents and injuries.

Notes

1. For instance, see K. F. H. Murrell, *Ergonomics: Man in His Working Environment* (London: Chapman and Hall, 1971); and Barry H. Kantowitz and Robert D. Sorkin, *Human Factors: Understanding People-System Relationships* (New York: Wiley, 1983).

2. See, for instance, Robert Gutman (Ed.), *People and Buildings* (New York: Basic Books, 1972); and Harold M. Proshansky, William H. Ittelson, and Leanne G. Rivlin (Eds.), *Environmental Psychology: Man and His Physical Setting,* 2nd ed. (New York: Holt, Rinehart & Winston, 1976).

3. Many of the recent social studies of technology cited in this and other chapters suggest the need for a new discipline to cope with the complexity of human/technology interactions. Even though there is now a recognized field of study in "human factors," either this field should be considerably expanded or an entirely new one should be created (we might call it "social factors in technology").

4. This expression is used by Edward Constant II in his *The Origins of the Turbojet Revolution* (Baltimore: Johns Hopkins University Press, 1980). Constant, however, tends to concentrate on the intellectuals in the process, rather than on the less imaginative but still vital members who build or who repair technological systems.

5. Those interested in users as reactors should consult Robert Sommer, *Social Design* (Englewood Cliffs, N.J.: Prentice-Hall, 1983). I also chaired a symposium that included some important contributions on this subject; see the proceedings of this symposium entitled *Organizations, Designs and the Future* (Ypsilanti: Eastern Michigan University, 1986), available from the Department of Interdisciplinary Technology, Eastern Michigan University, Ypsilanti, MI 48197.

6. See, for example, A. T. Welford, *Fundamentals of Skill* (London: Methuen, 1971).

7. Luis S. Kemnitzer, "Language Learning and Socialization on the Railroad," *Urban Life and Culture,* Vol. 1, No. 4 (Jan. 1973), pp. 363–378.

8. Jeffrey W. Riemer, "Becoming a Journeyman Electrician: Some Implicit Indicators in the Apprenticeship Process," *Sociology of Work and Occupations,* Vol. 4, No. 1 (Feb. 1977), pp. 87–98.

9. Donald Metz, *Running Hot: Structure and Stress in Ambulance Work* (Cambridge, Mass.: Abt Books, 1981), p. 81.

10. Chandra Mukerji, "Distinguishing Machines: Stratifications and Definitions of Technology in Film Schools," *Sociology of Work and Occupations,* Vol. 5, No. 1 (Feb. 1978), pp. 113–137.

11. Dennis Jenkinson, *The Racing Driver: The Theory and Practice of Fast Driving* (Cam-

bridge, Mass.: Robert Bentley, 1959), pp. 34–46. It is important to note that Jenkinson is a race driver himself.

12. On the skills and technology of picking pockets, see David Maurer, *Whiz Mob: A Correlation of the Technical Argot of Pickpockets with Their Behavior Patterns* (New Haven, Conn.: College and University Press, 1964).

13. Joe Barbenel, "Science Proves Music Myths Wrong," *New Scientist,* Apr. 3, 1986, pp. 29–32.

14. Harry Collins, "Building a TEA Laser," *Science Studies,* Vol. 4 (1974), pp. 165–186.

15. Robert Morris, Syracuse University, personal communication. In certain scientific circles it is generally held that the entrance of a theoretical physicist into a laboratory will cause apparatus to break. What factual basis this belief has I do not know.

16. Jasper E. Brody, *Tales of the Telegraph* (Chicago: Jamieson-Higgins, 1900), p. 65.

17. Barbenel, "Science Proves Music Myths Wrong," pp. 29–32.

18. Studs Terkel, *Working* (New York: Avon, 1975), pp. 297–302.

19. Tom Wolfe, *The Right Stuff* (New York: Bantam Books, 1980), pp. 16–34.

20. Walter Shapiro, "Sky Kings," *Washington Post Magazine,* Nov. 9, 1980, pp. 10–14. I am indebted for this citation to Edward Hall's book, *The Dance of Life: The Other Dimension of Time* (New York: Doubleday, 1984), which has stimulated much of the thought in this section.

21. John Stein, "Flow States," *Car and Driver,* May 1984, pp. 175–179.

22. Mihaly Csikszentmihalyi, *Beyond Boredom and Anxiety: The Experience of Play in Work and Games* (San Francisco: Jossey-Bass, 1982); and Mihaly Csikszentmihalyi and Isabella Selega Csikszentmihalyi (Eds.), *Optimal Experience: Psychological Studies of Flow in Consciousness* (New York: Cambridge University Press, 1988).

23. Chuck Yeager, *Yeager: An Autobiography* (New York: Bantam, 1986), p. 84.

24. John Keats, *The Insolent Chariots* (Philadelphia: Lippincott, 1958).

25. Csikszentmihalyi, *Beyond Boredom and Anxiety.*

26. Or so it seems from reading about the air controllers President Reagan fired; see N. R. Kleinfield, "The People Who Were Patco," *New York Times,* Sept. 28, 1986, Section IV, p. 4.

27. Terkel, *Working,* pp. 374–380.

28. Terkel, *Working,* pp. 17–22.

29. Claudia Deutsch, "What Do People Want, Anyway?" *New York Times,* Nov. 8, 1987, Section IV, p. 4.

30. See Lloyd Warner (Ed.), *Yankee City* (New Haven, Conn.: Yale University Press, 1963), p. 66, on the relationship between heirlooms and status.

31. Thomas Goldwasser, "Pumping Iron, Not Concrete," *New York Times,* Sept. 21, 1986, p. 4.

32. Katherine Bishop, "A Car Phone That Links People and Their Desires," *New York Times,* Apr. 24, 1988.

33. Jerry Adler et al., "The Wheels of Fortune," *Newsweek,* Nov. 4, 1985, p. 61.

34. Helmut O. K. Kirchner, "Fashions in Physics," *Interdisciplinary Science Reviews,* Vol. 9, No. 2 (1984), pp. 160–171.

35. See Ron Westrum, "Small Particles and Big Egos," *Social Psychology of Science Newsletter,* June 1987, p. 2.

36. I have learned by talking to friends that many owners of VCRs seem to have almost no idea how to get them to work; for example, how to record TV programs.

37. David Pauly et al., "Is Your VCR in the Shop? You Are Not Alone," *Newsweek,* Oct. 19, 1987, pp. 62–63.

38. Ed Hofmann, "My Turn: Thieves at the Auto Shop," *Newsweek,* Sept. 1987. This article is likely to bring out murderous feelings in all but the most blasé automobile owners. Nonetheless, the procedures recommended for cultivating a relationship with a repair shop are very sensible.

39. Craig Brod, *Technostress: The Human Cost of the Computer Revolution* (Reading, Mass.: Addison-Wesley, 1984), pp. 25–60.

40. Roger Becker, chief test driver for Lotus Engineering, feels that properly designed devices should not *surprise* the user. This is one of the criteria for the proper design of the beautiful, high-performance cars that Lotus engineers.

41. See, for instance, Georges Friedmann, *Industrial Society* (Glencoe, Ill.: Free Press, 1955); Robert Sommer, *Tight Spaces: Hard Architecture and How to Harmonize It* (Englewood Cliffs, N.J.: Prentice-Hall, 1974); and Ralph Keyes, "The Trade-Ins," in his *We, the Lonely People.*

42. More generally, see Ken Kusterer, *Know-How on the Job: The Important Knowledge of "Unskilled" Workers* (Boulder, Colo.: Westview Press, 1978).

43. G. de Terssac and J. Christol, "Division de Travail et Savoir Ouvrier," in *Culture Technique,* No. 8 (June 1982), pp. 141–145, translation Ron Westrum. Copyright 1982 by the Centre de Recherche sur la Culture Technique. Used with permission.

44. John A. Gordon, interview, Apr. 9, 1987.

45. Wolfe, *The Right Stuff,* p. 161.

46. Nehemiah Jordan, *Essays in Speculative Psychology* (London: Tavistock, 1968), pp. 213–215.

47. Benedict Crowell and Robert Forrest Wilson, *The Armies of Industry,* Vol. I (New Haven, Conn.: Yale University Press, 1920), pp. 362–382.

48. Barry G. Silverman, "Expert Intuition and Ill-Structured Problem-Solving," *IEEE Transactions on Engineering Management,* Vol. EM-32, No. 1 (Feb. 1985), pp. 29–33.

49. Frederick Rose, "Thinking Machine: An Electronic Clone of a Skilled Engineer Is Very Hard to Create," *Wall Street Journal,* Aug. 12, 1988, pp. 1, 14.

50. See, for instance, Ben B. Morgan, Glynn D. Coates, and Raymond H. Kirby, "Individual and Group Performances as Functions of the Team-Training Load," *Human Factors,* Vol. 26, No. 2 (1984), pp. 127–142.

51. During the hearings on the crash of Northwest Flight 255 near Detroit Metropolitan Airport in 1987, aviation psychologist John Lauber indicated that effective aircrews communicated more. In particular, crews in which communications were routinely acknowledged more were likely to make fewer errors; National Transportation Safety Board, *Hearings on Crash of Flight 255* (Detroit, 1987), p. 1109. Thanks to Tim Doyle for bringing this material to my attention.

52. Peter Vaill, "Toward a Behavioral Description of High Performance Teams," in Morgan McCall and Michael Lombardo (Eds.), *Leadership: Where Else Can We Go?* (Durham, N.C.: Duke University Press, 1978), pp. 103–124.

53. Martin Caidin, *Thunderbirds!* (New York: Dell, 1961).

54. Walter J. Boyne, *Phantom in Combat* (Washington, D.C.: Smithsonian Institution, 1985), pp. 71–80. I am told that the Soviet airline Aeroflot keeps its crews together to provide for growth of this kind of teamwork.

55. H. Clayton Foushee, "The Role of Communications, Socio-Psychological, and Personality Factors in the Maintenance of Crew Co-ordination," *Aviation, Space and Environmental Medicine,* Vol. 53, No. 11 (Nov. 1982), pp. 1062–1066.

56. George E. Cooper, Maurice D. White, John K. Lauber (Eds.), *Resource Management on the Flight Deck,* Proceedings of a NASA/Industry Workshop (Moffett Field, Calif.: NASA/Ames, 1979).

57. Testimony of Clayton Foushee, NASA/Ames, before the National Transportation Safety Board, hearings on the destruction of Northwest Flight 255 (1988), p. 1009.

58. The term "coordinate leadership" is my own, though the same phenomenon had already been termed "flickering authority" in Frederic G. Withington, *The Real Computer: Its Influence, Uses, and Effects* (Reading, Mass.: Addison-Wesley, 1969), pp. 199–203. Withington suggested that the computer's ability to store and process data—even in the era of mainframes—would allow junior as well as senior managers in the organization to check out novel hypotheses and plans. However, this potential contribution to generative rationality was not adopted.

59. Leonard I. Stein, "The Doctor-Nurse Game," *Archives of General Psychiatry,* Vol. 16 (1967), pp. 699–703.

60. Lincoln E. Moses and Frederick Mosteller, "Institutional Differences in Postoperative Death Rates," *Journal of the American Medical Association,* Vol. 203, No. 7 (Feb. 12, 1968), pp. 492–494.

61. Harold S. Luft, John P. Bunker, and Alain C. Enthoven, "Should Operations Be Regionalized? The Empire Relation Between Surgical Volume and Mortality," *New England Journal of Medicine,* Vol. 301, No. 25 (Dec. 20, 1979), pp. 1364–1369.

62. Joel Brinkley, "US Distributing List of Hospitals with Unusual Death Rates," *New York Times,* Mar. 3, 1986. These statistics, as one might expect, generated considerable discussion; see Martin Tolchin, "U.S. Challenged on Data on Hospital Death Rates," *New York Times,* July 11, 1988. For a book-length discussion of these issues, see Ann Barry Flood and W. Richard Scott, with Byron W. Brown, Jr., Donald E. Comstock, Wayne Ewy, William H. Forrest, Jr., and other members of the staff of the Stanford Center for Health Care Research, *Hospital Structure and Performance* (Baltimore: Johns Hopkins University Press, 1987).

63. Physicians who perform operations more often do seem to have lower patient mortality rates than their less-practiced colleagues; see Lawrence K. Altman, "Experience Tells in Surgery, New Study Says," *New York Times,* July 28, 1989.

64. Harold S. Luft, Sandra S. Hunt, and Susan C. Maerki, "The Volume-Outcome Relationship: Practice-Makes-Perfect or Selective Referral Patterns?" *Health Services Research,* Vol. 22, No. 2 (June 1987), pp. 157–182. For more on relative skill, see Ron Winslow, "Medical Clash: Big Hospitals' Moves into New Territories Draw Local Staffs' Ire," *Wall Street Journal,* Aug. 18, 1989, p. 1.

65. Robert Steinbrook, "Hospital Death Rates for Five Surgeries Vary," *Los Angeles Times,* Mar. 27, 1988, Part I, pp. 3, 35; Robert Steinbrook, "Heart Surgery Death Rates Found High in One in Six Hospitals," *Los Angeles Times,* July 24, 1988, Part I, pp. 3, 33.

66. Stephen M. Shortell, Selwyn W. Becker, and Duncan Neuhauser, "The Effects of Management Practices on Hospital Efficiency and Quality of Care," in S. M. Shortell and M. Brown

(Eds.), *Organizational Research in Hospitals* (Evanston, Ill.: Blue Cross Association, 1976), pp. 90–107.

67. "Judging Hospitals on Performance, Not Equipment," *New York Times,* Nov. 9, 1986, Section 4.

68. Paul Judge, "Building a National Profile: U-M Wants to Set Surgical Standard in Organ Transplants," *Ann Arbor News,* Feb. 1, 1987, Section A, pp. 1, 4; see also Ron Winslow, "Hospitals Rush to Transplant Organs," *Wall Street Journal,* Aug. 29, 1989, Section B, p. 1.

69. Gene I. Rochlin, Todd R. La Porte, and Karlene H. Roberts, "The Self-Designing High-Reliability Organization: Aircraft Carrier Flight Operations at Sea," *Naval War College Review,* Autumn 1987, pp. 76–90. This study is part of a larger project on high-reliability organizations; other parts of the study included electrical utility grids and air-traffic control operations.

70. Wernher von Braun, "Teamwork: Key to Success in Guided Missiles," *Missiles and Rockets,* Oct. 1956, pp. 38–42. On the concept of the technological maestro, see Arthur Squires, *The Tender Ship: Governmental Management of Technological Change* (Boston: Birkhauser, 1986), pp. 13–46.

71. Seymour Melman, *Profits Without Production* (New York: Knopf, 1983), pp. 40–53. See also David Halberstam, *The Reckoning* (New York: Morrow, 1986).

72. William J. Abernathy, Kim B. Clark, and Alan M. Kantrow, *Industrial Renaissance: Producing a Competitive Future for America* (New York: Basic Books, 1983).

73. For instance, note the remarks in *The Human Factor in Innovation and Productivity,* Committee on Science and Technology, U.S. House of Representatives, September 1981 (Washington, D.C.: U.S. Government Printing Office, 1981), pp. 701–727.

74. For instance, see Thomas Peters and Robert Waterman, *In Search of Excellence* (New York: Warner Books, 1982), pp. 292–307.

75. Joseph Trento, *Prescription for Disaster: From the Glory of Apollo to the Betrayal of the Shuttle* (New York: Crown, 1967).

76. Squires, *The Tender Ship,* pp. 13–46.

77. Myrna-Lynne Whitney, "Children Need Human Factors, Too," *Hazard Prevention,* Vol. 21, No. 1 (Jan.-Feb. 1985).

Adapting & Tinkering

W e have seen that the creativity shown by the designers of technology is sometimes impressive, but creativity can also be shown by a technology's users. Thus far we have discussed users' skills in coping with already-designed and manufactured technologies—we might call this the "black box approach." But users do not always accept a technology passively; sometimes they open the black box. Their understanding and experience may permit them to make creative new uses of it, including purposes unimagined by its designers. The extent of such changes can often be impressive.

Thus technological determinism is again a false view, for human beings are likely to shape technology's impacts on their lives by changing the function of the technology. Because the results of this adapting and tinkering can sometimes be interesting, useful, and creative, and can at other times be disturbing or dangerous, we must consider it more thoroughly. In this chapter we will explore how users "adjust" technology to fit their needs.

"What Is This Device for?"— The Ambiguity of Uses

Even though many people assume that the use for a product is evident at the time of its invention, this is often not the case. Many products, once brought to life, seem to be answers without problems. The inventor must thus seek to identify problems for which the invention is a solution. Some inventors seek an invention to answer a definite problem; other inventors seek a problem to which their invention can be applied. So an invention may come about either as the answer to someone's problem or as the application of a technology developed with another use in mind.[1]

The typewriter offers a splendid example of the uncertainties and ambiguities that surround technological innovation. When the typewriter was developed, its promoters did not understand that its largest market would be the business community.[2] Instead, they thought it would be important primarily to professional writers and other "literary people," for in the business community writing letters by hand was a matter of etiquette and was thought to help prevent forgery. Even the Automatic Telegraph Company turned the typewriter down—on the advice of a young employee named Thomas Edison, who thought he could invent a better version (he didn't). But when business expanded in the late 1880s, typewriters entered the office, as did female clerks.[3] Among the early adopters of the typewriter were court reporters, who needed the machine for accurate transcription of their notes.

This ambiguity about the nature of the users (and therefore the uses) of a technology is so important that an entire school of sociological inquiry has grown up around the issue of the definition or "social construction" of a technology.[4] A device may appear to be something different to different groups interested in exploiting it; depending on who takes charge of a given invention, it may evolve in quite different directions. Thus a bicycle may either be a piece of sports equipment or an alternative technology for personal transportation.[5] The technology's use depends on which social group "frames" the technology with its own concepts and thereby defines the benefits and problems the technology has to offer. Different groups may see what is seemingly the same artifact as part of very different systems of action.[6] What they do with it follows from what they envision it to be.

An interesting example of such contrasting points of view emerges from Boelie Elzen's study of the development of ultracentrifuges,[7] devices that use rapid rotation to separate substances of different densities.[8] In studying the way that the ultracentrifuge was developed, Elzen found that two men, Theodor Svedberg and Jesse Beams, had different approaches to its use. Svedberg was primarily interested in the ultracentrifuge as a way of solving problems in physical chemistry; Beams regarded the ultracentrifuge primarily as a device that might solve any number of problems other people might pose. Thus to Svedberg the ultracentrifuge was a part of colloid chemistry; to Beams, however, even though the ultracentrifuge might be "applied" to such problems, it was an interesting device in and of itself.

So technology is not simply "applied" to an obvious problem; sometimes the problem the technology solves must be found. Then again, once something is invented for one purpose, nothing prevents it from being used for another, often different, purpose. And sometimes people other than those for whom a device is developed become its primary users.

Colonel J. T. Thompson developed the machine gun as a military weapon, but his early attempts to interest the military failed. He then tried to interest the police; eventually he even approached private owners. But the primary users of machine guns turned out to be corporate security units (who used them to assist during labor disputes) and organized crime.[9]

An excellent example of the development of a technology as a potential "answer" to as yet undetermined questions was Harold Edgerton's creation of stroboscopic photography.[10] "Doc" Edgerton developed ultrashort flash photography as a result of his experiences as a young lineman on a powerline gang. When he was out on call during thunderstorms, he saw moving men spotlighted during flashes of light; the men appeared to be still. In 1930 he put this observation to work by developing a camera that would take pictures using extremely brief flashes of light. Originally he used this photographic technique to take pictures of electric motors, to see whether the parts were in synchrony. But since that time Edgerton has found many new uses for stroboscopic photography, including military reconnaissance, photographing physical events (from bullets penetrating objects to atomic explosions), and marine biology, in which he has collaborated with Jacques Cousteau.

The more closely we examine the users of technology, the more evident this ambiguity of use becomes;[11] that is, a device is basically a solution, but there may be more than one problem to which it applies. Many military weapons systems were designed for one mission but were ultimately used for another. The P-51 Mustang propeller plane, for instance, was designed primarily for tactical support of ground troops. Its most important use, however, was in escorting bombers.

In the Vietnam War, the threat of surface-to-air missiles meant that strategic bombers (for example, B-52s) could not be used for bombing North Vietnam. As a result, strategic bombing was done by tactical fighter/bombers, such as the F-4 Phantom, which were much faster and more maneuverable. The B-52, designed for the Strategic Air Command, is a heavy, slow bomber, a perfect target for anti-aircraft weapons. Thus the B-52 was used for *tactical* bombing in the south because its lack of speed was unimportant when missiles and other jets could not be used against it.[12] Another Vietnam surprise was the importance of aircraft cannon. With the advent of air-to-air missiles, machine guns and aerial cannon were presumed to be obsolete. There would not be any more dogfights, it was argued, because missiles would destroy opposing forces from miles away. Experience proved this to be incorrect. Terrified pilots described trying to dogfight against planes with cannons at close ranges (at which missiles were largely inoperative). Dogfights not only took place in Vietnam, but also in the Arab-Israeli Six-Day War. The cannons were subsequently put back on the planes.[13]

Similarly, the digital computer was originally designed, in all four of its incarnations, to do computations for the military (in ballistics, aviation, and cryptography). Yet its development led it to become one of the most versatile instruments of humanity. Today it is hardly an exaggeration to say that it is used far more often for word processing or video games than it is for the "computing" from which it got its name.

Computer systems generally show this ambiguity of use. Often firms that get computers must discover the uses to which the machines will be put. "Hackers" at universities quickly discovered that computers could be used to play Spacewar or Star Trek, as well as to carry out assigned computing tasks.[14] And according to a report by Frank Dubinskas, computer software often also has unanticipated uses. One "computer conferencing" system was a total failure in its designed use—linking two separated sites at which "knowledge engineers" were working.[15] The company librarian, however, discovered that it made an excellent "bulletin board" system. Dubinskas concluded that the "context of use" shapes the character of the artifact.

Why Users Adapt Technology

Designers of very large and complicated systems often lack control over users' applications of their systems. For instance, pilots of the Boeing 767 jet airliner find that using the plane as designed sometimes does not produce the desired results. Thus pilots have passed around an informal list of which circuit breakers to pull to get the plane to "fly better."[16] A near-crash of a 767 over the Rocky Mountains in 1983 may have involved this kind of behavior. The plane inexplicably lost power at 41,000 feet and begin to plummet, but at 14,500 feet the engines were restarted and a crash was barely averted. Ostensibly the engine stall was due to a storm; but according to a rumor, the incident took place following the de-activation of certain circuit breakers, which reputedly allowed the plane to fly better at high altitude. Upon returning to the cockpit an engineer noticed that the circuit breakers were out and put them back in, whereupon the plane went into the dive.[17]

Violation of operating rules is routine in many applications in which the rules don't make sense to the users. In some cases the designers have not communicated their full intentions. In other cases their intentions are inadequate for the realities users must face. For instance, a common problem with high-technology aircraft is checking the state of their electronics (avionics) equipment. The U.S. Navy developed an avionics checkout system known as VAST (Versatile Avionics Shop Test) to monitor the performance of electronic aircraft systems.[18] These VAST systems went on aircraft carriers, and thus they were used at too great a distance to allow surveillance by the designers. In spite of carefully worked out policies governing the procedures of VAST shops, the VAST workers tended to operate the equipment in a pragmatic way that was easier and made more sense to them. Rather than going by the book, they developed practices that they felt would best allow them to respond to the demands they felt were important. One example of their informal adaptations was the use of a "hanger queen." Because the mobile check-out units were finicky and unreliable, installations with VAST systems would frequently keep one system, whose components appeared to work well, in pristine order and off-line, to act as a source of spare electronic building blocks. This use of a "hanger queen" reduced the capacity of the shop to process incoming avionics units in the short run, but guaranteed its long-run functioning in another sense.

This systematic deviance from standard procedures on the part of the avionics technicians is one example of a more general phenomenon, which might be called the "existential" attitude. This attitude places a higher premium on the physical and job survival of the user than on following the "standard operating procedures." It is succinctly expressed by the remark, "Well, I work with it, and I don't care what they say, the blankety-blank thing doesn't work unless you give the sucker a good kick once in a while." In such cases the user employs the technology in whatever way is considered best by him or her to ensure "success" as operationally defined and thus to ensure continued survival in the job. Such uses may be very different from those intended by the equipment's designers.[19] The designers may understand very little about what the users' needs are because there is often a very lengthy channel or indirect process by which designer and user communicate.[20] Sometimes designers understand the equipment's users so poorly that their image of the users might almost be called a fantasy.

At the same time, users often approach a technology as a "black box" whose inner workings they do not understand at all. This situation is amplified in systems designed to be "idiot-proof" by limiting the options and variations available to users. The users then develop their own "theories in use" to be able to operate the technology. These theories are often satisfactory for normal operation, but they may break down under the pressure of crisis. Because the users do not understand the internal structure of the technology, it becomes impossible for them to tinker with or repair it.

Users develop not only novel uses, but also novel configurations, of a technology. Adaptations and variations of standard products are legion with automobiles, recreational equipment, cooking appliances, computer systems, and machine tools. The alterations may change the devices to make them either more effective or easier to use. These adaptations might simply reflect, on one hand, the very human tendency to adjust our environment and tools to our personal and operational needs. But on the other hand, they might suggest that often designers could better take into account users' needs. Everyday creativity is also important in regard to technological procedures. Recipes are often modified by intention or accident to accommodate the ingredients on hand as opposed to those that are supposed to be used. Carefully designed manuals or instructions issued with products are routinely ignored by users who develop their own procedures.[21]

Origins of Technological Adaptation

A very important source of training in ingenuity for Americans has been life on farms. Tinkering, adjusting, and repairing came naturally to people faced with complicated machinery, a low cash supply, and crops that needed harvesting. Because maintaining and repairing farm machinery was part of daily rural life, considerable skill was necessarily developed in adapting devices for special situations and in compensating for inevitable small accidents. This source of mechanical skills is now disappearing as farming has become heavily mechanized and capital-intensive, and as Americans have migrated to cities. At one time, however, farm life was a significant training ground—and in some cases actually a laboratory—for future inventors.[22] American women traditionally were taught to sew and to modify clothing to fit members of the family, but how well such skills have survived urbanization and the two-job family is not known.

The garage has also provided another potential training ground for mechanical skills. Car owners must often tinker with their vehicles to bring them into harmony with their needs. Teenagers often get their first engineering educations while taking apart the family car. Some segments of the population have a strong tradition of modifying automobiles for racing and display.[23] Working-class car owners often make considerable modifications to the suspension, lighting, and storage systems to accommodate individual tastes and preferences. Several subcultures exist among automobile owners who modify their cars by jacking them up or down to suit local fashions, by painting them exotic colors, by increasing the power of the engine, or by equipping them with radio communications suitable for a jet fighter. Garages thus also represent a potential incubator for inventors.[24]

Factory work provides further mechanical skills. People with skills derived from working in industrial plants often use them to good advantage in their personal lives—for example, in making extensive modifications to their homes, including additions that substantially increase the space of the dwelling.[25] Whether automated factories will continue to provide the same valuable hands-on education is not clear. Certainly today there seems to be little relationship between pushing buttons and setting dials and learning how the underlying machines work.[26] The traditional factory, by contrast, provided many opportunities for learning how to operate, fix, and design complicated machinery.

The Role of Tinkering in Technological Adaptation

As farms and garages dwindle in number and factories become more automated, tinkering with computers may be one of the few ways that Americans can test their "mechanical" skills. Just as teenagers in the 1950s learned mechanical skills by working on the family car, computer whizzes of the 1990s may learn their skills from early contact with the machines.[27] We might wonder, however, whether "do it yourself" experience with computers will provide an adequate substitute for "doing it yourself" skills with tools. Still, the impulses that drive the computer hacker and the workshop tinkerer or gadgeteer are the same.

This kind of tinkering can be applied to much larger and more expensive systems that require major engineering changes. One of the key features of many creative inventors is a willingness to take apart and improve the devices they encounter. A common feature of "skunk works" is a refusal to treat a device as a "black box" that must be left intact. Instead, such devices are often taken apart for inspection, and often considerable modifications may follow. The willingness to take devices apart and put them back together again in different ways is important, and it allows equipment to be very much improved.

In a similar vein, a study of safety teams at nuclear power plants found a relationship between how customized a team's procedures were and how successful it was in reducing accidents; the plants with fewer reported "incidents" had teams that had carefully tailored their approach to the plant's specific operating characteristics, but the teams that had simply carried out their duties "by the numbers" had more incidents.[28] Thus the willingness to open the "black box" and reconfigure the system to meet situational needs often results in increased efficiency.

But not all tinkering is helpful. Some innovations ought to be left alone, but instead they are "improved" into ineffectiveness. This kind of "improvement," as we have seen with the AR-15 Stoner rifle, is often detrimental to the effectiveness of the system. When innovations come from the wrong place on the organizational chart, those who have the formal responsibility for innovation often insist on dangerous modifications. Sometimes these modifications do not improve performance; instead they bring the device into line with "standard" procedures. Virtually anyone who has worked in research and development can speak from experience about how such modifications degraded the performance of a system. And suspicion of certain features—such as safety devices—as "unnecessary" may also lead to dangerous tinkering.

Some modifications by users may be recycled into the manufacturing process. Eric von Hippel has reported that in many industries, the devices purchased to perform major operations must be modified to fit the users' needs.[29] For instance, von Hippel found that in the scientific instruments business, 82 percent of the major improvements and 70 percent of the minor improvements in the sample studied had come from product users; in "process" industries such as oil, chemicals, and plastics, the proportions were 63 percent of the major improvements and 20 percent of the minor ones.[30] Such modifications are sometimes sufficiently important to compel the original manufacturer to redesign the product. It is especially noteworthy that in spite of such obvious need for further design, most of the manufacturers had no organizational way to respond to these new, customer-generated product improvements.

In the recent war between Iran and Iraq, the Iraqis showed an outstanding ability to modify the weapons sold to them by more developed nations. The missile attack on the USS *Stark,* for instance, involved two Exocet missiles fired from a French Mirage F-1 fighter. The Mirage F-1 is designed to carry only one Exocet; the Iraqis modified it to carry two. Similar tinkering allowed them to increase the range of the Russian Scud-B missile first from 175 to 400 miles and then to 560 miles.[31]

One aspect of tinkering that requires serious exploration is the adaptation of technologies in the workplace by those who work there. One of the realities of chemical process production is that skilled workers may have only a limited understanding of the transformation processes they supervise, because they are not engineers. But they may be very good at getting these processes to work by tinkering. In the previous chapter we noted the considerable amount of practical knowledge possessed by workers in the varnish business. But such seat-of-the-pants knowledge may be inadequate for dealing with major changes in critical processes.

Industrial processes may be altered to make them easier to carry out by workers who do not fully understand the consequences of such alterations. This can be very dangerous in chemical process plants, and it may be even more important in the nuclear industry. "Tinkering" can involve disconnecting safety devices or making changes that make operation of the machinery more hazardous. Virtually every investigation into technological accidents turns up some evidence of such hazardous tinkering.

The explosion of the ICMESA plant in Seveso, Italy, in 1976 involved exactly such a shortcut.[32] The plant in question manufactured trichlorophenol (TCP), which in turn was used to manufacture hexachlorophene, an antibacterial agent. Workers had changed one of the procedures for their own convenience. Ordinarily, chemical reactions should be carried to completion. However, workers had found it convenient over the weekend to leave uncompleted batches of TCP in the reactor. They had done this 34 times before, but on this occasion the result was disaster. The reactor overheated and then exploded, filling the air with dioxins, one of the worst poisons known. The results were horrifying: Birds fell dead out of the sky, leaves withered on trees, and people developed severe medical problems. The long-term medical effects of the Seveso explosion may be even more severe than those that have already been identified, for the area around Seveso is massively contaminated.

Similar tinkering goes on at nuclear plants, where again such adaptation of carefully designed (but not always reliable) technology can produce serious accidents. For instance, in the control rooms of some nuclear plants, there are over 2,000 gauges, many of which have alarms.[33] Some of these alarms go off all the time, thus producing a constant distraction. A natural response to this kind of problem is to disregard the alarms or turn some of them off.[34] The designs for the machinery, however, do not take into account such improvisation on the part of the workers. The result can be accidents because designers and workers have not collaborated on rectifying operating problems. In the chapter on technological accidents, we will see how such tinkering has contributed to important technological disasters.

Sometimes safety systems are disabled to make life easier for the workers. At age 16 Thomas Edison was working as a railroad telegrapher. To allow himself time for sleeping during slow periods, he rigged up a device that answered a call automatically. When the ruse was discovered, Edison got a severe reprimand from the company.[35]

Technologies that routinely present few problems may give workers a false sense of security, leading to careless or even reckless behavior. At the Peach Bottom Atomic Power Station, 65 miles west of Philadelphia, workers were "found sleeping, playing video games and reading magazines" with the reactors on at full power.[36] In one instance all the workers in the control room were found sleeping with the reactors on at full power. Part of the problem was that there was a shortage of personnel, so overtime was encouraged, and one way to incur overtime pay was to take the shift and sleep through it. This kind of reckless behavior is disturbing, when one considers the potential hazards reactors pose. This is a far cry from what we should expect with such dangerous forces as nuclear power, but familiarity sometimes breeds carelessness.[37]

Conclusion

A broad study of the "customization" of technologies would reveal some interesting behaviors. Hot-rodders constantly tinker with their cars, scientists improvise instruments by modifying off-the-shelf products, fishers are always devising new lures, and hackers constantly modify computer software. There is always, it seems, a way of making the product better by making certain small changes. So we file a little here, add a little something there, and presto! We have a new device.

We need to know more about how people adapt technologies to their own uses. People both change the things they get and use them for unanticipated purposes. We need to know more about this everyday creativity, partly because it might tell us something we need to know about better design and partly because it is something we probably need to teach people to do more. And finally, it's fascinating to see the variations!

Notes

1. An example of an invention that is an answer that once lacked a problem is the use of carpet adhesive materials as a replacement for latex paint. My colleague John A. Gordon noticed that many of the properties of plastic materials used for carpet backing would make them ideal as a replacement for latex paint, which tends to yellow with age. However, it was not until he became part of a "skunk works" group with another colleague and a student that this idea began to flower and he began active testing.

2. Cynthia Monaco, "The Difficult Birth of the Typewriter," *American Heritage of Invention and Technology,* Vol. 4, No. 1 (Spring/Summer 1988), pp. 10–21.

3. Thomas Whalen, "Office Technology and Socio-Economic Change 1870–1955," *IEEE Technology and Society Magazine,* June 1983, pp. 12–18, 29.

4. The works of many of this school's writers are brought together in W. E. Bijker, T. P. Hughes, and T. Pinch (Eds.), *The Social Construction of Technological Systems* (Cambridge, Mass.: MIT Press, 1987).

5. Trevor Pinch and Wiebe Bijker, "The Social Construction of Facts and Artifacts: Or How the Sociology of Science and the Sociology of Technology Might Benefit Each Other," *Social Studies of Science,* Vol. 14, No. 3 (1984), pp. 399–441.

6. Wiebe Bijker, "The Social Construction of Bakelite: Toward a Theory of Invention," in Bijker, Hughes, and Pinch, *The Social Construction of Technological Systems,* pp. 158–187.

7. Boelie Elzen, "Two Ultracentrifuges: A Comparative Study of the Social Construction of Artifacts," *Social Studies of Science,* Vol. 16, No. 4 (Nov. 1986), pp. 621–662.

8. An ultracentrifuge is a centrifuge that revolves faster than 50,000 revolutions per minute (rpm). Many dentists' drills, driven by compressed air, rotate in this speed range. By contrast, a steam turbine rotates in the range of 20,000 rpm, and an automobile crankshaft operates in the range of thousands of rpm.

9. John Ellis, *The Social History of the Machine Gun* (Baltimore: Johns Hopkins University Press, 1986), pp. 169–185. Many corporations had unregistered machine guns to deal with labor strikers or rioters in the pre–Wagner Act days (the Wagner Act prohibited corporations from interfering with union organizing through intimidation or threats of violence). The ownership of heavy-duty weaponry by corporations for use against their own workers in the 1930s also included gas guns. One corporation possessed 142 gas guns and 6,714 gas shells and grenades, according to the findings of the LaFollette committee.

10. Derek V. Goodwin, "The Sorcerer of Strobe Alley," *Science 82,* June 1982, pp. 37–45.

11. At Apex Engineering (a pseudonym), a firm associated with building sports and racing cars, a considerable disagreement arose over whether the thousands of parts used in the car ought to be catalogued in terms of how they fit into the car or in terms of their generic

properties. Thus a 2-inch bolt was not only a bolt, but was also a part of the car's frame. Whether its generic identity as a bolt or its functional identity as part of the frame was more important was the point at issue.

12. One result of the bombing raids by B-52s was terror. The B-52 bombing attacks were so numbing and deafening that they often terrified the Vietcong who survived them. Survivors have testified to the intense fear evoked by these attacks, which not infrequently caused involuntary defecation.

13. Lon O. Nordeen, *Air Warfare in the Missile Age* (Washington, D.C.: Smithsonian Press, 1985), pp. 9–50, 214–215.

14. Steve Levy, *Hackers: Heroes of the Computer Revolution* (New York: Doubleday, 1984), pp. 37–47.

15. Frank Dubinskas, "Collapsing the User-Designer Divide: Electronic Communication Software and the Social Construction of Advanced Technology," in Ron Westrum (Ed.), *Organizations, Designs, and the Future,* conference proceedings (Ypsilanti: Eastern Michigan University, 1986), pp. 184–208.

16. Personal communication with Earl Wiener.

17. An account of the near-accident is in Melinda Beck, with Connie Leslie, Jeff Copeland, Shawn Doherty, and Peter Rinzarson, "I Thought We Were Goners," *Newsweek,* Sept. 5, 1983, p. 20. The rumor was mentioned to me by an air-crash investigator. One of the reasons for suspicion is that the cockpit voice recorder was allowed to run on after landing, thus erasing records of key communications on the flight deck.

18. John L. Kmetz, "An Information-Processing Study of a Complex Workflow in Aircraft Electronics Repair," *Administrative Science Quarterly,* Vol. 29, No. 2 (June 1984), pp. 255–280.

19. Even when systems have been carefully designed for users, misunderstandings are still possible. The Sidewinder air-to-air missile had been carefully tailored for shipboard storage and use. Nonetheless, sailors on aircraft carriers thanked the designers for putting little "wheels" on the back fins, which could be used to roll the missiles along the deck, making the use of a cart unnecessary. The little "wheels" were actually gyroscopic stabilizers that had been carefully designed to keep the missile on course. Fortunately, they were sufficiently sturdy for this unintended use.

20. Obviously, military units need to be able to get prompt action from designers concerning proposed modifications based on combat experience. When Admiral Isaac Kidd, Jr. was chief of naval operations, very rapid engineering changes could be made through the "quick response capability" that he ordered each naval laboratory to develop. In Vietnam, soldiers using the M-16 rifle learned that the fastest way to deploy new ammunition clips was to tape one upside down onto the other. I have been told that this led to some clips actually being manufactured in this way, but I'm not sure if this is true.

21. There was considerable discussion about allowing the anti-acne drug accutane on the U.S. market because experiments with animals suggested it could cause birth defects. Drug companies argued that they could issue warning booklets with the drug and could educate doctors not to prescribe it to women likely to be pregnant. It is obvious, however, that often doctors and patients do not carefully read such instructions. After seven cases of birth defects associated with use of the drug were discovered (others were suspected), pressure began to mount to remove the drug from the market; see Gina Kolata, "Acne Drug's Relative Is Called Greater Threat of Birth Defects," *New York Times,* May 1, 1988, Section I, pp. 1, 13.

22. For examples, see Arnold Skromme, "Creative Students Are God's Neglected Children," booklet available from the author at 2605 31st Street, Moline, IL 61265.

23. See, for instance, Tom Wolfe, *The Kandy-Kolored Tangerine-Flake Streamline Baby* (New York: Pocket Books, 1970), pp. 62–89.

24. Thomas Lang, designer of the SWATH ship, built the prototype for his next major project in his garage. Lang's early tinkering with boats, assisted by his father, was influential in his later designs; see Thomas G. Lang, "SSP *Kaimalino:* Conception, Development History, Hurdles and Success," A.S.M.E. Paper 86-WA/HH-4 (New York: American Society of Mechanical Engineers, 1986). A humorous take-off on this propensity of innovators to work in garages is Steven Bleiberg, "Japan's Garage Gap," *New York Times,* June 26, 1988, Section E, p. 27.

25. Andy Beierwaltzes, a landscape architect, suggested to me that whereas the typical middle-class home owner is likely to respond to a major increase in income by moving to a "better" neighborhood, the typical working-class response is to build on another addition.

26. Robert Blauner, *Alienation and Freedom: The Factory Worker and His Industry* (Chicago: University of Chicago Press, 1964), pp. 124–142.

27. No firm data on this subject have come to my attention, but my impression is very strong that early contacts with computers are decisive to the development of skills. There seems to be what Goethe called "elective affinities" between programmers and their machines. See, for instance, Stephan Wilkinson, "Portraits in Software," *Technology,* Mar./Apr. 1982, pp. 31–37. In some cases (for instance, that of Stephen Wozniak), parental influence or occupation provides such early contacts.

28. Alfred Marcus and Isaac Fox, "Lessons Learned About Communicating Safety-Related Concerns to Industry: The Nuclear Regulatory Commission After Three Mile Island," paper presented at the 1988 Symposium on Science Communication: Environmental and Health Research, at the University of Southern California, Los Angeles, Dec. 1988.

29. Eric von Hippel, "Users as Innovators," *Technology Review,* Jan. 1978, pp. 31–39.

30. However, in other fields, such as polymers and plastics additives, very few changes are suggested by users; perhaps user tinkering is easier with mechanical products than with chemical ones.

31. Bernard E. Trainor, "Iraqi Missile with Extended Range Is New Peril to Iran's Cities and Oil," *New York Times,* May 1, 1988, Section I, p. 10.

32. This account is taken from Lee Niedringhaus Davis, *The Corporate Alchemists: Profit Takers and Problem Makers in the Chemical Industry* (New York: Morrow, 1984), pp. 199–204.

33. Peter Andow, "Failure of Process Plant Monitoring Systems," in J. A. Wise and A. Debons (Eds.), *Information Systems: Failure Analysis* (New York: Springer Verlag, 1987), pp. 233–240.

34. Cass Peterson, "Mission Fading for Aged A-Plant," *Washington Post,* Feb. 28, 1988, Section A, p. 4.

35. Matthew Josephson, *Edison: A Biography* (New York: McGraw-Hill, 1959), pp. 43–44.

36. See Philip E. Ross, "Pennsylvania Electric Executive Retires After Plant Closing," *New York Times,* Mar. 8, 1988; and Matthew L. Wald, "The Peach Bottom Syndrome," *New York Times,* Mar. 27, 1988, Business Section, pp. 1, 6, 7.

37. It is interesting to compare the reality of nuclear power operation with what was anticipated before nuclear power arrived. Robert Heinlein, a science fiction writer of some talent, wrote a story about a hypothetical nuclear reactor in 1940; see "Blowups Happen,"

in Heinlein's *The Past Through Tomorrow* (New York: Berkley, 1975). In this story, the reactor is monitored by two engineers. The engineers in turn are monitored by a psychiatrist. The careful selection and highly technical education of personnel in Heinlein's story and their constant monitoring by other professionals is an ironic contrast to the problems with drug use, goofing off, and sleeping reported in the Peach Bottom case. No airliner would ever be flown in the manner that we allow some of our nuclear reactors to be run. On some trans-Pacific flights, however, one can arrive on the flight deck to find the plane on autopilot and the entire crew sleeping; see Brian Moynahan, *Airport International* (London: Pan, 1978), p. 90. What will happen with the newer, more highly automated planes is anyone's guess.

Technological
Accidents

On March 21, 1986, Voyne Ray Cox, 33, was killed by a fatal dose of radiation from a Therac 25 linear accelerator machine. He had entered the East Texas Cancer Center in Tyler, Texas, to get radiation treatments as a follow-up to removal of a tumor. Instead, the million-dollar machine gave him three 25,000-rad doses, each dose 100 times stronger than the average treatment. When the machine was turned on, "Cox heard a frying sound and felt a pain like an electric shock shoot through his shoulder." After the third dose, Cox leaped from the table and ran for help. He died six months later. Another patient, Vernon Kidd, 66, died a month after receiving a similar overdose.

Checking the machine revealed no mechanical malfunction. What was at fault, investigation showed, was a small error, in the portion of the program that directed the electron beam. The beam had two modes. In electron mode the person would get the electron beam directly; in X-ray mode the beam's strength would be increased 100 times but would be aimed at a tungsten target, which would then give off a much milder dose of X rays. Under certain conditions, however, the patient would directly receive the high-intensity electron beam at X-ray strength. The "bug" in the program allowed the tungsten target to be retracted while the beam remained at high strength, thus delivering a lethal dose of radiation to the helpless patient. A programming error thus killed these two men, an error that was originally unnoticed and remained undetected in later testing.

How Vulnerable Are We to Technological Accidents?

In many respects the Therac 25 deaths represent a classic technological accident. A small mistake rendered dangerous a million-dollar, high-tech system. The programmers, remote in time and space from the results of their actions, had no idea that their mistake would be lethal. These characteristics apply to many systems in a high-tech society. Accordingly, it is important for us to consider how vulnerable society may be to such technological errors.

Accidents occur when actions or devices go astray. We attempt to do something but find ourselves suddenly blocked by a mistake, a failure, or an unexpected event. Accidents can have happy consequences, but usually they cause frustration or suffering. Because we live in a largely human-designed environment, most accidents involve technology. Sometimes people are the immediate causes of the accident, for mistakes can occur because of a variety of human frailties: a lapse of memory, too little sleep, carelessness, or too great a "mental workload." But mistakes can also be caused by inadequate equipment, faulty instructions, or even faulty software, as with the Therac 25. Then there is the simple fact that when systems are very complicated, some slipups are virtually inevitable.[1] The label "human error" may sometimes be applied to situations in which the real diagnosis should be "bad design."[2]

Even when people's actions are faultless, the technology itself can fail, for few technologies are perfect. Every device has a limited operational life or contains some "bugs" or imperfections, and in any case devices are likely to become less perfect with use.[3] The Greek physician Hippocrates spoke well when he said that "Life is short, art is long, the occasion is fleeting, experiment is perilous, and judgment is difficult." We are often the victims of our incomplete knowledge.

Working with technology is an inherently risky business. We can reduce the risks by learning and through technological progress, but perfection is an elusive and unrealistic goal. We should expect, however, to improve our devices with experience. With any technology we experience a "learning curve" that shows fewer accidents as the technology is improved. Thus the things we use are imperfect, and their failures and shortcomings cause us a good deal of trouble. The breakdowns that interest us most, of course, are those that cause injury or death.

Charles Darwin described the relative survival of species as the "survival of the fittest." Life is inherently a precarious business. A species endures largely because it manages to survive and reproduce more often than some other species that occupies the same biological niche. The weeding-out process Darwin made famous, "natural selection," occurs in many ways. One such method is accidental death. Unhappily, the victims nature selects for weeding out are not always those society would pick. Although many modern technologies are very much safer than the technologies people once used, accidents still occur. Home, industrial, and transportation accidents, for instance, still take hundreds of thousands of lives per year. But proportionately, many more people used to die from these accidents in former times. There are several reasons for our comparative safety.

First, more people today have less hazardous occupations. The massive change from farming to factory, store, and office work has moved us toward safer workplaces. Farming is a very hazardous occupation, given the proximity to large, powerful machines, sometimes dangerous animals, and the hazards of working outdoors. Factory work, although still dangerous, occurs in an environment largely controlled and supervised by human agents. Office work may be hazardous to health, but the risk of accident is low.

Second, things are designed better today. A bridge is less likely to fail, and a building is less likely to collapse, than once was the case. Consumer products are designed and regulated with safety in mind. Food no longer contains many of the noxious agents it once did, although many more than should still remain.[4] The passage of the Food and Drug Act in 1906 and subsequent legislation has done much to make what we eat safer.[5] The safety of drugs is monitored by government agencies. Patent medicines no longer contain the large quantities of opium and alcohol they once did, though many of them are still addictive.[6] Consumer products, such as tools, automobiles, and recreational equipment, are designed to prevent injuries to users, in part to spare their makers costly lawsuits and settlements.[7] In some states, stringent legal requirements have eliminated some dangerous products—firecrackers is one example—from the market entirely.

Finally, injuries sustained now are more effectively treated, thanks to first aid and medical science. From the band-aid to microsurgery, we have enormously improved and expanded the variety of means available for treating injuries. When we are injured, we can repair the injured body part and prevent serious injuries from becoming fatalities. Many injuries that were fatal in former times now require only a short hospital stay.

Thus today's injury and death rates due to accident are much lower than they once were, although they are hardly insignificant. In this chapter we will inquire into the character of technological accidents in modern society. We will do so by examining a number of case studies of large-scale "systems" accidents. These four case studies illustrate some of the unique dilemmas posed by modern technology, and they show that often very complicated and dangerous technologies may be given to institutions that are not equipped to handle them properly. As Charles Perrow has shown in his book *Normal Accidents,* there is a mismatch between the sophistication of the technology and the sophistication of the users.[8] Again, we seem to have third-generation machines and first-generation minds.

Four Case Studies

The *Challenger* Explosion. One of the saddest days in the history of America's space program was the explosion of the space shuttle *Challenger* on January 28, 1986. Six astronauts and a school teacher, Christa McAuliffe, died when the shuttle exploded in a ball of fire soon after takeoff. The flight was the twenty-fifth in a series of shuttle missions; it was the first fatal shuttle accident.[9] Subsequent research showed that one of the seals on the booster rocket had leaked, opening a pathway to the outside for burning gases. The "blowtorch" from the solid-fuel rocket penetrated the liquid-fuel tanks, leading to the explosion.[10] The seal had leaked because the booster was used at a temperature below its range of safe operation. The reason the launch occurred on a cold morning was that there were government pressures to have a launch at that time.[11]

The word "accident" creates the impression of an unforeseen event occurring from unexpected causes. Yet the investigations of technological accidents usually show that they are *not* unforeseen. Often investigations reveal that those who designed the system knew the risks and decided to take them. A risk means that there is some finite chance that something will go wrong; only when the chance is very small can the system be considered safe. In the case of the shuttle boosters, a reasonable estimate of the overall failure rate for solid-fuel rockets was about 1 in 50.[12] Because the shuttle used two boosters, this means that one would expect an accident about once very 25 missions. And if the booster failed, so would the entire system, killing all of the crew. That the twenty-fifth mission was a disaster is, of course, a coincidence. But a disaster was a perfectly reasonable expectation, given previous experience with solid-fuel rockets.

This leads to some obvious questions: How is it that so many missions were flown with this serious problem? How is it that segmented booster rockets, with joints prone to failure, were used on a very important space system with live crew members? The answers are disturbing, but they illustrate how technological accidents develop within a social context. Here the major factor to consider is the inner workings of NASA, which in 1969 had landed the first human on the moon. Understanding how this important government agency worked is crucial for understanding why the twenty-fifth shuttle mission failed.

Among the interesting revelations that emerged in the wake of the accident was the process by which it was decided to use segmented rockets made by Morton Thiokol, a Utah firm. Dr. James C. Fletcher, the head of NASA in 1973, was influential in getting the contract awarded to Morton Thiokol, which had strong support from state political officials and from the Mormon church, a powerful force in Utah. Dr.

Fletcher is a Mormon, and he also belonged to a Utah lobbying group called Pro-Utah.[13] It is thought that the insistence on a booster composed of segments was due to the wish to employ Morton Thiokol, which had decided that it would make and ship the boosters to Florida in segments. Morton Thiokol was rated fourth in competition with three other contractors. Another firm, Aerojet General (which was rated third) had wanted to make the boosters out of a single tube.[14] A single tube, of course, would have avoided any problem with the seals required for a segmented rocket.

Another strange aspect of the award was that Lewis Research Center, home to many of the world's experts on solid-fueled rockets, lost authority to oversee the boosters' development. Authority was transferred instead to the Marshall Space Flight Center; Lewis officials were not even asked to serve on the booster evaluation board. Two former Morton Thiokol officials, however, had high positions on this board. Protests were made about the decision by the passed-over contractors and by Congress, but Morton Thiokol still got the contract.[15]

But design was only one problem among many that the shuttle faced. Another serious problem was that NASA was inadequately funded for its shuttle missions. Major budget cuts had been made in areas related to safety. Between 1970 and 1985, for instance, quality control personnel at the Marshall Space Flight Center were cut from 615 to 88, a reduction of 86 percent.[16] The reductions in quality control and safety personnel were an ominous sign. The issue was not only quality control, but also decision-making influence. Safety concerns were not properly represented to top management during the launch procedure, which led to a recurrent pattern of the ignoring of technical warnings that otherwise would have kept NASA from launching shuttles.[17]

A third problem was the actual decision to launch. Inquiries carried out after the fact by the presidential commission revealed that NASA's decision making on the shuttle launch was seriously flawed. In addition to ignoring a number of previous problems with booster O-ring failures, NASA management had serious communication problems. One such problem related to the temperature of the rocket itself. The Morton Thiokol executives who made the decision to launch had much more specific and disturbing information than mere statistics on boosters. Roger Boisjoly, an engineer at Morton Thiokol, was responsible for the design of the joint that held together the segments of the solid rocket booster. Experiments had shown that the ability of the seals to seat properly was a function of temperature: The lower the temperature, the greater the probability of failure. By July 1985, Boisjoly had already given Thiokol management a memorandum stating that the problem with the booster seals was grave and that it might cause a "catastrophe." At that time he formed a special "seal task force" to cope with the seal problem.[18]

On the day of the *Challenger* launch Roger Boisjoly was extremely concerned about this problem,[19] for he feared the booster would fail if it were launched at the low temperature of the morning of the 28th. He knew the entire rocket assembly would be destroyed if there was a booster failure. On the night of January 27, when a NASA teleconference call to Thiokol about low temperatures occurred, Boisjoly stated that on the basis of his data, the seals would not be secure unless their temperature was at least 50° Fahrenheit. During the teleconference, Morton Thiokol management at first backed up this view and recommended against launching. But NASA officials asked Morton Thiokol to reconsider. Morton Thiokol asked for five minutes off-line while it consulted with its engineers. After the teleconference with NASA, Morton Thiokol took 25 minutes to ponder the decision.

The engineers were stunned at the pressure to launch. In the past they had always had to prove to NASA that launching was safe; now they were asked to show that launching would not be safe. The pressure from NASA led higher managers at Thiokol to reconsider and eventually to change their minds. The concerns of Boisjoly and other engineers were ignored. At one point, Robert Lund, the vice-president of engineering for Thiokol, was told by Gerald Mason, a senior vice-president, to "take off your engineering hat and put on your management hat," and agree to the launch.[20] Lund did so, and Morton Thiokol agreed to the launch. A similar problem involving ice formation on the spacecraft was also ignored.[21]

The events that Boisjoly feared took place just as he had indicated they might. The hot gas blowby and subsequent explosion in the primary motor shattered Americans' confidence in their space program.

The disagreements at Thiokol should have been communicated to NASA's top management; had this occurred, the decision to launch might not have been made. But officials in the communication chain decided not to transmit these disagreements,[22] and the decision to launch was made in ignorance of some major reasons against doing so. The communication system that surrounded the shuttle was inadequate to provide timely warning of problems, a common feature of major technological accidents.[23] Leadership at NASA may have been faulty.[24] It has been suggested that decision-makers at NASA were too far removed from the engineering issues to really understand the problems.[25]

The destruction of the *Challenger,* then, involved many factors: outside pressures, poor communications, chronic unresolved technological problems, and flawed decision making. The immediate cause of the accident was a hot gas blowby. The human system operating the shuttle, though, created the machinery for the accident, put it together, and set it in motion. The astronauts were killed not only by the effects of cold weather on the booster seals, but also by a human management system that was inadequate for its purpose and indifferent to its own faults.

Technological accidents, then, represent *people failures* as well as technology failures. People may fail in the technology's design, in managing it successfully, and in using it safely, as automobile accidents caused by drunk drivers constantly prove. But the same people who operate automobiles may also be at the controls of larger systems, such as railroad trains.[26]

The Amtrak-Conrail Collision. On January 4, 1987, a locomotive operated by the government agency Amtrak smashed into three linked locomotives operated by Conrail, another government agency.[27] Conrail trains carry freight; Amtrak trains carry passengers. The Conrail locomotives had pulled in front of the speeding Amtrak train when the collision occurred; both trains were speeding when they collided. The Amtrak train was going 23 mph more than its limit of 105 mph. The Conrail train was going twice as fast as the 30 mph indicated by signals that its engineer ignored. One hundred and seventy passengers were injured in the accident, and 16 died. The engineer on the Conrail train, Ricky Gates, was indicted for manslaughter. His blood and urine showed traces of marijuana. The warning whistle in his locomotive had been taped over to mute the sound.[28]

It would be easy to blame the accident on Ricky Gates. He was clearly negligent and he later admitted smoking marijuana, but other factors must be considered. The signals Gates drove through without slowing down would have slowed the Amtrak train automatically, for it had an automatic braking device that functioned independently of the engineer.[29] The Conrail train did not have such a device. That the Amtrak train had a braking device and the Conrail train did not reflects their respective cargoes: Society is more concerned about the safety of passengers than that of freight.[30] The Conrail train also had a broken radio, which had been temporarily replaced by an inadequate hand-held model. Both trains were speeding because of commercial pressures to keep on schedule.[31] And both trains were using the same tracks because it was cheaper to do so than to have separate tracks for passenger and freight lines.[32]

The accident occurred at a time marked by great controversy over the use of drug testing for train crews.[33] Unions had strongly resisted such testing, in spite of the involvement of drugs and alcohol in previous accidents.[34] Ricky Gates had previously had his driver's license suspended and was already under indictment on a drunk-driving charge.[35] Later, during testimony before a committee of the U.S. Senate, Gates admitted that beer drinking and other drug use by train crews was common.[36] Gates probably should not have been in the cab of a train.

Then there is the matter of the taped whistle—the disabling of a warning signal that had proven a nuisance.[37] But whistle-taping is common; the Federal Railroad Administration discovered 75 other taped whistles when they checked other trains.[38]

The accident, then, involved more than the collision of two trains. It occurred in a context in which there were commercial pressures for speed combined with a possibly drugged operator. A taped whistle muted the warning Gates should have heard. Lack of an automatic braking device and the use of a track traveled by both freight and passenger trains were important. High-tech systems, then, can be very vulnerable not only because of their design, but also because of the politics that inevitably surround their operation. Commercial pressures lead to high speeds; union pressures prevent drug testing; operators are willing to take chances; and economic considerations dictate multiple use of the tracks. In this high-tech accident, operators may have accidently "pulled the trigger," but "the gun was loaded" by cost pressures and interorganizational struggles.

A *Salmonella* Outbreak in Illinois. The Hillfarm Dairy of Jewel Food Stores is a typical example of high technology in the dairy industry. Using a completely closed production system, it produces 175,000 gallons of milk a day and distributes it to six states. The plant is impressive in its size and complexity:

> The plant itself sprawls over a city block. According to Illinois state dairy official Lewis Schultz, the plant pumps milk through a maze of 400 miles of stainless steel pipes, 638 air valves, a giant holding tank, and more than 600 large metal plates, which heat and cool milk during pasteurization.[39]

The dairy's size is a reflection of what economists call an "economy of scale": It is cheaper to produce some things in large quantities. But in March 1985, this dairy somehow became contaminated with *Salmonella* bacteria, and 14,000 people in the six-state area became sick with diarrhea, nausea, vomiting, and fever.[40] Public health officials took the plant apart, trying to determine what had gone wrong. *Salmonella* bacteria are often found in raw milk, but not in milk that has been partially sterilized by pasteurization.

The immediate causes of the accident were never determined, although the consensus was that somehow a faulty valve had mixed raw milk with pasteurized milk.[41] The important point is that very often the economic considerations surrounding the use of modern technology tend to increase the size of the systems on which we depend. As the system size increases, the impact of technological accidents also increases, and even a small part's malfunction can cause the entire system to fail. The Northeast power blackout of 1965, which affected a good portion of the East Coast, was caused by the malfunction of a single relay.[42] And when the system fails, the effects can be devastating, as at Bhopal, India, where a malfunction in a pesticide plant led to an unprecedented technological tragedy responsible for over 3,000 deaths.[43] The increasing size and complexity of our technology leads to greater risks and the potential for bigger disasters. The events at Chernobyl were such a disaster.

The Nuclear Accident at Chernobyl. One of the world's worst nuclear accidents took place in the Soviet Union on April 25, 1986, at Chernobyl, not far from Kiev, the capital of the Ukraine. As in the Conrail-Amtrak smashup, operators were negligent.[44] But Chernobyl #4 was a nuclear reactor filled with poisonous radioisotopes, and the mistakes made by its operators had worldwide repercussions. When the reactor blew up, it spread radioactive chemicals over much of northern Europe.[45]

The accident occurred during a test of the back-up electrical power system. The test required slowing the reactor down to a low level, but in slowing the reaction, the operators overdid it; they came close to shutting down the reaction. In an attempt to speed up the reaction, they disabled most of the reactor's safety systems by pulling out most of the rods that control the nuclear reaction. Even the Premier of the Soviet Union is not allowed to leave fewer than 15 rods in the reactor; the operators left 7. And they also reduced the volume of the reactor's cooling water, which served to moderate the reaction.

Once the operators began the test at 1:24 A.M., their fate was sealed. *Within a second* after the reaction accelerated, it zoomed out of control. The heat produced caused an explosion, blowing a 1,000-ton roof off the reactor and sending radio-isotopes shooting a half-mile into the sky.[46] A fire started and burned for days. Putting it out caused the death of more than 29 people (two died in the explosion and fire), including a dozen firefighters who died from exposure to radioactivity.[47] Chernobyl #4 was only yards away from another reactor in the complex, and only quick action by firefighters, who dumped boron, lead, dolomites, clay, and sand on it (water would have been useless), kept the fire from spreading. The emergency team also narrowly averted a meltdown of the reactor core into the Pripyat River, which flowed under the reactor. The Pripyat River provides the main water supply for Kiev, a city of 2.5 million people.

Clouds of radioisotopes poisoned crops and soils over hundreds of square miles of the U.S.S.R. In the area immediately downwind of the reactor, the land is considered to be uninhabitable.[48] Eighty thousand people were evacuated from the area, although some had to wait until several days had passed. Other "hot spots" created by radioactivity, even at considerable distances from the reactor, were also evacuated. Tens of thousands of people are expected to die prematurely of cancers caused by the accident. Other countries also experienced severe effects. Reindeer herds in Lapland were wiped out. Thousands of tons of vegetables in Italy were condemned, and children were advised not to drink milk. Food in Eastern Europe was condemned as unsafe to eat. The damages in Hungary alone were over $30 million.[49]

Soviet response to the accident was prompt but not entirely effective.[50] The highly centralized Soviet bureaucracy responded, but delays and red tape were common. The village of Pripyat, which is near the reactor, was not evacuated for 36 hours. The accident was not immediately admitted by the Soviet Union, which has a long tradition of secrecy regarding disasters. It was not until two days later, after Swedish scientists noticed an intense increase in radioactivity, that the Soviet government admitted to the West and to its own people that there had been a serious accident. Of course, this silence delayed measures that ordinary citizens might have taken to protect themselves from the fallout.

This accident was caused by the seriously mistaken judgment of three people, but it was allowed to happen by a system that had built the nuclear reactor in the first place, allowed engineers to carry out experiments at 1:24 A.M., had no containment vessel for the nuclear reactor in the event of a steam explosion, and had not properly set up plans for a major evacuation in the event of a disaster. If the firefighters had not been willing to sustain the lethal injuries they did, the second reactor would also have caught fire. These problems are worrisome to us in the United States because we have some 80 reactors of various types, some with and some without containment vessels. We also have had some near-disasters, such as those of Fermi I in 1966 and at Three Mile Island in 1979.[51]

The Anatomy of Technological Accidents

As a technology advances, the overall risk from accident is likely to decline because people learn how to use the technology safely. For a variety of reasons, though, it is worthwhile to concern ourselves with the hazards of technological systems. A number of tendencies in our technology threaten to make us more vulnerable.[52] First, as we become more dependent on microelectronic components, small malfunctions can translate into serious accidents. There may be virtually no relation between the size of the malfunctioning component and its impact on the overall system. A computer chip smaller than a thumbtack can send an airliner crashing into a hillside.[53] A "minor" programming error can result in lethal radiation doses, as in the Therac 25 example discussed above.

A second hazardous tendency results from combining systems into huge integrated complexes and operating units. The phrase "economy of scale" means that the larger a system, the more cost savings can be achieved through the use of common components. The dairy responsible for the *Salmonella* problem mentioned above is a good example; another is the centralization and high degree of interdependence of power systems that led to the Northeast blackouts.[54] Moving liquid natural gas in special supertankers is very inexpensive; it is also extremely hazardous. Already a number of major oil spills from supertankers have overwhelmed local ecological systems. But supertankers continue to be used because moving oil in this way saves money. Still, serious accidents are inevitable when two dozen people are put in charge of a ship that is a fifth of a mile long with a 70-foot draft, that is very difficult to steer, and that is very unforgiving when a mistake is made. It takes three miles to stop a 200,000-ton supertanker moving at 16 knots.[55] The ships are so big they can collide with and sink a trawler without the tanker's crew even realizing that anything has happened.[56]

A third hazardous tendency involves the system's human operators. Contrary to popular belief, highly automated systems are sometimes more vulnerable to human error than are manual ones. Small mistakes on the part of human operators can lead to big disasters in aviation, for instance.[57] Often a single person will be monitoring a huge installation, such as a process plant or a nuclear reactor; through an oversight or outright carelessness the lone controller may put hundreds of lives in jeopardy. The Conrail train involved in the smashup with Amtrak could well have been filled with explosives or toxic materials, in which case the loss of life would have been even greater, and train engineers are not the only operators who disable warning devices. The fault for the massive oil spill caused by the grounding of the *Exxon Valdez* has been laid at the feet of the tanker's captain.[58] We have already noted the factors that led to the accident at Chernobyl; similar errors in judgment might also deliver one of the "weapons" of a biological warfare laboratory to an unsuspecting target population.[59]

Sometimes technologies simply defy human analysis. The 1966 fuel-rod meltdown of the Fermi I reactor in Monroe, Michigan, involved first-rate engineers struggling with a situation they were unable to understand until long after the crisis.[60] Fortunately, the reactor was operating with a fractional load and was undergoing tests at this time. Fermi I was a "breeder" reactor, and therefore it was designed to contain far more radioisotopes than the usual light-water reactor. Had the reactor been at full charge, and had it been operated by the "all-thumbs" operators of the Three Mile Island accident, the results might have been horrifying. Similar problems in understanding the dynamics of the reactor, even after the accident, also became apparent at Three Mile Island. A large number of "surprises" were discovered when the reactor was dismantled eight years after the accident.[61] In another instance, engineers at the Tennessee Valley Authority's reactor at Savannah River were surprised to find cracks in stainless steel caused by hot water.[62] These unpredictable properties of ostensibly well-understood technologies are worrisome, and some specialists feel that nuclear power simply cannot be managed safely on a large scale.[63]

A fourth hazardous tendency is for managers of the technology to cut corners. In both capitalist and socialist countries, managers under pressure to keep costs down economize on safety devices because safety does not immediately affect operating efficiency. Chemical plants are often designed in a couple of days, and there may be little or no planning for catastrophic accidents. Chemical-process engineers sometimes must be able to make accurate designs of reactor vessels or distillation columns in a quarter of an hour.[64] At least partly because of these factors, major chemical accidents have been increasing over the last two decades.[65]

James Reason, a psychologist who has studied accidents in some depth, notes that hazardous conditions are often evident before accidents occur. He calls these conditions "resident pathogens" because they are lurking problems that increase the probability of an accident or intensify it should it indeed occur.[66] Resident pathogens were all too evident in the *Exxon Valdez*'s soiling of Alaska's coast. Exxon allowed safety devices to be out of service for long periods of time and minimized the need to keep safety teams on alert.[67] When the accident took place, not only was the equipment unready, but the company itself appeared unready to respond to the accident, both of which allowed the damage to intensify.

The final hazardous tendency involves the *stretch principle*. Called "risk homeostasis" by psychologists, this principle states that new safety measures sometimes encourage people to take new risks.[68] It is not that people like taking risks; rather, once people perceive that an activity is now "less risky," they tend to engage in that activity more. For instance, the miner's safety lamp, invented simultaneously in 1815 by George Stephenson and Humphrey Davy, allowed English miners to safely go near pockets of methane gas, previously a serious explosion hazard. We might think, then, that this invention would reduce the rate of death and injury in mines; this, however, was not the case. In the 18 years before the introduction of the lamp, 447 miners lost their lives in accidents in the counties of Durham and Northumberland; in the 18 years after, the number lost was 538. It has been speculated that use of the lamp, which was not *perfectly* safe, encouraged miners to work in more hazardous areas and that this accounted for the increased number of deaths.[69]

This kind of risk homeostasis occurs also with larger and more complicated technological systems. Although computer systems have the potential to increase safety, any increases in safety are often nullified because the technology encourages further risks. For instance, the Grumman X-29 jet plane is able to stay aloft only with the aid of computers. With its wings swept forward, the plane is aerodynamically unstable; it is able to fly only because the computer can adjust its canard control surfaces 40 times a second, a performance beyond human capability.[70] The control is managed by a digital computer that is backed up first by another digital computer and then by an analog computer. Should all three computers fail at speed, the plane would disintegrate in the air. The X-29 can be considered a model for other technological systems that are equally "far out on a limb," but if the computers fail on the X-29, only the pilot dies. If they fail at a nuclear power plant while the rods are up, what then?

This argument can be pushed too far. Obviously civilization itself depends on technologies that take us, so to speak, further out on a technological limb. The chains of dependence on agricultural, electrical, transport, and informational technologies are now taken so much for granted that only catastrophes force us to question them.

Many predictions of catastrophes due to large-scale technological failure seem overblown.[71] Large-scale electrical failure, for instance, can occur without catastrophic consequences, as it has during several major blackouts.[72] The failure of some systems, however, can be catastrophic, as was the case with the nuclear reactor at Chernobyl and the pesticide plant at Bhopal. If certain military systems fail, such as those associated with viral warfare, the word "catastrophe" may be an understatement.[73] "Learning by doing" can be a very dangerous way to learn, yet that seems to be the direction we have chosen with many hazardous systems.[74] This kind of experimental attitude may be even less appropriate with long-term ecological problems, such as stratospheric contamination, deforestation, or global warming, for in such matters feedback may come too late for us to take evasive action, or change may be irreversible.[75]

Avoiding Technological Accidents

Some social scientists contend that we have the problems associated with technological risk under control in the United States. In their book *Averting Catastrophe,* Joseph Morone and Edward J. Woodhouse argue that use of such high-tech systems as nuclear power and chlorofluorocarbons has been accompanied by the development of appropriate social controls.[76] How effective such controls have been, they say, has been shown by the infrequency of major catastrophes (their book was written after Bhopal but before Chernobyl). The strategies they discuss are:

1. Protecting against potential hazards

2. Proceeding cautiously

3. Testing the risks

4. Learning from experience

5. Setting priorities for studies of hazards

The accident at Chernobyl should make us consider how effective these controls have been. Although there was no immediate, large loss of life, many lives in the Soviet Union and nearby countries will be shortened by this accident. Hundreds of square miles of ecological disruption and continuing radiation dangers are worth noting. Just as luck has favored us in some nuclear accidents (Three Mile Island and Fermi I), it may turn against us again in the future, as it did at Chernobyl.

In a democracy, it is ultimately the responsibility of the citizens to decide what technological risks they will accept. But they can only decide responsibly if they are well informed. Unfortunately, society cannot depend on private interests to provide all the necessary information,[77] for the history of the nuclear industry clearly shows a long record of self-deception, soft-pedaling, underestimation, and outright lying. Nor can society depend on technical experts to be careful with the lives of others when strong economic or political interests are at stake.[78] Thus our safety depends on ensuring research, the free, public flow of information about the hazards of our technologies, and the existence of strong regulatory agencies to enforce decisions embodied in laws and administrative rulings. Government officials who weaken these agencies can be likened to the train drivers who tape over their warning whistles.[79] Safety is costly—but it is too important to ignore.

We must realize that many institutions are very capable of deceiving themselves about hazards. The design of many organizations does not encourage the free flow of thought about problems.[80] Hence, even when the organization is not intending to deceive, it may be deceiving itself. Gustaf Östberg, a Swedish technologist, has noted that often the most dangerous aspect of sociotechnical systems is the mind-set of those who manage them.[81]

Conclusion

A society's accidents reflect its character. In our society, the use of high technology is reflected in the sensitivity to failure of many of our systems. Failures will always occur; things made by people are by nature imperfect. We should not expect systems to function safely without extensive simulation, testing, and monitoring, but these are often absent or inadequate. Centralization, mathematical simulation, and computerized safeguards inherent in high technology encourage us to increase the size of risks. Still, systems do fail—through human error, bad design, or the intervention of the natural environment. We must take care to see that our failures are few and that the ones we have are noncatastrophic.

It is not at all clear whether our institutions are up to the test. The study of technological accidents and near-accidents suggests that the line between safety and disaster is often thin. Sometimes a single individual seems to make the difference, as when a human operator presses the right buttons in a crisis, or as was the case with James Asseltine's all-seeing presence on the Nuclear Regulatory Commission.[82] The current debate about chlorofluorocarbons was stimulated by a single scientist's research and advocacy;[83] in this case the margin of safety does not seem large enough. Although it is encouraging to see that individuals make a difference—it is always good to know that our actions count for something—an individual alone does not provide an adequate margin of safety considering the risks involved. Our ability to "see and avoid" should be kept well ahead of our ability to get into trouble. We will discuss this issue further in Chapter 17.

Notes

1. For horror stories in which human factors are responsible, it is difficult to top the 1200-psi steam propulsion system described by Charles Perrow in "The Organizational Context of Human Factors Engineering," *Administrative Science Quarterly,* Vol. 28, No. 4 (Dec. 1983), pp. 521–541. The original document is NPRDC SR 82-25, "Problems in Operating the 1200 Psi Steam Propulsion Plant: An Investigation," U.S. Navy Personnel Research and Development Center, San Diego, California, May 1982.

2. Charles Perrow has described such mistakes caused by "error-inducing systems" in his book *Normal Accidents: Living With High-Risk Technologies* (New York: Basic Books, 1984), p. 172.

3. Henry Petroski has written a splendid book giving a balanced portrait of these essential traits of designed and manufactured things; see *To Engineer Is Human: The Role of Failure in Successful Design* (New York: St. Martin's Press, 1965).

4. Gene Marine and Judith Van Allen, *Food Pollution: The Violation of Our Inner Ecology* (New York: Holt, Rinehart & Winston, 1972).

5. See, for instance, Oscar E. Anderson, Jr., *The Health of a Nation: Harvey Washington Wiley and the Fight for Pure Food* (Chicago: University of Chicago Press, 1958).

6. "Drug Use," Chapter 7, in Armand Mauss, *Social Problems as Social Movements* (Philadelphia: Lippincott, 1975).

7. Michael Brody, "When Products Turn into Liabilities," *Fortune,* Mar. 3, 1986, pp. 20–27; and George J. Church, "Sorry, Your Policy Is Cancelled," *Time,* Mar. 24, 1986, pp. 16–26.

8. Perrow, *Normal Accidents,* pp. 62–100.

9. Eliot Marshall, "The Shuttle Record: Risks, Achievements," *Science,* Vol. 231 (Feb. 14, 1986), pp. 664–666.

10. *Report of the Presidential Commission on the Space Shuttle* Challenger *Accident* (Washington, D.C.: June 6, 1986), 5 volumes.

11. Philip Boffey, "Rocket Engineers Tell of Pressure for a Launching," *New York Times,* Feb. 26, 1986, pp. 1, 17; see also *Presidential Commission on the Shuttle,* Vol. I, pp. 164–177.

12. Yet NASA's reports mentioned a failure rate for solid-fuel rockets of as low as 1 in 100,000; see Richard P. Feynmann, "Personal Observations on Reliability of the Shuttle," *Presidential Commission on the Shuttle,* Vol. II, pp. F-1–F-5.

13. William J. Broad, "NASA Chief Might Not Take Part in Decisions on Booster Contracts," *New York Times,* Dec. 7, 1986, Section K, pp. 1, 14.

14. Broad, "NASA Chief," Section K, p. 1.

15. Broad, "NASA Chief," Section K, p. 14.

16. Robert Pear, "Senator Says NASA Cut 70% of Staff Checking Quality," *New York Times,* May 8, 1986, pp. 1, 16.

17. Henry S. F. Cooper, "Letter from the Space Center," *New Yorker,* Nov. 10, 1986, pp. 83–114. This article is a very valuable source of information on the management of the shuttle program.

18. Roger Boisjoly spoke of his efforts to prevent launch of the *Challenger* 51-L flight in Ann

Arbor, Michigan, first at the University of Michigan's Collegiate Institute for Values and Science on November 3, 1987, and then in a lecture on "Ethical Decision-Making and the Space Shuttle *Challenger,*" given at the Chrysler Center for Continuing Education in Engineering on November 4, 1987; see also his "Ethical Decisions—Morton Thiokol and the Space Shuttle *Challenger* Disaster," American Society of Mechanical Engineers Paper 87-WA/TS-4 (New York: American Society of Mechanical Engineers, 1987).

19. Cooper, "Letter," pp. 91–92.

20. *Presidential Commission on the Shuttle,* Vol. IV, pp. 772–773.

21. Ed Magnuson, "A Serious Deficiency," *Time,* Mar. 10, 1987, pp. 38–42.

22. Michael Brody, "NASA's Challenge: Ending Isolation at the Top," *Fortune,* May 12, 1986, pp. 26–32.

23. Communications was a major factor in the Quebec Bridge Disaster of 1907, in which subordinates and the chief engineer were located too far apart to provide timely warning and action once problems began to appear; see John Tarkov, "A Disaster in the Making," *American Heritage of Invention and Technology,* Vol. 1, No. 3 (Spring 1986), pp. 10–17.

24. Hal Bowser, "Maestros of Technology," *American Heritage of Invention and Technology,* Vol. 3, No. 1 (Summer 1987), pp. 24–30.

25. Cooper, "Letter," p. 93.

26. And sometimes in charge of oil tankers that go astray, as in the case of *Exxon Valdez.*

27. William K. Stevens, "Starting at the Scene, Team Seeks Cause of Deadly Train Crash," *New York Times,* Jan. 11, 1987, p. Y-9.

28. David Gates and Susan E. Katz, "Legacy of a Railroad Disaster," *Newsweek,* May 11, 1987.

29. Stevens, "Scene," p. Y-9.

30. This special concern for the safety of passengers is one reason that marine accidents, which mostly involve cargoes, seldom result in the kind of safety improvements that air accidents produce; see Perrow, *Normal Accidents,* pp. 170–231.

31. William K. Stevens, "Metroliner in Northeast Gaining on Bullet Trains," *New York Times,* Nov. 3, 1985.

32. Reginald Stuart, "The Man on the Spot over Rail Safety," *New York Times,* Jan. 20, 1987.

33. In May 1989, the U.S. Supreme Court decided by a 7-2 margin that the Federal Railroad Administration could test crews for drugs if they were involved in a train crash; see "High Court Weighs Drug Tests," *Newsweek,* Apr. 3, 1989.

34. Reginald Stuart, "Rail Safety Campaign Is No Longer Spinning Wheels," *New York Times,* Nov. 3, 1985, Section E, p. 5.

35. Gates and Katz, "Legacy."

36. Laura Parker, "Drugs, Alcohol Widespread in Rail Industry, Ex-Workers Testify," *Washington Post,* Feb. 26, 1988, Section A, p. 19.

37. Unhappily, it is all too common to find that warning signals have been disabled by workers tired of "false alarms." This often occurs in nuclear power plant control rooms, where there are sometimes 2,000 or more meters, many of which include alarms; see Peter Andow, "Failure of Process Plant Monitoring Systems," in John Wise and Anthony Debons (Eds.), *Information Systems: Failure Analysis* (New York: Springer-Verlag, 1987), pp. 233–240. It can also occur at facilities that process hazardous substances. At one plant that manufactured uranium for nuclear bombs, inspectors discovered that many pollution

alarms had been deliberately disabled because they sounded so often; see Cass Peterson, "Mission Fading for Aged A-Plant," *Washington Post,* Feb. 28, 1988, Section A, p. 4.

38. Gates and Katz, "Legacy."

39. Marjorie Sun, "Desperately Seeking *Salmonella* in Illinois," *Science,* Vol. 228 (May 17, 1985), pp. 829–830.

40. Sun, "*Salmonella,*" pp. 829–830.

41. Telephone conversation with Susan Hark of the Illinois Department of Public Health's Information Service, May 27, 1987.

42. James R. Chiles, "Learning from the Big Blackouts," *American Heritage of Science and Technology,* Vol. 1, No. 2 (Fall 1985), pp. 26–33.

43. See Paul Shrivastava, *Bhopal: Anatomy of a Crisis* (Cambridge, Mass.: Ballinger, 1987).

44. This case description is largely based on Colin Norman, "Chernobyl: Errors and Design Flaws," *Science,* Vol. 233 (Sept. 5, 1986), pp. 1029–1031; and John F. Ahearne, "Nuclear Power After Chernobyl," *Science,* Vol. 236 (May 8, 1987), pp. 673–679.

45. John Greenwald, "Deadly Meltdown," *Time,* May 12, 1986, pp. 39–52.

46. Celestine Bohlen and Walter Pincus, "Anatomy of an Accident: A Logistical Nightmare," *Washington Post,* Oct. 26, 1986, Section A, pp. 1, 34, 35.

47. I think it is very likely that these firefighters realized that they were going to die from radiation exposure, and if so, they deserve to be remembered as heroes for being willing to die so that others might live.

48. Of the 50 million curies of radioactivity, about half of the ejected fission products may have been deposited within 30 kilometers of the reactor; see Norman, "Chernobyl: Errors," p. 1029.

49. Judy McKinsey, "Eastern Europe Farmers Feeling Chernobyl Fallout," *Detroit Free Press,* June 1, 1986, Section B, pp. 1–4.

50. Philip Taubman, "A Worst Case that Isn't a Scenario," *New York Times,* May 11, 1987, Section 4, p. 1.

51. Bill Kiesling, *Three Mile Island: The Turning Point* (Seattle: Veritas Books, 1980).

52. Ron Westrum, "Vulnerable Technologies: Accident, Crime, and Terrorism," *Interdisciplinary Science Reviews,* Vol. 11, No. 4 (December 1986), pp. 386–391.

53. Such components may be brought into being by negligent manufacture; see D. Sylvester, "The Dark Side of the Force," *Washington Monthly,* Feb. 1985.

54. See Amory B. Lovins and L. Hunter Lovins, *Brittle Power: Energy Strategy for National Security* (Andover, Mass.: Brick House, 1982).

55. Noel Mostert, *Supership* (New York: Knopf, 1974), p. 30.

56. In 1973 the South African trawler *Harvest Del Mar* was sunk by the Spanish tanker *Mostoles* without the tanker's crew becoming aware that anything had happened. Only later, when damage to the bow was discovered, was the trawler's sinking suspected; see Mostert, *Supership,* p. 32.

57. Earl Wiener, "Fallible Humans and Vulnerable Systems: Lessons Learned from Aviation," in Wise and Debons, *Information Systems: Failure Analysis,* pp. 163–182; see also James Reason, "Little Slips and Big Disasters," *Interdisciplinary Science Reviews,* Vol. 9, No. 2 (June 1982), pp. 179–189.

58. The *Exxon Valdez,* involved in the worst oil spill in American history, was under the

command of a tanker captain whose blood alcohol content was tested at .06 after the accident. The captain was not on the bridge at the time of the accident, and the ship, nearly 1,000 feet long, was being piloted by an unqualified third mate; see Philip Shabecoff, "Captain of Tanker Had Been Drinking, Blood Tests Show," *New York Times,* Mar. 31, 1989, pp. 1, 9. It is significant that Exxon had been monitoring the captain's drinking problems for four years before the accident; see John Cushman, Jr., "Exxon Monitored Drinking by Captain," *New York Times,* May 19, 1989, Section I.

59. A *New York Times* story on September 13, 1980, p. 7, entitled "Wrong Virus Cloned," tells how a prominent virologist, Dr. Ian Kennedy of the University of California, San Diego, cloned the genetic material of the wrong virus—a virus more dangerous than the one he intended to clone. He cloned the semliki virus, which can cause fever, headache, and even (rarely) death. Fortunately, the mishap hurt no one, and the design of the laboratory in which the mistake occurred included stringent protections. The incident does make one wonder, however.

60. John Fuller, *We Almost Lost Detroit* (New York: Ballantine Books, 1975). For a very different version of these events, see E. Pauline Alexanderson (Ed.), *Fermi: New Age for Nuclear Power* (Lagrange Park, Ill.: American Nuclear Society, 1979).

61. William Booth, "Postmortem on Three Mile Island," *Science,* Vol. 238 (Dec. 4, 1987), pp. 1342–1345.

62. Keith Schneider, "New Robot Seeks Cracks in Nuclear Reactors," *New York Times,* Jan. 31, 1989, pp. 19, 23.

63. Much of the problem, of course, comes about when private utilities try to save money by hiring semiliterate operators and maintenance people. Cheating on hiring exams for nuclear plant operator positions is common, according to Christopher Flavin, "How Many Chernobyls?" *World-Watch,* Jan.-Feb. 1988, pp. 14–18. But even in countries such as France, where reactors are carefully managed by the government, there are serious close calls; see David Dickson, "France Weighs Benefits, Risks of Nuclear Gamble," *Science,* Vol. 233, (Aug. 29, 1986), pp. 930–932.

64. Lee Niedringhaus Davis, *The Corporate Alchemists: Profit Takers and Trouble Makers in the Chemical Industry* (New York: Morrow, 1984), p. 195. See also the three-part series on chemical plant dangers in the *New York Times:* Stuart Diamond, "Problems at Chemical Plants Raise Broad Safety Concerns" (Nov. 25, 1985); Matthew Wald, "Industrial New Jersey Girds to Prevent Toxic Disasters" (Nov. 26, 1985); and Philip Shabecoff, "Tangled Rules on Toxic Hazards Hamper Efforts to Protect Public" (Nov. 27, 1985).

65. Fred Millar, "Braking the Slide in Chemical Safety," *New York Times,* May 11, 1986, Business Section, p. 3.

66. James Reason, "Resident Pathogens and Risk Management," presented to World Bank Conference on Safety Control and Risk Management, Washington, D.C., October 1988.

67. Ken Wells and Marilyn Chase, "Paradise Lost: Heartbreaking Scenes of Beauty Disfigured Follow Alaska Oil Spill," *Wall Street Journal,* Mar. 31, 1989, p. 1; and Keith Schneider, "Under Oil's Powerful Spell, Alaska Was Off Guard," *New York Times,* Apr. 2, 1989, pp. 1, 11.

68. The idea is that people will maintain a more or less constant level of risk, given that the level of risk is considered to be an acceptable one. Thus when safety devices are introduced into a situation with an already acceptable level of risk, people will venture, as it were, further out on a limb. The perceived level of risk, then, stays the same; see G. J. S. Wilde, "The Theory of Risk Homeostasis: Implications for Safety and Health," *Risk Analysis,* Vol. 2 (1982), pp. 209–225.

69. Samuel Smiles, *The Life of George Stephenson, Railway Engineer* (Boston: Ticknor and Fields, 1858), p. 130.

70. John Tierney, "The Real Stuff," *Science 85*, Sept. 1985, pp. 24–35.

71. See, for example, Roberto Vacca, *The Coming Dark Age* (Garden City, N.J.: Doubleday, 1974).

72. This is not to say that future blackouts could not cause catastrophes; the extent of damage depends on what capabilities are lost in such a blackout; see James R. Chiles, "Learning from the Big Blackouts," *American Heritage of Technology and Invention,* Vol. 1, No. 2 (1985), pp. 27–30. The 1977 blackout on the East Coast caused damages estimated to be between $300 million and $1 billion.

73. Accidents with military systems necessarily present larger hazards because these systems are designed to kill large numbers of people. Nuclear weapons and biowarfare present the biggest, but by no means the only, hazard of this type. The potential for the use of these weapons by terrorists is, of course, very serious.

74. Abigail Trafford and Andrea Gabor, "Living Dangerously," *U.S. News and World Report,* May 19, 1986, pp. 19–22.

75. See Chapter 17 of this text for a more complete discussion of this topic.

76. Joseph G. Morone and Edward J. Woodhouse, *Averting Catastrophe: Strategies for Regulating Risky Technologies* (Berkeley: University of California Press, 1986), pp. 121–135.

77. John Gofman and Arthur R. Tamplin, *Poisoned Power* (New York: New American Library, 1971).

78. See, for instance, Michael Uhl and Tod Ensign, *GI Guinea Pigs: How the Pentagon Exposed Our Troops to Dangers More Deadly Than War* (New York: Wideview Press, 1980); Leslie J. Freeman, *Nuclear Witnesses: Insiders Speak Out* (New York: Norton, 1981); Paul Brodeur, "Annals of Asbestos," *The New Yorker;* and Paul Brodeur, "Annals of Chemistry: In the Face of Doubt," *The New Yorker,* June 9, 1986, pp. 70–87.

79. Several U.S. administrations have had a strong influence on safety; however, the Reagan administration's impact has been to weaken many of the agencies responsible for our protection; see, for instance, James Rathlesberger (Ed.), *Nixon and the Environment: The Politics of Devastation* (New York: Taurus Communications, 1972); Joan Claybrook and the Staff of Public Citizen, *Retreat from Safety: Reagan's Attack on America's Health* (New York: Pantheon, 1984); and Jonathan Lash, Katherine Gillman, and David Sheridan, *A Season of Spoils: The Reagan Administration's Attack on the Environment* (New York: Pantheon, 1984).

80. See Ron Westrum, "Organizational and Inter-Organizational Thought," Proceedings of World Bank Conference on Safety Control and Risk Management, October 1988; see also Tim Clark and Ron Westrum, "Paradigms and Ferrets," *Social Studies of Science,* Vol. 17, No. 1 (Feb. 1988), pp. 3–34.

81. Gustaf Östberg, "Some Reflections on Attitudes and Mind-Sets Concerning Pressure Vessel Failure in Nuclear Reactors," *International Journal of Pressure Vessels and Piping,* Vol. 34 (1988), pp. 315–330.

82. Eliot Marshall, "NRC's Political Meltdown," *Science,* Vol. 237 (July 10, 1987), pp. 123–124.

83. Brodeur, "Annals of Chemistry."

The Distancing Effects of Technology

In previous chapters we have largely considered the general categories of technological action. In this chapter we will take a slightly different approach and discuss one of the potential impacts of high technology—psychological distancing. We will explore the way in which electronic communication and similar technologies have shifted the ways in which we relate to one another. Electronic communication is usually considered to link people; here we will learn how it can also divide them. In a similar way, the "convenience" technologies are changing the character of our relations with one another. This chapter's thesis is simple: Technology that puts physical distance between people distances them psychologically as well, putting important human values at risk.

When people interact face-to-face, they are able to get the full spectrum of other people's responses. They can tell—from words, posture, gestures, and facial expressions—how another person feels.[1] This kind of face-to-face contact allows us to understand other people's feelings, positive and negative, and to identify with them. It is easier to give affection face-to-face, and it is more difficult to inflict suffering. At least in principle, this personal contact provides a sensitive link. When this link is replaced, however, empathy declines. As other people recede in physical distance, signals reminding us that they are human become fainter. We are more likely to treat another as an abstraction, a number, a case, which provides for both a greater degree of stereotyping and fantasy and a greater callousness in regard to others' suffering.

A particularly striking example of the value of physical closeness is a mother's physical contact with her baby shortly after its birth. In American society, it was once customary to isolate babies born in the hospital in a nursery, where they would be free of contaminating germs. Recent studies have shown, however, that this removal of the child from the mother can strongly affect their bonding. Mothers allowed direct interaction with their children 30–45 minutes after birth, and those allowed longer interaction thereafter, show more interest in and affection toward their babies.[2] A father, too, needs bonding with his child, and interaction with the baby shortly after birth helps such bonding. Use of the sterile nursery may have solved some problems, but it caused new ones.

A danger in modern society is the impact of institutions that separate people. Such physical distancing may be becoming more common, harming our ability to be close to others and to feel part of a community with them. In this chapter we will explore a few of the ways in which distancing occurs.[3]

The Implications of Communications Technology

Communication is the transmission of signals among people. Among primitive peoples most communication takes place face-to-face; what remote communication exists is usually acoustic—it uses sounds, such as beating on drums, sounding horns, or pounding on the earth.[4] With the development of civilization came optical communication through semaphores, animal messengers, and mail systems.[5] With high technology we get telephony and satellite television. This ability to communicate over ever-longer distances now makes it possible to communicate with people on the moon, or even on other planets. Developments in technology have thus allowed people to coordinate, to exchange ideas, and to influence each other without face-to-face contact.

In some respects these technologies have allowed us to reach out to others. Farmers' lives were made much richer and less isolated with the inventions of the telephone, the automobile, and the radio. But along with the resultant "global village," we also have the "global organization." Just as communication innovations have allowed individuals to reach out, they also have allowed the long arm of the organization to extend farther as well, which has ended local isolation, quickened the pace of life, and brought new pressures to bear on all of us. With the telegraph, the telephone, the computer, the pager, the cellular phone, and now the fax machine, personal privacy has continuously eroded, and we now are plugged in to the "global organization" at all times.

The innovations we will consider here are largely those involving electricity: the telegraph, telephone, radio, and computer. The social impact of these media is so great that we will confine ourselves here to only one aspect of this impact—how communications media influence social and psychological distancing. In our approach to this subject, let's consider one of the most striking examples of the social impacts of a communications medium.

You are sitting at home one night, and the telephone rings. The person on the other end of the line says that she represents the (fictitious) Acme Bank and Trust Company, and that they are interested in giving you a credit card. But first you must answer a few questions, which you do, and in due time the credit card arrives in the mail. You now have a credit account for $1,500 with this bank, which you have never seen and never will see. Nor will you ever meet the person who, by calling you on the phone, has made you a credit-card holder able to spend the bank's money on loan. That something like this can actually happen is, in social terms, a minor miracle. In many societies, it would seem incomprehensible.

Traditionally, lenders extend credit only to people they have known for a long time. Even in societies advanced enough to have banks, loans were granted only after a personal interview and only to known members of the community, or to members of another community who could provide local references. In such societies, business deals take place only face-to-face, usually between people who know each other, and sometimes only when they have known each other for a long time. The issuance of a credit card over the phone is unthinkable in such a context.

But consider why, in the context of modern society, it is not only thinkable, but natural. In modern society information circulates freely. By making a few telephone calls about you to credit bureaus, credit-card companies, and banks, the would-be lender can get a wealth of information about you, information that in terms of reliability is in fact far superior to what could be gathered during an interview. The person granting the loan may be a conservative "straight arrow" kind of person, and you may be a "far-out deviant," but if you have paid your bills on time you have a good credit rating, and as a result you can get a credit card.

An illustrative incident occurred while I was ordering a pizza from a friend's house. (Please note that the ability to make the order in this way is a gift of the telephone.) Just after I placed the order, I was astonished to hear the person on the other end say, "OK, 2510 Edgerow, that's a single-story red brick, right?" His information came not from an encyclopedic knowledge of the neighborhood, but from a computer data base available at the push of a few keys. That a pizza firm should have this kind of information is staggering. The rationale is obvious: Pizza deliverers should know what kind of situation they are getting into, and in any case they need to be able to find the house in the first place. But the implications are disturbing. Who else can get access to this information, and what else can they do with it?

From these examples it is not surprising to discover that every day hundreds of billions of dollars change hands electronically in ways that are almost impossible to control and that can be subject to serious tampering.[6] The current international economy is very much dependent on a flow of electronic signals representing dollars, marks, francs, yen, and so forth. The basis for the transfers is that money, already a symbol, is represented by other symbols. The process of abstraction has already reached a frightening level, and it is likely to continue as new computer systems and financial innovations are put into place.[7] When the British created a new stock exchange that operates largely by computer, the old one was virtually deserted within a week.

Furthermore, we increasingly find that people prefer to do business with machines rather than with people, as an increasing interest in self-service testifies.[8] Automated bank tellers, shopping by mail and computer, and pump-your-own gas stations have cut into service occupations—or rather, the high cost of employing service personnel has stimulated the replacement of the person by the robot. And the apparent preference for interacting with machines extends into people's private lives. When single people arrive home in the evening, often their wish is to be alone with their machines: their VCR, their compact disc player, their answering machine, and their microwave oven. The personal support system is becoming increasingly mechanical. Machines have a predictability that is often appealing to the stressed-out white-collar worker who toils in a large organization. The machine yields what people seldom do: complete control, at least until the machine breaks down.

The Effects of the Stretch Principle

George Kirkham, a professor of criminology who gave frequent lectures on the nature of police work, was once challenged by one of his students, himself a police officer. The student pointed out that Kirkham had never personally been out on patrol, and therefore he could not really appreciate the kinds of problems officers faced. Kirkham accepted the challenge, went to the police academy, trained, and went on patrol in the ghetto.[9] Not surprisingly, Kirkham discovered that police work, and the human reactions to it, looked very different from the street than they did from the classroom. The police behavior that Kirkham had previously viewed as paranoid now seemed perfectly understandable, given the stresses to which police officers were subjected on the street. Kirkham had discovered a very basic principle: Being there was very different from reading about being there. But the real issue this example raises is how Kirkham could be allowed to teach about something he had never personally experienced. The educational system had obviously allowed a kind of participation at a distance, a distance that allowed Kirkham to ignore the real situations his lectures covered.

Technology also allows participation at a distance. Technology allows us to do things more safely, quickly, and easily. Yet often technological change results in only a temporary improvement in such matters as safety. The problem is that as a technology improves our capabilities, we experience pressure to depend on it—the technology can be used, therefore it must be used. New safety devices lower risks, but they also allow us to try things that are riskier. We now have "fly-by-wire" commercial airliners, in which control surfaces are actuated not by mechanical connections, but by electronic signals. Furthermore, planes can be flown almost entirely by autopilot.[10] We might wonder how long we will continue to use pilots. Thus the stretch principle—that increased capabilities lead to novel risks—allows us to go further out on a limb.

This point was very forcibly brought home to me recently while attending a conference on "Safety Management and Risk Control." After a long and somewhat daunting catalogue of technological disasters and near-disasters, the chairman of the conference, an eminent British manager, in fact a former Managing Director (equivalent to our "chief executive officer"), got up to give his reactions to the conference. One of his observations went something like "We managers are used to managing at the edge. Now that we have a better idea of where the edge *is*, we can manage closer to it." His attitude is by no means unique.

Here's the basic way the stretch principle works: Without continuous technological change, institutions might well reach a kind of equilibrium, and our lives would stabilize due to balancing pressures from existing social institutions. Technology, however, eases the pressure from one set of institutions: It allows us to handle the same demands with less effort. As a result, we tend to expend more energy on those institutions whose demands are not relaxed by a given innovation. So, for instance, when there are conflicting demands between work and family, we tend to use technology to help meet the demands of one of these spheres. Although technology may allow only a *minimal* response to the situation's demands, it allows us to devote ourselves more fully to the institution that technology will not help. Too often it is the family that gets the technological first aid, while work gets a more complete commitment.

An important result of the stretch principle is that communication allows us to stretch our personal influence as far as wires or radio signals will allow. We can operate remote vehicles on the moon (240,000 miles away) and even change the course and attitude of satellites as far away as Jupiter. But in doing this we do not necessarily get good or timely information on the effects of our actions, for remote control can also imply a remoteness from the effects. Consider what Albert Speer wrote about Adolf Hitler's long-range supervision of his armies:

> The communications apparatus at headquarters was remarkable for that period. It was possible to communicate directly with all the important theatres of war. But Hitler overestimated the merits of the telephone, radio, and teletype. For thanks to this apparatus the responsible army commanders were robbed of every chance for independent action, in contrast to earlier wars. Hitler was constantly intervening on their sectors of the front. Because of

this communications apparatus individual decisions in all the theatres of war could be directed from Hitler's table in the situation room.[11]

Thus Hitler made decisions about the Russian front that made sense in Berlin, but not in the subfreezing weather in which the orders would have to be carried out. During the Vietnam War, politicians in the United States often dealt with groups and numbers that were little more than figments of imagination; Vietnam was too far away, its culture too foreign, and its politics too complex for them to grasp. Yet they continued to make decisions, to order bombing and strafing sorties against people and defoliation of an environment that they had never seen, and never would see.

In a more recent instance, the guided-missile cruiser *Vincennes,* patrolling the Persian Gulf, shot down an airliner with 290 passengers aboard, under the illusion that it was being approached by a jet fighter armed with missiles. Civilians were killed from nine miles away, victims not only of war, but of a technology that allows people to kill others they cannot even see, using detection systems that can detect airplanes but cannot determine their nature or the intentions of their occupants.[12]

As Jacques Vallee has nicely put it, we are "getting out of our sphere."[13] We acquire the ability to communicate over great distances, but we do not have the benefit of being there—of feeling what it's like or seeing the face of the person we are talking to—and we do not have to face the consequences of the words we say or the missiles we launch.

The Social Effects of Technological Distancing

The family is affected by technological stretching, and one of the impacts of "convenience technology" may well be the destruction of community. The coming of the telephone began the unraveling of social processes. Although the telephone helped increase the sense of community for far-flung farmers and those unable to travel, for many others it meant replacing direct personal contacts with telephone calls. Robert and Helen Lynd's study of Muncie, Indiana, showed that with the advent of the telephone, visiting tended to fall off, and a half-hour chat replaced a whole morning's conversation.[14] Whereas at first the telephone supplemented face-to-face conversations, in time it began to replace them. Indeed, people discovered that a large variety of face-to-face contacts could be replaced by telephone calls.[15] With this replacement, furthermore, society began to "stretch out," and people became willing to accept physical separation so long as contact could still be maintained by telephone. But telephone contact is not the same as being there, and it creates a different kind of society from one in which people interact in person.

Convenience technologies, a development of the 1950s, made further inroads. One of the most important changes was a reduction in joint activities due to control of our physical environment. For most of human history, people have been dependent on external cycles—night and day and the seasons. Even in tropical climates, there were limitations on when we might do something, whether because of the presence of absence of water, light, or other species, or because of climatological conditions. As people have developed more elaborate technologies for creating an environment that is independent of these external conditions, the necessity to conform to these external cycles has decreased as well. But an important effect of conforming to these external cycles was the creation of community. Because people could do certain activities only when the external environment permitted, of necessity everybody had to do them at the same time. As technology developed, however, this necessity for synchronous action—working, eating, sleeping, or playing—declined. Today, because of devices that permit us to schedule activities almost at our leisure, this synchronization is no longer as important.

Let's consider for a moment the implications of synchronization. Even though I live in a small city (Ann Arbor, Michigan), I still experience traffic jams. I know that if I leave work around 5 p.m., it will take longer to get home than if I left earlier. In nearby Detroit, of course, the situation is usually much worse, and in Los Angeles it is a nightmare. People leave work at the same time because business hours are synchronized. And shortly after people leave work, they also begin to think about eating dinner; if we go to a restaurant at this time, the place will be jammed. Synchrony can mean more than doing things at the same time; it can often mean that people will do things together. Even watching television, which recent polls suggest American adults do an average of three hours a day, is often done together because a program is on at a given time.

Synchronization presents some problems for us, for often several people want to use the same facilities—the highways, the restaurants, the bathroom—at the same time. But synchronization also allows a sense of community to develop. When we do such things as eating dinner, watching television, or working together, we are forced to take account of one another, to become aware of one anothers' concerns. This makes us feel part of a group, family, or a wider community. Such interaction can also produce friction, but conflict is often a necessary part of life with others.

High technology, however, allows a different system. With a microwave oven, dinner can be prepared at one time and consumed at another, often by different people.[16] The necessity to arrive home at "dinner time" and eat dinner with other family members no longer exists. Similarly, with video recorders we do not have to watch television programs when they are broadcast, for they can be recorded and then watched whenever an individual so desires. Telephone messages can be recorded and listened to when people arrive home or when they call in to collect their messages via electronic cueing. American homes now are filled with such electronic gadgets (see Table 14-1). All this is a tremendous help to busy people—to the two-career family or couples who are on different physiological schedules ("early birds" vs. "night owls"). The convenience of such systems is enormous, but there is a cost. Whereas meals can be prepared, set aside, and rewarmed, human interaction cannot be.[17] The vital social function of interacting with others face-to-face, in bodily proximity, begins to disappear.

**Table 14-1   Electronic Systems Owner-
ship in American Homes**
(Estimated percentages of
households as of June
1988)

Radios	98%
Television	98%
Color TV	94%
Audio systems	90%
Microwave ovens	66%
Videocassette recorders	56%
Home computers	21%
Telephone answering machines	20%
Stereo-adaptable color TV	12%
Alarm systems	10%
Compact disc players	10%
Camcorders	5%

SOURCE: New York Times, *Dec. 18, 1988. Copyright © 1988 by the New York Times Company.
Reprinted with permission.*

Thus, technology allows us to do things we could not do before, we stretch our limits to what the technology will allow. Now that we have jet planes, we can have a married couple that lives 250 miles apart and rationalizes that because they can get together on weekends, they can both enjoy careers, even if the careers are physically separated. What this does to marriages is often not very good. Similarly, because the family doesn't have to eat together, the pressure on either spouse to stay late at the office, or for the children to arrive late, is often accepted. Activities that were formerly done together now can be done separately; although this may be a great convenience, it is also a great hazard to our emotional health and to the health of our relationships. All this takes place because of the stretch principle.

The family is particularly at risk in such a situation, for so much of our technology is likely to be used to the family's detriment. If food can be microwaved, other demands are likely to take precedence. If telephone messages left on the recording machine can be retrieved from afar, do we really need to go home at all? Children's needs, too, can be deferred until it is more convenient to meet them; out of this notion grew the concept of "quality time." A recent newspaper article discussed a service through which children left home alone can "phone a friend" who will talk to them sympathetically. We thus become more distant from one another.

Technology also provides excuses for noninteraction. Because technology allows relationships to stretch, other institutions' demands may *force* relationships to stretch. And they often do. "Professional" occupations frequently require late hours at the office, trips to remote places, and paperwork at home after hours. Such jobs may even force people to move to another state at the company's whim. Technology allows relationships to stretch in time and space as institutional demands put pressure on them. Unfortunately, the stretches that result may degrade a relationship's quality.

Subtly, too, these technologies allow the destruction of community because they encourage far-flung operations and far-flung relationships. Studies of frequently moved managers' families have shown that these people have little involvement with local institutions and have expendable relationships with friends, including those of the children.[18] We seem to have moved into an era in which we have intimate relationships with organizations and distant ones with our families. The complex organization is now becoming a focus of loyalty for employees who consider its goals to be as important as those of a community or even a nation. But this loyalty is rarely returned by the corporation, which often hires and fires them as economic convenience dictates.

Distancing and Responsibility

Theodore Taylor is a very unusual person.[19] For many years he designed some of the most efficient atomic weapons in the U.S. arsenal. He is a brilliant technologist, and he has now turned his attention to designing "appropriate technologies" instead of weapons of war. But what makes Taylor particularly intriguing is that he has always been a pacifist. Even while designing nuclear bombs, he was hoping that they would end the possibility of war, and apparently his weapons design work was violently at odds with his pacifist inclinations. The split in Taylor's soul between his values and his activities is worth considering, for it mirrors, in exaggerated form, the split between our moral standards and the activities in which we frequently find ourselves engaged. For Theodore Taylor, a confrontation between his activities' potential victims and his values never took place. Thus he could *distance* himself from the consequences.

When someone puts a revolver to another person's head and pulls the trigger, all of us can agree that he or she has committed murder. But suppose someone working in a factory is making pitons—metal pins used by climbers to hold ropes on a mountainside—and this person ignores a defective unit, which then gets shipped out. The piton breaks in use, and a climber falls to her death as a result. Guilt here is less evident. Could the maker have anticipated that the defective product would *kill?* Did he or she know the victim? Even if the answer to these questions is no, there is still a serious crime here—perhaps not murder, but at least criminal negligence.

Suppose the product involved is a brake that goes on a school bus, and when the bus's brakes fail at speed, a busload of schoolchildren dies. Here the offense seems more serious, but only because of the number of people involved. What we see here is a characteristic change in the perception of guilt, both on the part of those responsible and of society at large. This difference in perception is due to the removal of the victims from those responsible by time, space, and anonymity. The killer who kills face-to-face offends us more; but equal or greater damage may be done by those whose victims are removed by time, distance, and lack of contact.

In a sense these problems have existed since the social division of labor first began to mean that one person had to use the products of another person's work. But high technology poses this problem of relation between agent and victim with special force. It presents us with a much higher standard of responsibility, for the results of our actions may be almost endlessly ramified. We encountered this phenomenon when we discussed how a "minor" software bug resulted in the death of two medical patients due to a radiation overdose. Obviously, the use of such radioactive substances as radium and plutonium, with half-lives in the tens of thousands of years, changes the standards of responsibility.[20] An abandoned medical machine or factory becomes a frightening time bomb, waiting for unfortunate victims to set it off.[21] Toxic-waste dumps may similarly harm people many years in the future, often, we might note, as a result of the most careless actions. Technology, then, can create serious dangers at the same time that it creates psychological distance between agents and victims. Although we may feel responsible for effects that we can see, we may feel less responsible for those effects that are distant in time, space, and familiarity.

Technocrime. In the chapter on sponsors we noted that quite often manufacturers do not have a sense of responsibility for a product that has left their hands. The law may hold them accountable later, but this threat is rarely great enough to discourage everyone, especially not those on the brink of bankruptcy. After all, careless or criminal workers will not see the victims of their negligence—"It's someone else's problem now" is a common sentiment. It is easy to fail to connect the release of a batch of defective products with far-off plane crashes and poisonings.[22]

The complexity of modern systems means that a system's safety and integrity may rest on the quality of very small parts. Yet often the practices by which we supply even the nuts and bolts of our systems mean that the systems are unsafe. The Navy recently tested many of the bolts in the stocks of their shipyards and found that 58 percent were unable to pass the tests they were designed to pass.[23] Similarly, half the nuclear reactors in the country were discovered to have some substandard bolts in them.[24] If we cannot depend on the bolts that hold our systems together, how can we depend on the systems? Yet the people who supply the bolts see themselves as sharp businesspeople, not as criminals.

During the 1970s the Ford Motor Company received a great deal of negative publicity for a memo that had been influential in shaping decisions about the Ford Pinto. The Pinto gas tank, it had been discovered, had a certain vulnerability to rupture. The question was: How much would it cost to fix it *versus* not fixing it? Ford decided to cope with this problem by carrying out a cost-benefit analysis. The cost (including loss of market share) of fixing the problem was determined to be greater than the projected legal costs resulting from the deaths of the 200 or so victims whose families could be expected to sue the company. So, rather than incur the expense of redesigning and retrofitting a technological solution, Ford did nothing. People did die in Pinto crashes—how many is not entirely clear—and many victims' families sued. An article in *Mother Jones* magazine revealing the existence of the cost-benefit study angered many. Ford paid out some $50 million in settlements and had to face a trial on three counts of "reckless homicide."[25]

Certainly, choosing to do nothing about the Pinto gas tank was a questionable decision, but its basis is understandable. Every safety improvement has a cost.[26] If a manufacturer made a completely crash-proof car, no one could afford to buy it. So where, then, is the trade-off between safety and cost? It is certainly true that before Ralph Nader's consumer advocacy the automobile industry was less sensitive than it ought to have been about safety.[27] Although Edwin Land developed polarized glass specifically for glare-proof automobile headlights, he was never able to get the automobile industry to adopt it for that purpose.[28] And even as "safe driving" is emphasized in public service announcements, automobile advertisements have generally suggested very different conduct.[29] But safety necessarily costs money. How safe must a car be before minimum standards are met? And who sets the standards? Such were the problems with which Ford grappled. Its decision seems remarkably cold-hearted, but what if it had not carried out the calculation at all? What was wrong was not the calculation, but the spirit in which it was done. We need to encourage firms to consider the consequences of their actions, and it is also important that they consider other issues besides cost in the decisions they make.

Societal Norms for Distancing. A very reasonable, all-purpose standard was proposed by Roger Boisjoly, a rocket engineer, in discussing his experience with the space shuttle *Challenger*. He said, very simply, that he would not sign off on any piece of technology that he would not feel good about letting his wife and children use.[30] We know this standard has been violated in the United States during the atomic bomb experiments in the 1950s. Scientists associated with the experiments allowed soldiers to be exposed to doses of radiation they claimed were safe, but to which they were unwilling to subject themselves. The result is that large numbers of these soldiers are now dying of leukemia and other cancers. The scientists had miscalculated, but they had made someone else take the risk.[31] To further avoid responsibility, the government refused to allow the victims to sue the scientists and officers who had placed them in jeopardy.[32]

Technology allows a separation of what people do from the consequences of doing it. During World War II the Nazis sent a huge number of people to their deaths in gas chambers, one of the worst mass murders of all time.[33] After the war many of the perpetrators of these shocking crimes, despite their attempts to distance themselves from their actions, were imprisoned, tried, and executed. But a similarly appalling number of German and Japanese civilians were killed by Allied bombing raids. In one of the raids on Tokyo, an estimated 130,000 civilians perished in the firestorm.[34] In the raid on Dresden, Germany, the second wave of bombing was intentionally timed to coincide with the presence of fire-fighting and rescue vehicles on the streets; these too were burned up in the raid.[35] The atomic bombs dropped on Hiroshima and Nagasaki were yet another part of this campaign of destruction, and the airmen who carried out this frightful mission were seen as heroes.[36] The "free fire zones" and scorched earth campaigns in Vietnam provide still more evidence that technology permits pilots to behave callously toward civilians because they are separated by distance.[37]

Thus society judges slaughter at a distance differently from slaughter face-to-face. The anonymity of the victims allows people to feel less guilt for killing, but the victims still die. Chuck Yeager, recalling orders to strafe randomly in Germany during World War II, remembers whispering to a fellow pilot, "If we're going to do things like this, we better sure as hell make sure we're on the winning side."[38] Gunning down civilians from an airplane seems wrong, but to some it doesn't seem as wrong as gunning them down face-to-face. In a similar way the perpetrators of technological crimes feel less guilt because they do not see their victims face-to-face. The village whose water is poisoned because of a toxic chemical dump; the hundreds of people who die of cancer after a radiation accident; those who fall victim to a faulty computer chip or a structural failure—these victims do not see those responsible for their deaths, and those responsible do not see their victims. But *society* needs to acknowledge that crimes are in fact being committed and should act accordingly.

Computer Hacking
as Technocrime

The problems of social responsibility that high technology can cause were dramatically revealed during the 1980s by the rise of teenage "computer hackers." We can define a hacker as a person who through expertise with software is able to bypass security measures and use computer systems for personal amusement. Even though hacking can have serious consequences, this motive—amusement—separates the hacker from the computer criminal, who expects to get financial benefits from illegal access.[39] Activities like hacking have a considerable history—they began long before computers—but the modern tradition of hacking began at the Massachusetts Institute of Technology.[40] Students there in the 1960s wanted to use M.I.T.'s minicomputers at times and in ways that were not officially permitted. The students' response was to find ways of entering the locked rooms where the computers were kept and use them anyway.[41] Thus computer hacking began as a form of electronic "joy-riding" by which one might steal "time" on large computers, typically those that were part of university time-sharing mainframe systems.

The next form of hacking to become popular was making free long-distance telephone calls by electronically cheating the phone system. Its beginning occurred when a hacker named John Draper discovered that a small whistle, included as a prize in Cap'n Crunch breakfast cereal, would generate a pure 2600 cycle-per-second tone. This tone, when blown into a telephone at the beginning of a long-distance call, would cause the billing code to disappear. Effective enough in itself, this discovery was followed by the invention of electronic aids to such "phone phreaking." The tone generators used for such purposes were known as "blue boxes" from the color used in A.T.&T. advertisements. An article in *Esquire* magazine in 1971 spread information about Draper (since known as "Captain Crunch") and his activities, multiplying the number of perpetrators. Two college students, Steven Wozniak and Stephen Jobs, made and sold such blue boxes years before they embarked upon the more important business of developing the Apple II computer, which was to make them both millionaires.[42] A.T.&T. investigated the problem of the blue boxes and prosecuted their makers, but no doubt some "phone phreaking" still continues.

With the development of personal computers in the 1980s, however, younger hackers appeared.[43] These hackers were often teenagers, not college students, and hacking was done via telephone modem, which allowed hackers to access computers at remote sites in banks, schools, and government bureaus, and to create serious problems. Whereas earlier hackers had mostly tackled computers used in university research programs, the new hackers were tapping into systems used for everyday decision making, and their tampering involved bank accounts, school grade records, and even hospital data banks.[44]

The possibilities for abuse were enormous, and in some cases the welfare of others was seriously damaged. One teenage group, known as the 414's (their area code), became especially notorious. After they penetrated a computer at Sloan-Kettering Cancer Institute, a computer operator became suspicious and notified the FBI, which began to monitor the activities of the group. As the investigation proceeded,

> it became apparent that the 414's had probably been involved in at least sixty other computer break-ins involving corporate and governmental computers in the United States and Canada. Among the computers violated were those of the Security Pacific National Bank in Los Angeles and the Los Alamos National Laboratory.[45]

Other groups have caused millions of dollars of damage to university, corporate, and governmental computer systems. In one case a group used their personal computers to alter credit ratings, insert obscenities, and attempt to "crash" the entire California telephone system.[46] Because much of the operation of the telephone system can be changed remotely, hackers who get the access codes can alter the functioning of giant switching devices from their home computers.[47] Given the variety and sensitivity of the communication and data transfer taking place through the telephone system, the possibilities for intentional and unintentional sabotage are ominous.

The development of electronic "bulletin boards" made the problems with teenage hackers even worse.[48] Access codes, credit-card numbers, unlisted telephone numbers, and other sensitive information would be displayed on the 3,000 or so national bulletin boards into which hackers could tap. Many bulletin boards are available around the clock; some are used for activities as diverse as exchanging philosophical commentary and meeting sexual partners.[49] The possibilities for mischief and crime multiply once information is exchanged between people who never meet and may therefore feel no sense of responsibility.

Hackers often use such monikers as "Phantom" or "Stainless Steel Rat," and some even attempt to sabotage each other. When a critical article on hackers appeared in a national magazine, the author was "targeted" by hackers who wanted to punish him. His telephone and credit-card numbers were prominently displayed on an electronic bulletin board with instructions to callers to use them constantly.[50] One of the biggest problems related to hackers is that many of them are teenagers who have little experience with the damage that misinformation can cause. The combination of juvenile attitudes and high-tech capabilities can be frightening. Excuses such as "I just wanted to see what I could do" sound rather hollow when a mother with a sick baby or someone reporting a fire is unable to use the telephone system because a hacker has just crashed it.

The next challenge for many hackers is tapping into signals from satellites.[51] For many this involves only passively listening in, but for others it can involve using unoccupied channels for illicit communications, such as those involved in drug trafficking. Clearly, possibilities for serious interference exist. The signals of some television broadcasts from satellites are intentionally encoded, and only those who have paid subscriptions can get access to the required decoders. One unhappy customer, John R. MacDougall, an electronics engineer, decided he was going to protest the practice of encoding. With a 30-foot radio "dish" he overpowered a Home Box Office broadcast signal, and for nearly a minute he broadcast his own defiance in the guise of "Captain Midnight." Half a million viewers were surprised to find a message on their screen that began "Goodevening HBO." For military, business, and government applications, a large amount of data is transmitted via satellite, and interference with it could cause serious problems. Already accidental interference is a problem; intentional sabotage could cause even more serious consequences.

Employee sabotage has already caused serious problems for business firms. If demented employees feel compelled to come to the office and shoot their former employers—or even, as in one recent case, crash an entire airliner—to get revenge, then one can imagine the exquisite forms of computer revenge possible. Unpaid consultants and disgruntled employees have been known to leave "logic bombs" on their firms' computers, and in some cases these efforts have wiped data banks clean after the employee has departed.[52] Computer "worms" and the even more dangerous "viruses"—programs that can reproduce themselves and will do so with every new computer with which they come in contact—are capable of spreading damage through an entire computer network.[53]

At one time viruses were largely a college prank; and their possibilities for more serious harm were hypothetical. Now these once theoretical problems have become realities. Universities are now routinely finding that students are using viruses as pranks, even in mainframe computers. But many of the other instances are much more ominous.

In one of several serious incidents, some 40 percent of the records at an Eastern medical center were erased by a "virus," which then proceeded to infect other, interconnected, computer systems.[54] In 1987 a virus spread through computer data systems in Israel over a two-month period. The program was intended to shut down computers all over the country on May 13 (Israeli independence day is May 14). The virus was discovered largely by a fluke.[55]

Such computer viruses, because they can multiply virtually endlessly, producing harm far beyond their original targets, are the epitome of crimes in which the victims are anonymous. Although "physician" programs to clean viruses out of a system exist, the problem is unlikely to go away soon.

Big Brother at Work— Distancing and the Invasion of Privacy

Much of people's fear of computers concerns governmental invasion of citizens' privacy.[56] Our government has expanded its ability to use computers for tracking criminals.[57] Businesses also use sophisticated computerized tactics to target, monitor, and prod consumers to buy.[58] But an important aspect of computers' new capabilities is their ability to monitor those inside as well as outside organizations.[59] In some ways the use of computers has enhanced the jobs of workers,[60] but computers are likely to aid employers more than employees. With the aid of high technology, supervision has become closer; computers have made it easier to watch over people. In addition to providing great geographical reach for supervisors, technology now allows much more direct and detailed monitoring, raising fears of a society like that described in George Orwell's *1984*.

Innovations in communications have always affected commerce. The advent of the telegraph and telephone allowed managers to keep closer tabs on employees. Before the telegraph, for instance, foreign commercial agents and salespeople on the road had to be given more discretion, for it was impossible to communicate with them in a timely way. The coming of the telegraph quite dramatically curtailed the power of the British viceroys in India by making it easy for Whitehall to supervise them.[61] With the coming of the telephone, even owners could supervise their managers and employees from a distance. This could create new pressures on those supervised:

Special "periods of service" from coal mining companies to their New York offices were quite popular. In a few cases their officials brought me (not for publication) the suggestion that the "periods" should not begin so early in the morning. It was hinted that the "Old Man" up at the mines (who was not infrequently the owner) began to call them up at 8:00 A.M. and it was embarrassing, sometimes, to have to explain that 9:00 or 9:30 was a usual starting time for office hours in New York.[62]

Modern branch banking was virtually created by the telephone. Once, the manager of a branch was typically a member of the owning family or had a financial share in the bank's ownership. With the advent of the telephone, however, it was possible to supervise branch managers to such an extent that they need only be mere employees.[63] And soon, they were.

At the same time, of course, the size of firms in general expanded enormously, due in part to the capability communications gave people to manage remote operations.[64] The stretch principle was operative, and businesses stretched out as far as their communication capabilities—the telephone lines—would let them. In the 1940s the invention of the computer put yet another supervision method in the hands of top management.

But with the advent of the computer, the ability to scrutinize employees grew enormously. All kinds of data could now be collected and fed into central data banks,[65] and the performance of the employee became the data generated by the computer. Soon the computer became the way to manage the entire business, especially for managers of the new conglomerate firms who knew little about the product but could always read the quarterly profit-and-loss statement. Furthermore, as computers became smaller and were given to employees to use, the machines could provide even more information on employee performance. More than one introduction of a computer system into an office failed because employees sabotaged a system they suspected, not without reason, was going to be used to monitor them.[66]

The largest group being monitored is the legion of clerical workers and computer operators, largely women and often minorities, who use computers every day. Workers have discovered that management in some companies can monitor every keystroke, a problem that is receiving increasing attention. The problem is likely to be most acute in "white-collar factories" in which highly repetitive—and therefore highly measurable—operations take place. Alan Westin, an expert on the protection of privacy, has estimated that 20–35 percent of the clerical work force is affected by such monitoring measures.[67] Such scrutiny can cause extreme stress in employees and has a serious potential for abuse.[68] Management makes claims that such monitoring is helpful to spot problems, but we might wonder if having supervisors "electronically breathing down the neck" of workers is truly very efficient.

Some offices have broken up the factory-like work design of data processing after finding out that it is actually more time-consuming, expensive, and conducive to errors.[69] Some evidence suggests that human-centered offices may not only be more pleasant to work in, but may actually be more efficient at getting the job done. If this is true, the computer may not only be a nuisance; it may be a threat to sound operation.[70] But for many operations—the back room of a bank, for instance—repetitive, boring jobs may be difficult to avoid.

Other parts of the work force can be monitored, too; grocery-store checkout clerks, for instance, can be watched by TV cameras to see if they are cheating when their friends are in line. Cameras can reveal when employees are loafing on the job, formerly a prerogative of the back-room employee.

And now performance can be measured in yet other ways. For instance, consider what is happening to truck drivers, whose machines often cost $100,000, to say nothing of their cargoes. Because truckers frequently exceed speed limits and drive for longer stretches than they should, they are prone to serious accidents, and their rigs and cargoes are in jeopardy. It has been argued that driver fatigue is the primary cause of some 41 percent of truck crashes.[71] With rigs that weigh 20 tons, a very formidable potential for harm exists. So now the truckers may well be monitored; what could be simpler than giving the trucks "little black boxes" like airplane flight recorders?[72] The boxes keep track of miles driven, speeds, and stops.[73] These small computers, costing about $1,000 apiece, can often be read directly into mainframes at the trucking headquarters, and the data can then be processed for use by various groups in the company, including management, operations, and maintenance. For the trucking companies, and perhaps for their overworked drivers as well, the computers may be useful. And overworked truckers should not be exceeding the speed limit with huge cargoes on roads filled with passenger cars. But such practices also raise problems: What kind of supervision is legitimate? Do we want to live in a society where "big brother" is watching over us on the job like this?

Conclusion

This chapter has considered only some of the ways that technology can provide communications media that separate as well as unite people. As we introduce more technology into our lives, as society becomes more "convenient," we must ask ourselves some questions about what is being lost. We have not discussed television—a major problem in itself—nor have we speculated about what future devices are likely to bring. The trends we have considered here need more examination, for we are increasingly giving up our relations with people in favor of our relations with machines. We must examine what is being traded to achieve these gains in convenience and in instant (and impersonal) gratification. The disturbing thing is that there seems to be relatively little resistance to these technologies that are transforming our lives and altering our communal existence. I must confess that it bothers me when I see people walking around with earphones on their heads. Where will it end?[74]

Notes

1. Desmond Morris, *Manwatching: A Field Guide to Human Behavior* (New York: Abrams, 1977).

2. Marshall H. Klaus and John H. Kennell, *Maternal-Infant Bonding* (New York: Mosby, 1976), p. 6.

3. For earlier treatments of this topic, see Ralph Keyes, *We, the Lonely People* (New York: Harper & Row, 1973); and Philip Slater, *The Pursuit of Loneliness: American Culture at the Breaking Point* (Boston: Beacon Press, 1970).

4. Consider, for instance, the interesting communication among Serbian shepherds achieved by earth-pounding, discussed by Michael Pupin in his book *From Immigrant to Inventor* (New York: Scribner, 1960), pp. 15–16; see also John F. Carrington, "The Talking Drums of Africa," *Scientific American,* Vol. 225 (Dec. 1971), pp. 90–94.

5. For instance, see Geoffrey Wilson, *The Old Telegraphs* (London: Phillimore, 1976).

6. Stewart Brand, "The World Information Economy," *Whole Earth Review,* Winter 1986, pp. 88–97.

7. David E. Sanger, "Wall Street's Tomorrow Machine," *New York Times,* Oct. 19, 1986, Business Section, pp. 1, 30.

8. Claudia H. Deutsch, "The Powerful Push for Self-Service," *New York Times,* Apr. 9, 1989, Section 3, pp. 1, 15.

9. See George Kirkham, *Signal Zero* (New York: Ballantine Books, 1976).

10. Richard Sandza, "The Phantom of the Cockpit," *Newsweek,* July 17, 1989, p. 61.

11. Albert Speer, *Inside the Third Reich* (New York: Macmillan, 1970), p. 304.

12. Stephen Engelberg, "Downing of Flight 655: Questions Keep Coming," *New York Times,* July 11, 1988, pp. 1, 4.

13. Jacques Vallee, *The Network Revolution: Confessions of a Computer Scientist* (Berkeley, Calif.: And/Or Press, 1982), pp. 12–26. I have found this to be a most thought-provoking book on the "computer revolution."

14. Robert S. Lynd and Helen M. Lynd, *Middletown: A Study in Modern American Culture* (New York: Harcourt, Brace, 1956), p. 275.

15. See Sidney H. Aronson, "The Sociology of the Telephone," *International Journal of Comparative Sociology,* Vol. 12 (Sept. 1971), pp. 153–167; see also his "Bell's Electrical Toy: What's the Use? The Sociology of Early Telephone Usage," in Ithiel de Solla Pool (Ed.), *The Social Impact of the Telephone* (Cambridge, Mass.: MIT Press, 1977), pp. 15–39.

16. The damaging impact of the microwave on convivial eating is brought out in Dena Kleiman, "Fast Food? It Just Isn't Fast Enough Anymore," *New York Times,* Dec. 6, 1989, pp. 1, 24.

17. I shudder to think that mechanical surrogates for sex are the next step, but insiders assure me that such products are already in the developmental stage.

18. Diane Rothbard Margolis, *The Managers: Corporate Life in America* (New York: Morrow, 1979), pp. 143–259.

19. For more about Theodore Taylor, see John McPhee, *The Curve of Binding Energy* (New York: Ballantine Books, 1974).

20. It is very sobering to realize that today archeologists are considering how to mark sites of radioactive dumps because they will be dangerous so far in the future that we cannot count on the earth's inhabitants recognizing any of our current languages or symbols; see Maureen F. Kaplan, "Using the Past to Protect the Future: Archeology and the Disposal of Highly Radioactive Wastes," *Interdisciplinary Science Reviews,* Vol. 11, No. 3 (Sept. 1986), pp. 257–268.

21. Leslie Roberts, "Radiation Grips Goiania," *Science,* Vol. 238, (Nov. 20, 1987), pp. 1028–1031; and David E. Pitt, "New York Radium Supplier Faulted on Safety by Legislative Committee," *New York Times,* Oct. 4, 1987, p. 18.

22. Arthur Miller's drama *All My Sons* made this point very forcefully. For current examples, see David Sylvester, "The Dark Side of the Force," *Washington Monthly,* Feb. 1985; and Steven Strasser, Kim Willenson, and Peter McAlevey, "A Farewell to Faulty Arms," *Newsweek,* Sept. 24, 1984, pp. 30–31.

23. Associated Press, "Committee Finds Faulty Bolts Plague Military and Industry," *New York Times,* July 31, 1988, p. 18.

24. Ben A. Franklin, "Defective Bolts Are Found in Half of Nuclear Reactors," *New York Times,* June 19, 1988, p. 16.

25. Lee Patrick Strobel, *Reckless Homicide: Ford's Pinto Trial* (South Bend, Ind.: And Books, 1980); Richard T. DeGeorge, "Ethical Responsibilities of Engineers in Large Organizations: The Pinto Case," *Business and Professional Ethics Journal,* Vol. 1, No. 1 (Fall 1981), pp. 1–14; Malcolm E. Wheeler, "Product Liability, Civil or Criminal—The Pinto Litigation," *A.B.A. Forum,* Vol. 17, No. 2 (Fall 1981), pp. 250–265; and Ford News Release of Sept. 26, 1977, available from the Office of the General Counsel, Ford Motor Company, Detroit, Mich.

26. Dekkers L. Davidson, "Managing Product Safety: The Ford Pinto," Harvard Business School Case Number 383-129 (1984).

27. Jeffrey O'Connell and Arthur Myers, *Safety Last: An Indictment of the Auto Industry* (New York: Random House, 1966).

28. Mark Olshaker, *The Instant Image: Edwin Land and the Polaroid Experience* (New York: Stein and Day, 1978), pp. 25–26; and Peter Wensberg, *Land's Polaroid: A Company and the Man Who Invented It* (Boston: Houghton Mifflin, 1987), pp. 63–64.

29. Jeffrey O'Connell, "Contradictions in Automobile Advertising," *Columbia Journalism Review,* Fall 1967, pp. 21–28.

30. Roger Boisjoly, "Ethical Decisions—Morton Thiokol and the Space Shuttle *Challenger,*" ASME Paper 87-WA/TS-4 (New York: American Society of Mechanical Engineers, Dec. 1987), p. 11.

31. Thomas H. Saffer and Orville E. Kelly, *Countdown Zero* (New York: Putnam, 1962), pp. 19–52.

32. David G. Savage, "Justices Widen Federal Immunity to Lawsuits," *Los Angeles Times,* Jan. 12, 1987.

33. See Heinz Hohne, *The Order of the Death's Head: The Story of Hitler's SS* (New York: Coward-McCann, 1970).

34. Martin Caidin, *A Torch to the Enemy* (New York: Ballantine Books, 1960), p. 153.

35. Dieter Georgi, "The Bombings of Dresden," *Harvard Magazine,* Mar.-Apr. 1985, pp. 56–64.

36. Howard Zinn, *The Politics of History* (Boston: Beacon Press, 1970), pp. 268–269.

37. Slater, "Kill Anything That Moves," in his *The Pursuit of Loneliness,* pp. 29–52.

38. Chuck Yeager and Leo Janos, *Yeager: An Autobiography* (New York: Bantam, 1985), p. 80.

39. Thomas Whiteside, *Computer Capers* (New York: New American Library, 1978); and Donn Parker, *Crime by Computer* (New York: Scribner, 1976). The average size of a reported theft by computer manipulation is $500,000, which makes computer crime one of the most lucrative forms of theft; see Milton Maskowitz, Michael Katz, and Robert Levering (Eds.), *Everybody's Business: An Almanac* (San Francisco: Harper & Row, 1982), p. 422.

40. Steve Levy, *Hackers: Heroes of the Computer Revolution* (Garden City, N.Y.: Doubleday, 1984), pp. 3–25.

41. The lock-picking aspect of hacking may be considerably older than the computer genera- tion. For instance, while my father was a graduate student in chemistry at University of California, Berkeley, in the 1930s, he had the reputation there of being able to get into any room in the building, locked or not. He explains that many students felt that if you weren't able get the chemicals you needed by getting into rooms after hours, you proba- bly didn't belong at Berkeley. Similarly, when Bill McLean (see Chapter 6) was at Cal Tech, lock-picking of other students' rooms (with appropriate telltale clues left behind) was equally in vogue.

42. Doug Garr, *Woz: Prodigal Son of Silicon Valley* (New York: Avon, 1984), pp. 47–56.

43. August Bequai, *Technocrimes: Computerization of Crime and Terrorism* (Lexington, Mass.: Heath, 1987), pp. 29–44.

44. William Marbach, Madlyn Resener, John Carey, Richard Sandza, Michael Rogers, Jennet Conant, and Susan Agnest, "Beware: Hackers at Play," *Newsweek,* Sept. 5, 1983, pp. 42–48; Judith Cummings, "Computer-Falsified Grades Raise New Issue at Colleges," *New York Times,* Feb. 10, 1985, Section I, p. 22.

45. Bequai, *Technocrimes,* p. 40.

46. Bequai, *Technocrimes,* pp. 34–35.

47. John Markoff and Andrew Pollack, "Personal Computer Users Penetrating Nation's Phone System," *Ann Arbor News,* July 24, 1986, Section B, p. 13 (reprinted from the *New York Times*).

48. Bill Landreth, *Out of the Inner Circle* (Bellevue, Wash.: Microsoft Press, 1985). This book, written by a former hacker, discusses the psychology of computer hacking.

49. Richard Sandza, "The Night of the Hackers," *Newsweek,* Nov. 12, 1984.

50. Richard Sandza, "The Revenge of the Hackers," *Newsweek,* Dec. 10, 1984, p. 81. This ar- ticle reported hackers' responses to Sandza's article published just a month earlier.

51. Michael Rogers and Brad Risinger, "Uplinks and High Jinks," *Newsweek,* Sept. 29, 1986, pp. 56–57.

52. Peter J. Ognibene, "Computer Saboteurs," *Science Digest,* July 1984, pp. 59–61.

53. Lee Dembart, "Attack of the Computer Virus," *Discover,* Nov. 1984, pp. 90–92.

54. William D. Marbach, Richard Sandza, and Michael Rogers, "Is Your Computer Infected?" *Newsweek,* Feb. 1, 1988, p. 48.

55. Vin McLellan, "Computer Systems Under Seige," *New York Times,* Jan. 31, 1988, Business Section, pp. 1, 8.

56. Gary T. Marx and Nancy Reichman, "Routinizing the Discovery of Secrets," *American Behavioral Scientist,* Vol. 27, No. 4 (Mar.-Apr. 1984), pp. 423–452.

57. Elaine Shannon, "Taking a Byte Out of Crime," *Time,* May 25, 1987, p. 63.

58. Robert Samuelson, "Computer Communities," *Newsweek,* Dec. 15, 1986, p. 66; Anonymous, "Mishmash of Laws Gives Consumers Little Defense Against Computer Calls" (UPI story), *Ann Arbor News,* Apr. 23, 1986, Section D, pp. 1, 2.

59. Michael J. Smith, "Electronic Performance Monitoring at the Workplace: Part of a New Industrial Revolution," *Human Factors Society Bulletin,* Vol. 31, No. 2 (Feb. 1988), pp. 1–3.

60. Computers, especially personal computers, can provide jobs that require greater knowledge and reasoning skills; see, for instance, Mitchell Fleischer and Jonathan A. Morell, "The Organizational and Managerial Consequences of Computer Technology," *Computers in Human Behavior,* Vol. 1, No. 1 (1985), pp. 83–94. In this article, the use of personal computers was shown to decentralize decision making in an insurance company, just as a generation before, mainframes had tended to centralize the process.

61. John Cell, *British Colonial Administration in the Mid-Nineteenth Century: The Policy-Making Process* (New Haven, Conn.: Yale University Press, 1970), p. 43.

62. Angus Hibbard, *Hello-Goodbye: My Story of Telephone Pioneering* (Chicago: McClurg, 1941), p. 84.

63. Fritz Redlich, "Bank Administration, 1780–1914," *Journal of Economic History,* Vol. 12, No. 4 (1952), pp. 438–453.

64. This point, I feel, has not been sufficiently addressed in Alfred Chandler's otherwise masterful book, *The Visible Hand: The Managerial Revolution in American Business* (Cambridge, Mass.: Harvard University Press, 1978). But see Joanne Yates, "The Telegraph's Effects on Nineteenth Century Markets and Firms," *Business and Economic History,* Vol. 15 (1986), pp. 149–193.

65. Frederic Withington, *The Real Computer: Its Influence, Uses and Effects* (Reading, Mass.: Addison-Wesley, 1969), pp. 181–192.

66. See, for instance, Gary Albrecht, "Defusing Technological Change in Juvenile Courts: The Probation Officer's Struggle for Professional Autonomy," *Sociology of Work and Occupations,* Vol. 6, No. 3 (Aug. 1979), pp. 259–282.

67. William Booth, "Big Brother Is Counting Your Keystrokes," *Science,* Vol. 238 (Oct. 2, 1987), p. 17.

68. Susan Gilbert, "What Are the Limits on Electronic Monitoring of Workers' Behavior?" *Christian Science Monitor,* June 4, 1987, p. 27.

69. Jonathan Schiller, "Office Automation and Bureaucracy," *Technology Review,* July 1983, pp. 32–40.

70. S. L. Smith, "Information Technology: Taylorization or Human-Centered Office Systems," *Science and Public Policy,* Vol. 14, No. 3 (June 1987), pp. 159–167.

71. William E. Schmidt, "Sharp Rise in Truck Crashes Prompts Action Across U.S.," *New York Times,* Dec. 7, 1987, pp. 1, 15.

72. Previous solutions have included the tachograph, a recording device that used a paper disk to keep track of truck speeds; see Anonymous, "A Backseat Aid for Truck Drivers," *Business Week,* June 19, 1978, pp. 110G–110I.

73. Robb Deigh, "The Little Computer-Overseer Threatens Trucking Tradition," *Insight,* June 1, 1987, pp. 20–21.

74. In the 1950s Ray Bradbury's science fiction novel *Fahrenheit 451* predicted much of the technology we are now seeing. Bradbury was disturbed when transistor radios came out and teenagers could be seen driving around with radios plugged into their ears; this new development reminded him too much of his book.

PART FIVE

Monitoring Technology: Shaping the Present & the Future

Social Control
of Technologies

Previous chapters have pointed to a very serious difficulty: How do we control the problems associated with the technologies we use? In this chapter we will focus on the issue of control. We will examine a variety of social institutions whose purpose is to control technological problems. This examination will show that many of our current institutions are inadequate for controlling the technologies we use. They often lack essential knowledge, legitimacy, and power, and often they must deal with powerful sponsors, complex technologies, and obsolete laws. In this situation they cannot function, as they should, to monitor, regulate, and protect. In the face of difficulty they often fail.

But the institutions are still valuable; even if such institutions of control are inadequate, they are not useless. In fact they frequently do have impacts, and sometimes decisive ones. And they can be very much improved, if we are willing to make the effort to do so.

To develop more adequate institutions of control is an important goal for any society, for control is the counterpoise to creativity. Just as we want a society with strong creative institutions, we want one with strong control institutions. If the creative institutions are too powerful, society is likely to be overwhelmed by poorly assimilated innovations. If the control organizations are too strong, society is likely to be throttled by its own protective mechanisms. Finding the right balance is difficult, but it is also crucial.

Control institutions' effectiveness hinges on three things: (1) getting the relevant information, (2) having the intellectual manpower to process the information correctly, and (3) having the legal or social power to compel action toward desired behaviors. Institutions of control must have information, talent, and power if they are 'o benefit society. We might regard them as safety devices built into the social fabric. Our health, freedom, and ultimate survival depend on these institutions. Thus it is very important for social scientists to see how these institutions work and, whenever they are not adequate, to improve them.[1]

In this chapter we will examine a number of these institutions, but we will only scratch the surface of a very deep subject. We will begin by looking at a case study, and then we will consider some aspects of social control of technology, such as ethics, regulation, whistle-blowing, and confidential feedback systems.

When Sponsors Are Careless: A Case Study of Technology Out of Control

To understand the kinds of pressures control institutions must face, we will first examine an incident in which virtually all the safeguards set up to protect citizens broke down in the face of a network of careless sponsors. In 1973 some very dangerous chemicals were widely disseminated across Michigan.[2] These chemicals were polybrominated biphenyls (PBBs), the toxicity of which is measured in parts per million.[3] Although large doses of PBBs produce manifestly negative effects on cows and humans, it is believed that PBBs in small amounts can cause cancer. Most of the people living in Michigan during this period were exposed to PBBs, and they will always retain these accumulated chemicals in their fat tissue, for PBBs cannot be digested and excreted. In fact, they come very near to being indestructible because they have a half-life on the order of thousands of years. How did this poisoning take place?

PBBs are used as a fire retardant. They go into such products as plastic dashboards so that if a dashboard is exposed to fire, it will not burn. But PBBs are also very toxic, so the decision to manufacture them involved a certain degree of peril.

This decision was made by a small company called Michigan Chemical Corporation, a subsidiary of a larger corporation called Velsicol, which in turn was a subsidiary of the still-larger Northwest Industries. When Michigan Chemical decided to manufacture PBBs, they were known to be toxic to mammals. Two larger firms, Dupont and Dow Chemical, had decided not to manufacture PBBs because of their toxicity; they were afraid of lawsuits should people be poisoned by PBBs.[4] Michigan Chemical, however, decided to take the risk.

At the time that Michigan Chemical was making PBBs, they also began to make another product, magnesium oxide, that had a similar appearance. The PBBs were marketed under the name "Firemaster," the magnesium oxide under the name "Nutrimaster." Magnesium oxide was fed to cattle because it both helped to prevent a serious disease called wheat pasture fever and increased the butterfat content of their milk. The magnesium oxide was mixed into cattle feed (at about 8 lbs. per ton) by Farm Bureau Services, a corporate subsidiary of the Farm Bureau, a cooperative organization.

Both Nutrimaster and Firemaster were shipped in brown bags with black stenciled lettering. On May 2, 1973, a mix-up occurred. A forklift operator for Michigan Chemical included at least 10–20 50-lb. bags of Firemaster in a shipment of salt to Farm Bureau Services. The exact amount is impossible to determine, but roughly a ton of PBBs were thought to have been shipped. Because the name "Firemaster" looked very similar to "Nutrimaster," the PBBs got mixed into the cattle feed. And because Farm Bureau Services distributed cattle feed over a wide area, many farms received contaminated feed, and large numbers of cattle got sick and died. In the meantime, however, their milk and meat were consumed by human beings, and PBBs became spread throughout the food chain in Michigan. Within a few months, the population of Michigan had become seriously contaminated with PBBs.

Let's pause at this point and consider the chain of events so far. A small company decides to produce a hazardous product. Through a human error the product is distributed and fed to animals whose meat and milk people eat. Because of the highly interconnected commercial institutions in American society, this contaminated food is quickly spread over the state. Thus corporate decisions can put large populations at risk, and only the threat of lawsuits exists to deter such actions. Human error combined with mass production and distribution can turn this risk into a serious accident.[5] (We should also note that Michigan Chemical disposed of another 160 *tons* of PBBs by burying some of it and by dumping some of it into a nearby river. Judging from the damaging effects of the original ton of PBBs, this company has clearly bequeathed a dangerous legacy to the people of Michigan.)

But the accident was only the beginning. As word of sick cows spread, the authorities at first refused to acknowledge that anything was amiss. The Farm Bureau, the state Department of Agriculture, the U.S. Public Health Service, the federal Food and Drug Administration, and finally Michigan State University all showed relative indifference to the problem. Only the initiative of private citizens, including a remarkable veterinarian, Alpha Clark, and a farmer named Rick Halbert, finally brought the problem into public view.

Many farmers had to kill their animals. Some were compensated by the government; some were not. Well over 22,000 cattle died or were destroyed because of the poisoning, and millions of chickens were killed.[6] Some of the dead cows in turn contaminated the farmland where they were buried. No one—not Farm Services Bureau workers, not farmers, and not Department of Agriculture representatives—was used to working with substances toxic at the parts-per-million level.

Many farmers sold cows that seemed to be diseased. At first, it was only a food production or an environmental problem, but what about the contaminated people? They needed to be examined, diagnosed, and treated, but the authorities were very reluctant to do so. Michigan has powerful beef and dairy industries, and it had a governor who definitely did not want to see Michigan meat and dairy products quarantined. So getting proper attention paid to the problem was not easy. Many of the appropriate authorities proved to have direct or indirect ties to Michigan agricultural interests, and so they were reluctant to stir up trouble. The problem still persists, and the authorities are still very reluctant to admit that serious contamination remains.

This episode was a widespread technological problem that touched the lives of almost everyone living in the Lower Peninsula of Michigan. It was caused by corporate decision, by human error, and by the commercial structure of mass production and distribution, and it was made worse by conflicting interests, economic pressures, indifference, fear, and ultimately a willingness to place private interests above public ones. Without the dedicated and resourceful work of private citizens, the problem might have gone undetected much longer. The PBB event is worth studying because, in a world filled with toxic chemicals, there will be other such accidents. Every strategy for protecting society against the dangers of technology must take into account the reality of what powerful organizations did in this case.

In considering the PBB case, then, we must note that the hazards posed by risky technologies are shaped by the institutions that are their sponsors. In a larger sense, the PBB case is literally "business as usual." Because so many of the technologies in our society are managed by private firms, we must try to understand something of the internal rules by which such firms operate if we are to understand the nature of the hazards we face. Whenever risky technologies are managed by profit-oriented firms on the economic margin, we should not expect that scrupulous attention will be devoted to safety issues. As we will see in Chapter 17, those who make decisions about technologies often do not take a very long-term view of potential consequences. We must encourage foresight and provide a broader motivation for these executives if the safety of our lives and of those who follow us is to be ensured.

Ethical Awareness: A Sense of Social Responsibility

Thus far in this book we have talked primarily about behavior and relatively little about values. But ethical values are important because they shape what people do. In principle, the primary responsibility for ensuring that technologies are sound, useful, and safe rests with every individual in society. Yet as Americans we tend to take it for granted that everyone naturally protects his or her own self-interest, even to the detriment of the public good. In many respects this is a self-fulfilling prophecy; because we assume that people will act selfishly, they feel no obligation to act otherwise. Yet people can be trained to think and act responsibility with respect to technologies, and we need to train them to think this way. If people can be encouraged to design things in a responsible manner, if organizations can be made to care about the impacts of the decisions they make, if people reflect on the impacts of the actions they take, then the technologies we deploy will be less dangerous to us. Laws are designed to protect us from people who lack a sense of responsibility, for there will always be people willing to hurt others out of malice or indifference. But laws alone are not enough; we cannot rely on them to be a panacea. Without broader social acceptance of responsibility for technological impacts, many of the problems we now face will become worse.

Learning Responsibility. Our society could do a great deal more to increase this sense of responsibility. The first task is to teach people that they are responsible for other people. This value is too often neglected in the education of children; but in a fragile world, it is very much needed. Societies get the kind of people they admire; we nourish what we cherish.[7] The heroes of our society often seem to be those who act with reckless disregard for the welfare of others. Television in particular seems to celebrate gun-toting heroes who seek quick solutions much more often than it celebrates those who cultivate more socially relevant, long-term projects. We ignore and neglect those whose quiet contributions have increased our well-being and safety.

A second task in increasing a sense of responsibility is to identify the effects of people's actions. There are two parts to this task: The first is giving people general, abstract knowledge about the world in which they live so that they can anticipate what their actions will do; the second relates to experiential learning—that is, pointing out the consequences of their actions as they occur. The factory manager who dumps toxic chemicals into a river does not stop to think that his or her children will have to drink the water, but often they will. People who drive large, powerful automobiles often have no conception of how their actions make the energy problem worse; or if they do, they may be indifferent to the problem. When the price of natural gas drops, builders of houses no longer feel it necessary to build designs that include sufficient insulation that helps retain heat. There is a lack of constructive imagination here, the inability to imagine consequences. Ignorance of basic science and of ecological issues encourages people to be irresponsible.

The Role of Feedback in Responsible Behavior. Giving people direct feedback on their actions as they occur is also important. People are less willing to injure others when they have feedback on their own actions. When people are made to feel ashamed whenever they are careless with technology, they are less likely to be careless. And people are more willing to engage in dangerous actions when they remain unaware of the dangers. The results of badly conceived technological actions must be made clear to those who engage in them.

Amasa Stone, a bridge engineer in the nineteenth century, designed a railroad span at Ashtabula, Ohio. When the bridge failed on a stormy night in 1876, a train with over 85 passengers in it crashed and burned. A jury investigating the accident found that the bridge was an experiment "which ought never to have been tried." Amasa Stone took his own life.[8] Suicide is not a cure for bridge failures, but Stone's action illustrates a sense of responsibility often absent in today's corporate world. The managers of Morton Thiokol, who authorized the launch of *Challenger* over the objections of their own engineers, could have used a dose of this personal responsibility.

Dean Wilson, a systems engineer working in Colombia, was concerned about mistakes that people made with drugs in a local hospital. He discovered a very high rate of failure in administering to patients the correct dosage of prescribed drugs. He identified a number of steps that intervened between the doctor's prescription of the drug and its final delivery to the patient. As he studied the process, he began to display the failures in graphic form, so that people could see where mistakes were made in the communication chain between doctors and patients; for instance, nurses often made mistakes copying the prescriptions.[9] Even though no punishment was given to personnel who made mistakes, the mere existence of feedback caused a behavioral change: As they realized that their actions were being recorded and monitored, they became more careful about what they did, and the accuracy of the prescriptions improved enormously. The same technique was then used to motivate personnel to change the bed linen promptly, with equally good results.[10] The mere existence of feedback, then, will often be enough to shift behavior toward more responsible action. When people are aware of the results of their actions, they act differently.

Social Pressures and the Health Care System.

An awareness of others' needs may encourage a greater sense of responsibility concerning technology than even litigation can. A sense of responsibility is far more important than liability suits in getting people to act wisely in relation to technology. Enforcers of the law cannot watch over every action; they can never succeed in spotting all violations. Nor can the legal system prevent crime as easily as it can punish criminals. Breaking the law is a tried-and-true American tradition, but when taking risks with technology offends an inner sense of fitness, it becomes very much less likely. The most immediate problem, of course, is to instill this sense of responsibility in corporate leaders. Officers of corporations will do things to aid their firm that they would be ashamed to do for themselves in private life.

Groups that act without due regard for others also invite revenge. When a social group seems to act without due regard for the welfare of others, people remember and strike out against it when the opportunity arises. It is possible that one of the reasons that liability suits against physicians have become so popular, and settlements so large, is that people resent the attitudes of physicians.[11] When people sit on a jury judging a physician who intentionally or accidentally has injured a patient, people tend to see the situation as a chance to strike back. It seems extremely unlikely that this kind of behavior would have been directed against the family doctor of the old days, who visited the patients' homes. But the typical contemporary physician is a businessperson who treats patients less as people than as cases. The older emphasis on service has been replaced by attention to business efficiency. The patient who gets only a five-minute examination for a medical problem that has been bothering him or her for weeks is sometimes shocked. Many physicians schedule such hasty interactions with patients that they have "traffic lights" in their examination rooms to tell them when time is up. How did this situation come to be?

Considering the origins of the behavior of the contemporary American physician, we can make some interesting observations. We find that a variety of social decisions, made at different periods of time, have shaped the current business-oriented practice of medicine.[12] Restriction of the number of medical schools as a "quality measure" reduced the supply of physicians available to practice. As the supply of physicians has dwindled, the price of their services has increased. Other measures designed to maintain the quality of services have also increased costs. Increasing the requirements for training, certification, and licensing meant both increased training costs and access to medical school only for those who could afford the training. Insurance costs have led to "defensive medicine," including the performance of tests simply to provide protection against liability suits. Now we have a relatively small number of highly trained physicians who often must act as rapid-fire decision-makers in their brief contacts with patients.[13] Finally, to add insult to injury, much of the information gathered from patients is collected on computerized, paper-and-pencil forms.[14] Although this is admittedly efficient and may even allow superior care, it does little to improve the patients' view of the system's responsiveness. So patients have come to view physicians less as public servants than as entrepreneurs who hold a monopoly in the practice of medicine. When sitting in judgment over one of these business-oriented practitioners, juries tend to award the maximum damages.

Social forces, then, have placed physicians and patients into patterned roles that neither finds comfortable. The patient wishes to be cured; the physician wishes to provide care at a reasonable cost. Yet both are forced by the society around them to interact in an almost mechanical way—the patient tries to be satisfied with the scant time the physician can afford to respond to his or her problems, and the physician is rushed and conscious of time and financial pressures. The physician's accountability may ultimately be defined in court; insurance premiums of $30,000 to $100,000 annually remind the physician of this possibility. The patient, however, would prefer that the physician's attention be directed instead to the relief of pain and suffering. An attitude of mutual trust and responsibility seems very difficult in this atmosphere. Yet both physician and patient are faced with the same technological problem: What procedures will best help the patient cope with the complaint that brought him or her to the physician's office in the first place? In this task, both parties could cooperate, but the ethic of the marketplace makes this difficult.

In medicine, as elsewhere, values have consequences for behavior. Values also continue to be shaped by institutions that carry out technological tasks. In the United States we have shaped a medical establishment that is expert at high-tech operations. This same establishment, in turn, pays relatively little attention to routine care or to public health. The infant mortality rate in the United States is far above those of many other developed countries. Our care of the poor is shocking. Those who require emergency medical care outside the hospital may be shocked to discover how poorly trained and paid are the personnel who come to aid them.[15] This, however, is a natural consequence of a system that emphasizes the importance of a relatively small number of very highly trained (and therefore extremely expensive) physicians. Only if we are willing to dig in and change medical institutions can we expect the priorities of American medicine to change. Otherwise, we will continue to have costly machines of mind-boggling complexity in the same system that neglects such elementary aspects of perinatal care as the education and nutrition of mothers.

Bioethics and the Dilemmas Posed by New Technologies. Not only does technology reflect values; it poses choices as well. One impact of new technology in medicine is the creation of ethical dilemmas. A dilemma is a situation involving a choice between unsatisfactory alternatives. In some of the problems we encountered with technological accidents, there were obvious measures to take to avoid the problems that occurred; in other words, there was a course of action that was clearly superior. A dilemma, however, forces a choice between two courses of action, each of which is unsatisfactory in some way.[16] A good example is the decision about the extent to which resources should be devoted to help a terminally ill patient.

Physicians have always done their best to save patients' lives. But often in the past there was little they could do for very sick or severely injured patients. As technology has developed, however, more methods have evolved. Now people who previously would have died can be kept alive.[17] But there is a price, often a very steep one. Much of the new high technology to avert death is used for older patients who are often on the brink of dying and in considerable pain. High sums of money are expended keeping such patients alive. In one case over $200,000 was expended on an 89-year-old nursing-home resident over a 77-day period during which he remained unconscious. He was never aware of the hundreds of tests and medical procedures performed on him.[18] Although letting someone die is never a pleasant prospect, in this case the resources used to save this dying man might well have been better used elsewhere.

The same dilemma exists with decisions about the newborn. Modern technologies allow us now to save babies born prematurely, sometimes as early as the seventh month of pregnancy. Yet such life-saving measures are expensive; they can cost more than $300,000.[19] And the children saved may be permanently impaired by their premature births. The money might better be spent on preventing premature births in the first place. Yet American society simply will not spend much money to support prenatal care for poor mothers. For instance, from 1978 to 1984, federal support for key sources of maternal care dropped by a third.[20] Thus, at the same time that our society is developing high-tech ways to save premature babies, it is in effect encouraging their occurrence by cutting back funding for preventive measures.

The decision about who should be saved and who should be allowed to die is only one of a variety of ethical dilemmas caused by new technology in medicine. Let's consider some others.

1. *Organ transplants.* Over the years transplantation of organs from one patient to another has become more successful, but the cost is high. There are many

more people who need organs than there are potential donors,[21] and states vary in their policies regarding organ transplants. The well-off can better afford replacements for failing organs; often the only hope for a poor child needing an organ is a charity drive.[22] Obvious inequities in this situation make the dilemma more dramatic, but differences in financial resources in general result in very different health statuses and life-spans for those of high and low incomes.

2. *New reproductive technologies.* Whereas some technologies allow patients to prevent pregnancy, others, such as in vitro ("test tube") fertilization, allow the infertile to have children.[23] These technologies are expensive, and many are still in the experimental stage, given their low success rates. Ethical dilemmas are also associated with sperm banks and the matter of absentee paternity. In the not-too-distant future technologies will be able to alter the genetic makeup of babies as well as assess it. Should parents be allowed to select their babies' hair color, height, athletic ability, or intelligence? What about parents who want a child that will be less intelligent than they are?

3. *Prenatal monitoring.* Increased information provided by fetal monitor technology (such as ultrasound) will allow more-informed decisions about whether to allow a baby to be delivered normally or by cesarean section, or whether a fetus should be allowed to continue development at all. An informed decision, however, is still shaped by the preferences and biases of those who make it. In India, for instance, female fetuses are much more likely to be aborted than male ones.[24] Thus increased information not only provides greater freedom of action for prospective parents; it also allows prejudices more scope.[25]

4. *Expensive technologies.* One of the major ethical issues is the rationing of medical care as "heroic" technologies make it possible to save the life of almost anyone—for a hefty price. The high cost of heroic treatments forces decisions about how scarce resources should be allocated.[26] For instance, the development of "intensive care" centers has meant greatly increased medical care costs. Critics have charged that these centers produce too little improvement for too much money. Yet physicians and patients alike have grown to depend on them and feel that they are an important tool.[27] New diagnostic technologies for the physician's office are also expensive. In a time of "defensive medicine"—when physicians often wish to protect themselves from lawsuits—tests and operations are sometimes ordered when their medical value is doubtful; their main function sometimes seems to be to protect the physician.[28] The cost of such tests is a significant threat to affordable medical care.[29]

5. *The role of medications.* Although drugs have done much to ease pain as well as improve performance (for example, steroids), a number of disturbing questions arise. Many drugs developed for medical purposes have "recreational" uses and have led to serious problems of addiction and abuse. Some tranquilizers are very much overprescribed by physicians. As more is learned about the chemistry of the nervous system, increasingly difficult decisions will be faced about mind-altering substances. Similarly, drugs that affect growth, appearance, and resistance to illness are likely to produce more powerful effects and thus are likely to generate more serious dilemmas.[30]

The development of new medications and medical techniques thus provides us with a host of new issues. To cope with these novel problems, a new field called "bioethics" has developed.[31] The development of new technology, then, is forcing even philosophers to examine many basic premises about life and human action.

Values and norms shape behavior in society, but as we have seen, values and norms are embodied in particular institutions—such as business and medicine—whose dynamics affect how technology is used. It would be splendid if such institutions could be counted on to take responsible attitudes toward the use of technology. Even if they did, however, this would not be sufficient to guarantee our safety. An additional safeguard is the use of law and regulatory agencies to provide scrutiny and to decide on codes of practice. Regulatory and enforcement agencies are institutions also, of course, so we must consider their dynamics as well.

The Regulation of Technologies

Regulation is one of the principal tools of social control of technology. Protection against hazards is only one of the functions of regulation. Other functions are the promotion of social fairness (for instance, in the setting of freight rates) and the setting and maintenance of standards. Issues of fairness are outside the scope of this study, but standards are important, and we need to take a moment to discuss them.

The Importance of Standards. The setting of standards is an activity with a history that extends back into antiquity. This practice provides for ease of commerce, fairness in trade, technical compatibility, and not least of all a chance for government to control and tax.[32] When technological standards exist, activities are facilitated because of the availability of systems that interconnect, of spare parts, and of knowledgeable repair persons. Simply considering what automobile repair would be like if there were 100 automobile companies provides a sense of the value of standardization. This problem was forcibly brought to public attention in the United States during the Baltimore fire of 1904, when fire companies from neighboring cities found that their hoses would not connect to local fire hydrants.[33] Experiences such as the Baltimore fire have promoted standardization in a variety of fields, but the development of every new technology poses new problems with standards.

In the electronics industry different computer operating systems, software programs, disc sizes, and so forth often make interconnection difficult. Problems also exist with such systems as the following:

Computer helpers. Automated programs for designing software are largely in-compatible. Computer Automated Software Engineering (CASE) products that are designed by different firms don't interface. This prevents the market from growing, for without common systems, the products of different firms cannot interconnect.[34]

High-definition television. The United States is currently proposing to use a standard that is different from those of other countries but that will be compati-ble with the current domestic television sets. This will present problems for the quality of broadcasting and for selling American television sets abroad.

Videotex systems. In the United States, several different and incompatible sys-tems of videoshopping, information queries, and so on exist. In France there is only one system, the "Mini-Tel." Thus in the United States, very little is happen-ing in this field, whereas a lot is happening in France.

Videoconferencing systems. In the United States, three incompatible systems exist for coding-decoding videoconferencing signals. Thus it is not possible to interconnect the three different systems for long-distance videoconversations.

Although setting standards is an important activity for technology users, it is also an exercise of power that helps some sponsors and their technologies and harms others. Sometimes regulation is done by industry bodies in ways that seem sus-piciously motivated by vested interests.[35] A recent dramatic case involved a voluntary standard-setting body of the American Society of Mechanical Engineers, in a case re-ferred to as *A.S.M.E., Inc.* v. *Hydrolevel Corporation.*[36]

The Hydrolevel Corporation had developed a new version of low-water fuel cut-off for boilers, used to prevent them from running dry and exploding. Standards for fuel cutoff devices are set voluntarily by an A.S.M.E. committee, rather than through governmental action. Hydrolevel's form of fuel cutoff was criticized (on A.S.M.E. sta-tionery) as being unsafe by two members of a committee in charge of the boiler and pressure vessel code. This criticism was sufficient to make the company's products seem undesirable to potential buyers, and it went out of business. The two commit-tee members, however, were employed by one of Hydrolevel's major competitors, and their abuse of power resulted in a lawsuit against A.S.M.E., which the U.S. Su-preme Court found guilty of "anti-competitive activities"—restraint of trade. In the wake of this case, a much closer examination of the possible abuses of power by private standard-setting bodies has taken place, and companies as well as profes-sional associations have become much more aware of potential liability for their actions.[37]

However standards are set, they are likely to be strongly influenced by powerful sponsors. Standards for technology are shaped by monopolies, dominant producers, major customers, and industry pressure groups. When I.B.M. introduced its personal computer (the "I.B.M. PC"), it was widely adopted, not because of its superiority, but because I.B.M. had set industry standards, and thus personal computers would have to interface with other I.B.M. machines. Similarly, when electrical transmission sys-tems were being set up, there was a major "battle of the currents" to decide whether the systems would run on direct or alternating current.[38]

The amounts of money involved in these and similar battles are often enormous, and the fate of giant sponsors is bound up with them. Often not only the major sponsor, but also hosts of related industries, are involved; thus appliance manufacturers would be involved, for instance, in changes to electrical systems. Scientific and technical talent is used to develop "scientific" rationales for why one or the other system should be adopted. Enormous public relations campaigns are used to influence public opinion.[39] Important social institutions are enlisted on one side or the other. Free and nearly-free services are provided to thousands of customers to influence their choices.

Political considerations are often important in standardization. Civilian transportation and communication systems are often designed to interface with military systems or to provide easy access to military vehicles. Civilian technologies may be required to use designs that allow them to be converted to military use. The American superhighways, for instance, were designed with overpasses that would accommodate the passage under them of trucks bearing intercontinental ballistic missiles. Technologies also provide access within and among countries.[40] A common railroad gauge allows free movement within a country, and incompatible gauges can impede movement between countries. Incompatible radio or television broadcasting systems can provide barriers to the flow of information. The sizes of trucks allowed on local roads will influence the amount of produce that can be shipped in from another state.[41]

Setting standards, then, is likely to involve powerful sponsors and vested interests. Whether standards are set by private or public bodies, pressure is likely to be applied by groups that will benefit from favorable treatment. Even though the public can benefit from uniform standards, who sets the standards and why is an important issue.

Protecting Against Technological Hazards.

Regulation is also important for controlling hazards. We have seen in the case of PBB contamination in Michigan how public health can be threatened by a seemingly minor decision by a small company. This episode points out the basic difficulty with the regulation of technologies: Such small decisions are made daily by the hundreds of thousands, yet our ability to keep track of them, let alone screen them for problems or dangers, is dangerously small.[42] Furthermore, the sponsors of dangerous technologies are often large and powerful, not small and insignificant. To take merely one example, the biotechnology industry in the United States can be expected to grow rapidly over the next decade. Yet the agencies charged with monitoring this growth are far too weak, and the laws governing the introduction of new organisms into the system are largely obsolete.[43] We are going to have problems.

A large number of government organizations in the United States exist to monitor hazards posed by technologies.[44] At the federal level, probably the most significant of these is the Environmental Protection Agency, created in 1969. The EPA is responsible for protecting the environment against physical and chemical abuse. Other organizations, such as the Office of Technology Assessment, the National Transportation Safety Board, the Nuclear Regulatory Commission, the Food and

Drug Administration, and the Occupational Safety and Health Administration play important roles in protecting us from serious technological hazards.[45] And there are many state and local versions of such organizations.

Regulatory organizations anticipate problems and process information about current situations. Then they either present this information to other governmental bodies (as does the Office of Technology Assessment) or take action themselves (as does the Food and Drug Administration). Rather than acting in isolation, these organizations act as parts of a web of governmental and private action. Their effectiveness is shaped by the organizations' public support (constituencies), by the legal tools given them to take action, and by the human resources they can bring to bear on the problem.

Problems in Technology Regulation. Even though regulation is indispensable, it is difficult in a rapidly changing society. Both laws providing regulations and regulatory organizations are likely to follow the introduction of technological hazards, not precede them. Regulation is likely to arrive late on the scene and with knowledge that is almost always less adequate than it should be. Further, the ability to enforce regulations depends on adequate personnel. For a variety of reasons, the number of inspectors is likely to be less than the number needed. Because commercial pressures are strong, there is a tendency for industrial firms to "push the envelope" and take marginally acceptable risks. The larger the amounts of money involved, the stronger these pressures are likely to be. Finally, regulation is clumsy. Laws do not change as rapidly as the society in which regulation operates. On one hand, this produces a tendency toward rigidity and bureaucracy; on the other, it results in a dangerous irrelevance of much of the regulatory process, sapping the confidence of inspectors and encouraging cynicism on the part of those regulated.

Manpower. Often a lack of manpower is a serious barrier to effectiveness. These organizations can enforce regulations only so far as they have the eyes to observe and the legal talent to prosecute. Often these are insufficient. For instance, one expert on trucking regulation reported that the most important factor he has detected in reducing truck crashes is the number of roadside inspections, which obviously is a function of the number of inspectors.[46] According to a recent article, the Nuclear Regulatory Commission's Office of Investigation has a total of 45 employees, including 32 investigators. Yet this office must oversee more than 120 operating nuclear reactors.[47] Thus a small number of people is responsible for policing one of the most dangerous—and powerfully sponsored—industries in the United States.

Given the existence of the kind of nuclear problems we have, such a small investigation force is not an adequate safety net.[48] And the Reagan administration reduced many of the staffs of bureaus that monitor technology, on the principle that less government—that is, less regulation—will be better for the country.[49] The result is that many groups whose safety depends on such inspections are endangered. Workers depend on such organizations as the federal Occupational Safety and Health Administration; but with fewer inspectors, OSHA's ability to inspect has declined.[50]

Regulatory Conflict. Another problem for regulatory agencies is that their funding and support come from elected government officials—from the U.S. Congress. Elected officials, however, are often very responsive to potential campaign sponsors, which often include large industrial interests and powerful government agencies like the military services. Very often, then, government agencies feel themselves pressured both by their express mandates (to clean up the environment, for example) and by the wishes of elected officials who are supporters of big business. Action on important technological control problems is often paralyzed by this conflict.

The use of antibiotics in cattle feed illustrates a conflict of this kind. Cattle feeders have frequently dosed their cattle with subtherapeutic amounts of antibiotics. A subtherapeutic dose is less than the amount necessary to cure the animal if it is sick. Of course, the animals to which these doses are fed are not sick. But these drugs, for reasons that are not entirely clear, help the cattle grow larger and fatter. At the same time, the use of the drugs in this way leads to an environment in the cow's body that is favorable to disease-resistant strains of bacteria. Because the cow's body is an environment in which only antibiotic-resistant strains of bacteria can thrive, the cow acts as a selective breeder of those strains, a process akin to natural selection. These bacteria, in turn, are no longer killed by the antibiotics, a very dangerous situation when the diseases they cause also affect humans. Recently there have been outbreaks of such antibiotic-resistant diseases among humans.[51]

Lobbying by Sponsors. How did this situation come to exist? It came about because the sponsors of this technological practice are very powerful. The pharmaceutical and dairy industries are large and powerful, and both have strong vested interests in the use of antibiotics in cattle feed.[52] These industries have lobbied effectively against legislation that would prohibit the practice. So far this lobbying has meant that antibiotics can still be fed to cattle.

Similar lobbying has taken place by companies that manufacture chlorofluorocarbons (CFCs). CFCs are believed to damage the ozone layer of the stratosphere, which protects life on earth from cancer-producing ultraviolet rays.[53] CFCs are currently manufactured in quantities of a million tons a year worldwide. Yet producers of CFCs have consistently delayed regulatory action and continued to produce large quantities of CFCs.[54] Their strategies have included large "public information" campaigns and extensive scientific studies aimed at countering scientists who believe that CFCs are dangerous. CFC use has actually increased since the U.S. banned their inclusion in aerosol sprays in 1978.

At present, there is pressure for worldwide action on the production and use of CFCs.[55] Even the chlorofluorocarbon industry has conceded that scientific evidence for potential danger from CFCs is stronger than it once was willing to admit. Again, however, what the CFC situation shows is that powerful economic interests will try to shape not only laws, but also public perceptions about the safety of a technology. The effectiveness of such efforts in the CFC case should make us think very carefully about how effective regulation can be when economic interests are strong.

(One aspect of the CFC business, though, is worthy of reflection. The executives of the chlorofluorocarbon industry, and their spouses and children and neighbors, all live in the world that their actions threaten. Why were they not more concerned and responsible about this world? How much did their concern for making money and assisting their organizations shape their perceptions of the world around them? It is important to remember that when the effects of a person's decision are remote from the person in space, time, or social network, many moral constraints are absent. It must be recognized that people are often willing to do dangerous things with technology because the effects are not immediately present. Words and slogans conceal from the mind the potential effects and give the individual an altered sense of morality. In his work on "groupthink," Irving Janis has shown how people's environments can make them feel good about doing dangerous things.[56] This observation shows the inadequacy of the individual conscience in the current ethical climate. We must develop institutions capable of being conscientious when individuals are not.)

Regulatory Rigidity. Another problem with regulation is the inertia of the processes of law- and rule-making. Over time there may be a slow buildup of overly specific rules. As time passes, some rules become obsolete, but laws are added on rather than revised in toto. As a result, enforcers often are stuck with a set of unreasonable and unworkable rules that they are expected to enforce.[57] If they allow deviations, they may be subject to disciplinary action or subsequent congressional scrutiny. This situation is made worse in the United States because regulators are supposed to act like police. In other countries, there is more effort to work with industries instead of trying to police them. The Japanese, for instance, see regulation as a form of negotiation rather than one of enforcement.[58] To an American, constructive problem-solving with industries—working *with* rather than *against* them—sounds suspiciously like corruption.[59] Often the losers in such a rigid approach, however, are those the rules are supposed to protect. The aim of regulatory agencies is to protect citizens, but sometimes their practices get in the way of doing this. Even though it is certainly the champion of workers in some cases, OSHA, for instance, has been excessively rigid in other cases.[60] But in the end, rigidity may be another symptom of insufficient manpower for enforcement.

Judicial Inadequacy. The courts are also part of the regulatory process, and in the past they have been an important tool for coping with changing technologies.[61] How adequate they may be at present, however, is another matter. Courts do decide some very important conflicts of interest over technology,[62] yet training in law is often inadequate for dealing with the complexities of science and technology. Increasingly, scientific competence may have to be represented on the judicial bench as well as through expert witnesses.[63] A more serious problem is that such new technologies as in vitro fertilization create moral and ethical problems that courts are asked to resolve, often in advance of either adequate technical knowledge or appropriate social debate.[64] Here the courts may be asked to cope with problems before even philosophers have had a crack at them. This places too much burden on an already overloaded system.

The Role of Zealots, Crusaders, and Pressure Groups

Regulation by itself is never enough. In a healthy society there must be ways to embody concern about technology apart from the officially constituted agencies. In American society one of the greatest forces for reform lies in the ability of citizens to organize. The ease with which Americans form voluntary associations makes possible one of the strongest tools for coping with technological hazards. Such groups form an important counterpoise to corporate interests, and they are frequently the incubator from which future government regulation comes.

We must not underestimate, of course, the ability of bureaucratic officials to act as "moral entrepreneurs" in relation to technology.[65] A moral entrepreneur is a person who tries to change social norms to some new standard. One such moral entrepreneur was Harvey Washington Wiley (1844–1930), known during much of his life simply as "Doctor Wiley."[66] Wiley, chief chemist at the U.S. Department of Agriculture, fought a long battle to get many of the unnecessary and noxious additives out of food. With his extensive research and his forceful, articulate public statements, he caused the purity of food to become a national focus. He was one of the major sponsors of the Pure Food and Drug Act of 1906, and he was also a powerful activist against addictive drugs in medicines.[67] His "poison squad" of 20 young men who personally tested the effects of food ingredients was an important public-relations tool for the causes that Wiley championed. Although Wiley was assisted in his struggle by crusading organizations, his ability to act as a governmental champion of such causes was important.

Similar moral crusading against the dangers of using tobacco has been carried out by C. Everett Koop, the former Surgeon General of the United States. Koop is almost the model of the bureaucratic reformer. He surprised those both inside and outside of government, not only with his independence of thought, but also with his willingness to do battle with powerful interests, including those associated with President Reagan, who had appointed him.[68] He took strong stands on smoking, AIDS, and abortion, and he gave government a stronger role in the regulation of technology than many who formerly held his position.

Wiley and Koop have many counterparts who are outside the government. A related human resource, and perhaps an even more important one, is the technology critic who assembles information about corporate or official abuses and then publicizes them. Ralph Nader is probably the best known of the current critics, but there are many others, particularly in environmental organizations.

At the turn of the century, a number of social critics, who Theodore Roosevelt labeled "muckrakers," exposed serious commercial and industrial abuses—in patent medicines, food manufacture, and industrial safety.[69] One of the most effective of these was Upton Sinclair, whose fact-filled novel *The Jungle* exposed abuses in the meat-packing business.[70] The muckrakers were interested in much more than technology. They also attacked monopolies, political corruption, sweatshops, child labor, prostitution, and many of the other evils of a rapidly urbanizing society dominated by large corporations and political machines.

The muckrakers were part of the larger "Progressive Movement" that worked for the creation of a government more responsive to citizen interests and more protective of the individual's rights.[71] Many of the muckrakers were newspaper reporters and free-lance writers who made their livelihoods from the sensational stories that unhappily were only too easy to find. They were supported in turn by a variety of "citizen action" groups, such as the Consumer's Union, which put its label only on goods manufactured under decent working conditions.[72] Many of these groups pioneered in the collection of statistics on such technological issues such as public health and injuries at work, an activity that later became a governmental function.

In the latter part of the twentieth century, such citizen action organizations are still with us. These groups have been strongly criticized as alarmist, elitist, or "bleeding hearts," yet their function is vital. Critics of smoking, fluoridation, abortion, alcohol use, and nuclear power have been influential. One of the values of such organizations in the environmental field has been to provide an external watchdog of the actions of industry and government.[73] In other countries where such critics are fewer and much less influential, this protection is absent. The absence, until recently, of similar groups in the Soviet Union and Eastern Europe has meant that the pressures for environmental protection have not been as strong there. In France the government has been able to maintain greater secrecy regarding the problems with its nuclear program because of a weaker environmental movement.[74]

Women have played a key role in environmental action as critics of technology. One important critic was Rachel Carson (1907–1964), a biologist and best-selling author of books on nature. When she became concerned about the dangers pesticides posed to wildlife and the environment, she researched and wrote *Silent Spring* (1962), a book that can be said to have launched the current environmental movement.[75] This book was violently attacked by the agricultural-chemical industry and many university scientists associated with it.[76] But the concerns raised by *Silent Spring* have since been proven to be well-founded and have resulted in laws and restrictions on pesticides and other agricultural chemicals. Mary Sinclair was a major force in the cancellation of a nuclear reactor under construction by Consumers' Power in Midland, Michigan.[77] In Monroe, Michigan, Jennifer Puntunney, executive director of the Safe Energy Coalition, has been an active critic of the Fermi II nuclear plant, which once had its license suspended after a premature nuclear reaction was started accidently on July 2, 1985.[78] Private action of this kind plays a major role in protecting society against technological dangers.

Private groups can also act as advocates for and distributors of "alternative technologies." Few hippie communes survived the 1960s, but "health food" stores, alternative medicine, and passive solar technology are very much alive. The existence of alternatives is important, for they challenge the record of "official" science, medicine, and technology. Such private ventures help shape norms, provide employment for dissident intellectuals, and increase the variety of products and services available to consumers. It is worth remembering that birth-control technologies were pioneered by such private groups.[79] Inevitably, some of the alternative technologies are dangerous, but many of the official technologies are dangerous, too.[80]

Much of the information used by these groups comes from intellectuals outside of government. Some of these critics are in universities, but many are simply freelance writers. We have already mentioned Rachel Carson in this connection. Probably the best-known critic of this kind today is Ralph Nader, whose name is closely linked with criticism of corporate neglect of safety. Nader began public advocacy through magazine articles in 1959, but his first major work was a book on the automobile industry titled *Unsafe at Any Speed* (1965).[81] As a result of this book, he was investigated by detectives hired by General Motors, who harassed him and tried to gather information to defame him.[82] Nader discovered this surveillance, and he took General Motors to court and was paid a large settlement that helped contribute to his continuing activities in automobile safety and other areas.[83] Nader's current organization, Public Citizen, investigates a very broad spectrum of complaints and problems, ranging from antitrust problems to pollution.

Moral entrepreneurs and citizen action organizations are thus an essential part of the struggle to control technology. Probably the most important function of these groups is simply to get people aroused. Keeping citizens interested in technological issues is a major force for technological education. If industries and government bureaus were the only ones concerned with technological hazards, we would be in a much more vulnerable position than we are now; the proponents of technology do not worry enough. The active citizens of a society constitute an important resource for thought and action. Their function is not simply to protest and complain, but also to actively examine directions, issues, and hazards. We need to make sure that citizens are kept alert and aware.

Addressing Problems: The "Open Free Loop"

One of the most serious nuclear hazards in the United States was posed by faulty reactors of the Tennessee Valley Authority.[84] The TVA is a regional government agency that supplies power to seven states. In the 1960s it decided to start building nuclear reactors to provide power it had formerly generated at its famous dams and coal-fired plants. But problems arose. Poor workmanship and supervision and pervasive corruption led to public outcry. Of the 17 nuclear reactors planned by the Authority, only 5 had been completed by 1986, and all of them were shut down because of safety problems. Construction on the others had virtually halted while attempts to repair or replace shoddy work took place instead. Employees who reported violations were harassed and intimidated. One employee was murdered in a parking lot, probably because of her knowledge of a theft ring. A remarkable firm, Quality Technology Company, was called in to try to improve information flow and to get the reactors operating again.

The story of Quality Technology Company is unusual.[85] Owen Thero, the president of Q.T.C., had worked for 24 years as a manager of quality assurance and testing for General Electric. Thero's responsibility was to see that nuclear reactors produced by GE met the standards spelled out in the contract. In testing the reactors, Thero often found it necessary to reject things built by the company and send them back to the drawing board. The company's system, fortunately, had institutionalized this process of "take your lumps and redo it." So Thero's suggestions for improvements were accepted and acted upon.

Still, Thero found that no matter how good was the product that General Electric shipped, its quality was often compromised by its users. Utilities were very careless with installation and maintenance of the reactors. The utilities, Thero felt, "just didn't seem to care" how the equipment was installed, maintained, and used. The only thing that seemed to matter was getting the reactors "on line" so that they would make money for the utility. The utilities often cut corners in installation, skimped on maintenance, and overused the equipment by running it at maximum power.

Thero was frustrated by this situation, and with the aid of business associates he created an organization that could respond to utilities' increasing problems with regulatory agencies. Thero's bias, he freely admits, was to get reactors working—but safely. Quality Technology Company was "custom-made" to work on nuclear utility problems. The organization was set up to process worker complaints and provide constructive solutions. Q.T.C. was successful in doing this at the Waterford III reactor, and it was then hired to work on problems with the TVA's reactors. Thero looked forward to a bright future solving the problems of the nuclear industry.

In 1985, Quality Technology Company went to work for the TVA. Q.T.C. interviewed about 5,000 TVA workers; some 40 percent had complaints, of which a sizeable fraction (93 percent) were safety-related. We have already mentioned the crisis precipitated by these problems. Because it also received reports of very severe crime and drug problems, Q.T.C. provided confidentiality for workers, an "open free loop" in which safety and other problems noticed by them could be addressed without the threat of severe punishment. Because TVA previously had harassed many workers, opening this new channel of communication was the only way in which many problems could be brought to light. But Q.T.C. was about to experience its own problems.[86]

Admiral Steven White was brought in as manager of nuclear power for the TVA.[87] It was generally expected that White would make all necessary changes to get the TVA reactors relicensed within two years, the length of his contract with TVA. Critics charged that many of the changes would be cosmetic and that many of the problems would be "papered over." In the process of taking charge, White found that Q.T.C. was interfering. According to White, Q.T.C. was charging too much for the services they offered. According to Thero, the real intent was to close down an organization that could provide criticism of White's assertions that TVA's problems had been solved. Q.T.C. was dismissed, and the "open free loop" was closed. Furthermore, according to Thero, Q.T.C. was blacklisted by the nuclear industry, and the company could no longer find contracts to work. Q.T.C.'s personnel went from 70 to one in a very short period. As of January 1989, Q.T.C. consisted of Owen Thero and an office manager.

It is important to observe here that Q.T.C. was not a whistle-blowing organization. Its intent was not to provide public notice of problems, but rather to help organizations solve problems by providing an "open free loop" of communication outside ordinary lines of authority. In principle, then, it might have been even more useful than ordinary whistle-blowers, because there is no guarantee that once a whistle is blown, any action will be taken. The important feature of Q.T.C. was involvement in solving the problems identified. This commitment to solving problems is a valuable social innovation for two reasons. First, problems are more likely to surface in an environment in which they are part of an ongoing problem-solving process. Rather than generating images of "crime and punishment," they become part of a dialectic of "find and solve." Second, whistle-blowing itself may do little to effect action to change the situation. Quite often organizations caught doing something wrong can create enough confusion about the motivations and charges of the whistle-blower that the problems are forgotten. This is a common strategy used by organizations threatened by this kind of exposure. The public's attention span is much shorter than the organization's.

For this reason, it may well be that we should consider companies like Q.T.C. to be one of a variety of new instruments for coping with technological complexity. It links the observations and concerns of workers with a mechanism for investigating and acting on their complaints. Ideally, organizational systems like that of Q.T.C. can be used not only as external aids, but also inside organizations. Use of "hotlines," such as the one recently installed at NASA, are also good, but they do not have the anonymity or "action bias" of the Q.T.C. system. The NASA system, apparently based on a similar program at the FAA, has workers telephone a number at Batelle Memorial Institute in Columbus, Ohio. This allows callers to bypass normal channels, but it does not provide for the integration of problem-solving with the rest of the organization, as Q.T.C.'s system does.[88]

One of the best ways in which technologies are controlled is within organizations themselves. Often those who have ethical concerns can team up with others in the organization to promote scrutiny of dangerous organizational practices. Because this kind of lobbying is an internal matter, it is less likely to promote the sense of betrayal and external conflict caused by whistle-blowing. Roger Boisjoly was a rocket engineer employed by Morton Thiokol in the early 1980s. While designing the solid-rocket boosters for the space shuttle program, Boisjoly became aware of the precarious nature of the joints holding the segments of the boosters in place. These joints were made tight by rubber seals, and cold temperatures, Boisjoly discovered, made these seals unable to prevent leakage of the hot gases generated by the burning grain of the rocket. So he developed a "seal team" to lobby for either replacing the seals or not using them under cold conditions. This team, however, was unable to prevent the launch of the space shuttle *Challenger* on the cold morning of January 28, 1986. The seals leaked and the *Challenger* was destroyed.[89]

Whistle-Blowing

As we have seen in the previous examples and in other chapters, sponsorship of a technology does not necessarily imply a sense of responsibility toward it. Some organizations will always be willing—for profit, reputation, or market share—to put others at risk. Such unethical activity is likely to occur in spite of the most stringent regulations, even in organizations ostensibly dedicated to the public good.[90] In such a situation the protection of the public may well hinge on the actions of rare individuals willing to make these organizational crimes public. This activity, commonly known as "whistle-blowing," is indispensable to a society vulnerable to the risks high technology brings.[91] Such a society, whose activities intimately depend on minute or complicated devices or on enormous systems whose workings may seem arcane to the average person, and which makes use of technologies of potentially great destructiveness, needs every safeguard conceivable. And one such safeguard is the knowledgeable insider who knows that something wrong is going on.

In 1971 three engineers who were involved in the construction of the Bay Area Rapid Transit System thought they had detected a serious system problem.[92] They had reason to believe that the computers—the nervous system of the electric train complex—would not be able to handle the complicated interactions that were bound to arise. If the computers could not cope with the inevitable complexity, accidents—perhaps serious accidents—could endanger trains full of people.[93] Through anonymous memos and eventually through making the problems known to a board member, the engineers caused the computer problems to become public. As a result, they were eventually fired from the project. Their concerns were justified, however, as it was later discovered that the computers would temporarily lose track of trains, thus creating "phantoms" that could nonetheless be involved in real accidents. In another case, a crystal oscillator that had escaped checkout malfunctioned, and at the Fremont station on October 2, 1972, a train hit a sandpile positioned to stop trains in an emergency.

A. Ernest Fitzgerald was a cost analyst for the Department of the Air Force when he discovered enormous cost overruns on the C-5A Galaxy transport, then being developed by Lockheed Aircraft. He reported these overruns, which amounted to over a billion dollars, to a congressional committee chaired by Senator William Proxmire. Fitzgerald had previously received awards for saving the Air Force money. The C-5A's sponsors in the Defense Department, however, were extremely powerful; they included the President of the United States, Richard Nixon. Fitzgerald was investigated (nothing unseemly was found) and then transferred to work on minor projects such as bowling alleys in Thailand. When his reassignment did not stop his investigations—he found things to investigate even in Thailand—his job was abolished (as an "economy move") and he was blacklisted in the aviation industry.[94]

During the 1980s a series of problems occurred at the nuclear reactors operated by the Tennessee Valley Authority. By June 8, 1986, all 5 of the authority's operating reactors had been shut down, 8 of the 17 planned reactors had been cancelled, and 4 more were put on indefinite "hold."[95] The problems that led to these shutdowns had been known for years. Concerned employees had tried to point out the problems to their supervisors, only to be told to back off. Some 2,000 "concerns" had been reported involving safety problems in construction and operating "shortcuts." More strenuous efforts to call public attention to the problems led to harassment and intimidation of the whistle-blowers.[96] As we have seen, Quality Technology Company was finally brought in to hear workers' complaints because the TVA's official channels were no longer trusted by the Nuclear Regulatory Commission.

Whistle-blowers have little in common except the courage to act on their moral convictions. Some of them report on unsafe nuclear power plants, others on cost overruns of military systems, others on falsified studies or hidden white-collar crimes.[97] A great many of them are involved with technological problems. What is common to nearly all of them is that they are severely punished.[98] When firms or industries discover that an employee has chosen the public over the private interest, the employee is nearly always dismissed and later blackballed by the rest of the industry. Even the night watchman who discovered Richard Nixon's "plumbers" bugging the Watergate has had difficulty finding a job since.

Obviously, however, a high-tech society needs these people who have a strong concern for public safety. As a result, many states have now passed laws or, better still, created agencies, to protect whistle-blowers. For instance, Michigan passed such a law—the Michigan Whistleblowers' Protection Act—on March 31, 1981. According to press reports, the calls to a pollution hotline increased from 1,755 for the first eight months of 1980 to 2,094 for the same period in 1981. Michigan's law was unique at the time in that it was intended to cover the employees of private firms who wished to report illegal activities, whereas the laws of other states covered only public employees.[99]

For the most part, though, these measures are very inadequate. Corporations and official bureaucracies have an enormous range of weapons that can be used against whistle-blowers, from career stagnation to murder.[100] These weapons can be used without due process or public scrutiny, unlike the protective measures that require legal action, public funds, or government support.[101] It is instructive to realize that A. Ernest Fitzgerald had an extremely strong case, no detectable vices, and enormous public support. Yet he was out of a job for four years, and he was not reinstated fully until 13 years later.[102] His current suggestion to the many whistle-blowers who call him for advice is to think twice about going public. William Bush, a retired (and formerly demoted) NASA whistle-blower, has kept a computer file on other whistle-blowers; there are already 8,500 entries in this file. He is equally dubious about the fate of whistle-blowers:

When individuals phone him with dark secrets he exhorts them to keep quiet unless they're independently wealthy. "I want to emphasize this one thing," he says. "Whistle blowing is dangerous. I've seen people bloodied. And it's not going to get easier to do. Nobody wants a snitch." [103]

The fate of whistle-blowers in our society is troubling. In spite of their willingness to risk career and security for the public good, whistle-blowers are largely neglected by society. Instead of being treated as heroes—an appropriate designation in some cases—they are treated as pariahs. In part this is true because we are an organizational society that is used to cooperation with organizations rather than resistance to them. Offenders against organizations may occasionally be regarded sympathetically, but typically they are neither treated well nor readily given new jobs. We should regard this situation with some concern, for our lives may depend on the single, ethical individual willing to call attention to hazards that others are more interested in covering up.

Whistle-blowing is essentially a last-ditch measure. It should only be used when everything else has failed. It cannot solve the problems of most organizations, and many abuses are not corrected even when brought to public attention. Samuel Florman notes that "a system that relies on heroism is neither stable nor efficient." [104] I agree. It would be far better if organizations built in suitable mechanisms for handling dissent, such as ombudsmen, "organizational troubleshooters," or internal "devil's advocate" systems. [105] Increasingly, organizations are coming to appreciate the value of these systems, which often can be used to improve the value of products and can provide timely warning of coming crises. These built-in devices for encouraging and channeling dissent make the organization able to improve itself. Unhappily, whistle-blowing will always be necessary in some cases.

Conclusion

Examining the mechanisms we have discussed here for social control of technologies is a sobering activity. When we recognize the powerful influence of sponsors, the slow response of government and voluntary associations, and the unpredictability of the technology itself, a serious problem is evident. It is much easier to bring a technology into the world and get it into full-scale operation than it is to control this same technology, mitigate its ill effects, or stop it completely if it is dangerous. With the technology comes, of course, a community of technically skilled people who know the technology, will tinker with it, fix it, improve it, and often defend it at great length in industry publications and before government committees. [106] Against this massive scientific and technological power, regulatory agencies have very limited resources. Sometimes the personnel who manage such agencies come from the industry itself and share its values and outlook. They often will do little to stem serious abuses without a public protest. And even when regulators are antagonistic to the technology, their power often remains limited. Ultimately, they may be able to stop the technology by shutting it down or making it illegal. This kind of resistance, however, is a last resort, and unless a technological alternative is present, such action may prove politically unacceptable. Voluntary associations and whistle-blowers can often only put roadblocks in the way of the juggernaut. The most difficult thing to do is to civilize the technology, to make it congruent with social needs, and this requires foresight. How is such foresight to be acquired? We will tackle this issue in the next chapter.

Notes

1. A very useful approach to this matter is the one adopted by William Foote Whyte in his presidential address to the American Sociological Association; see W. F. Whyte, "Social Inventions for Solving Human Problems," *American Sociological Review,* Vol. 47, No. 1 (Feb. 1982), pp. 1–13.

2. The account in this text is largely based on Joyce Egginton, *The Poisoning of Michigan* (New York: Norton, 1980), and on conversations with my friend Claudia Capos, who investigated the PBB situation in Michigan as a reporter for the *Detroit News;* see "PBB: The Problem That Won't Go Away," *Michigan: The Detroit News Magazine,* Sept. 12 and 19, 1982. These articles drew a pointed response from the governor's office. An aide to Governor William Milliken, Robert Berg, wrote a stinging eight-page letter criticizing Ms. Capos's methods and facts. The letter was published on the editorial page of the *News,* and only an extremely strong rebuttal from Ms. Capos, published the following week, saved her reputation from damage.

3. A substance's toxicity indicates how concentrated a danger it represents; that only parts per million—only few millionths of an ounce—of PBBs are considered toxic to humans is an indication of how dangerous the ingestion of these chemicals can be.

4. Egginton, *The Poisoning of Michigan,* p. 92.

5. Compare this situation with that described in Marjorie Sun, "Desperately Seeking *Salmonella* in Illinois," *Science,* Vol. 228 (May 17, 1985), pp. 829–830.

6. Curtis K. Stadtfeld, "Cheap Chemicals and Dumb Luck," *Audubon,* Jan. 1976, pp. 110–118.

7. It is interesting in this regard that lately we seem to cherish the environment more. Public opinion polls have shown a strong increase (from 45 percent to 78 percent) from 1981 to 1989 in the number of Americans who feel that the environment must be protected "regardless of cost"; see Roberto Suro, "Grass-Roots Groups Show Power Battling Pollution Close to Home," *New York Times,* July 2, 1989, pp. 1, 12.

8. Elting E. Morison, "The Master Builder," *American Heritage of Invention and Technology,* Vol. 2, No. 2 (Fall 1986), pp. 34–40.

9. The handwriting of physicians is notoriously bad, and sometimes it threatens the health of their patients. In view of the importance of what they write, it is disturbing that in

controlled studies in Australia, physicians' handwriting has been found to be significantly worse than that of others; see Lawrence Altman, "Warning: Your Physician's Penmanship May Be Hazardous to Your Health," (*New York Times* News Service), printed in the *Ann Arbor News,* Mar. 13, 1986.

10. Dean Wilson, personal communication, about 1986.

11. See Aric Press, with Ann McDaniel, Tessa Namuth, Ginny Carroll, and Mark Starr, "The Malpractice Mess," *Newsweek,* Feb. 17, 1986, pp. 74–75. The general crisis in insurance liability has led to a number of studies and analyses, and our understanding of this subject remains far from satisfactory. For contrasting views, see Robert Hunter, "Taming the Latest Insurance Crisis," *New York Times,* Apr. 13, 1986, Business Section; Peter Huber, "Injury Litigation and Liability Insurance Dynamics," *Science,* Vol. 238 (Oct. 2, 1987), pp. 31–36; and Scott Harrington and Robert E. Litan, "Causes of the Liability Crisis," *Science,* Vol. 239 (Feb. 12, 1988), pp. 737–741.

12. See a fine review of these decisions in Gregg Easterbrook, "The Revolution in Medicine," *Newsweek,* Jan. 26, 1987, pp. 40–74.

13. I have been surprised to find that some physicians feel that an oversupply of doctors exists. Because this is seldom the view of patients, I must conclude that these physicians consider the fate of their incomes to be more important than the fate of their patients.

14. Thanks to the stretch principle, the use of computerized "expert systems" for diagnosis may become more commonplace, in some cases replacing, not just supplementing, physicians' intelligence; see William Marbach, Jennet Conant, and Mary Hager, "Doctor Digital, We Presume," *Newsweek,* May 20, 1985, p. 83.

15. Donald Metz, *Running Hot: Structure and Stress in Ambulance Work* (Cambridge, Mass.: Abt Books, 1981), pp. 57–82.

16. See, for example, Lester B. Lave, "Conflicting Objectives in Regulating the Automobile," *Science,* Vol. 212 (May 22, 1981), pp. 893–899.

17. Formerly, many relatively minor wounds that were the result of assaults, duels, and warfare were commonly fatal because they led to infection. Techniques for treating such wounds are now so much more effective that mortality rates have been seriously reduced.

18. Ronald Kotulak, "Never-Say-Die Policy Prompts Rationing Call," *Chicago Tribune,* June 15, 1986, pp. 1, 12.

19. Barbara Kantrowitz, Pat Wingert, and Mary Hager, "Preemies," *Newsweek,* May 16, 1988, pp. 62–69.

20. Terence Monmaney, "Preventing Early Births," *Newsweek,* May 16, 1988, p. 70.

21. Julie Wiernik, "A Critical Shortage," *Ann Arbor News,* Apr. 13, 1986, Section G, pp. 1, 2. This article noted that between 20 and 25 of the 580 people in Michigan who needed a kidney transplant that year would probably die before donor organs became available.

22. Timothy Egan, "Oregon Cut in Transplant Aid Spurs Victims to Turn Actor to Avert Death," *New Year Times,* May 1, 1988, p. 12.

23. Sandra Blakeslee, "Trying to Make Money Making 'Test-Tube' Babies," *New York Times,* May 17, 1987, p. 6. This article notes that since 1981, in-vitro fertilization has produced 800 babies—and 150 clinics—in the United States.

24. Jo McGown, "In India, They Abort Females," *Newsweek,* Jan. 30, 1989, p. 12.

25. Tamar Lewin, "Despite Criticism, Use of Fetal Monitors Persists," *New York Times,* Mar. 27, 1988, Section 1.

26. B. D. Colen, *Hard Choices: Mixed Blessings of Modern Medical Technology* (New York: Putnam, 1986); Kotulak, "Never-Say-Die Policy," p. 1.

27. See Bryan Jennett, *High Technology Medicine,* 2nd ed. (Oxford: Oxford University Press, 1986). This is a valuable introduction to this problem, with detailed discussions by both critics and defenders. See also Matt Clark, Mary Hagar, Daniel Shapiro, Elisa Williams, Sue Hutchinson, and Erik Calonius, "Trauma in the Emergency Room," *Newsweek,* Feb. 16, 1987, pp. 76–77.

28. See, for instance, the Associated Press story, "Study Critical of Antistroke Operation: Some Patients Harmed by Carotid Endarterectomy, RAND says," *Washington Post,* Mar. 24, 1988.

29. Glenn Kramon, "Good Medicine, Better Business," *New York Times,* May 15, 1988, p. 6.

30. Pamela Abramson, "Genentech's Drug Problem: The Perils of Marketing a Synthetic Hormone," *Newsweek,* Nov. 25, 1985, p. 70.

31. See, for instance, Doris Teichler-Zallen and Colleen D. Clements, *Science and Morality: New Directions in Bioethics* (Lexington, Mass.: Heath, 1982), pp. xv–xviii.

32. John Perry, *The Story of Standards* (New York: Funk & Wagnalls, 1955).

33. Achsah Nesmith, "A Long, Arduous March Toward Standardization," *Smithsonian,* Vol. 15 (Mar. 1985), pp. 176–194.

34. Barnaby J. Feder, "Computer Helper: Software that Writes Software," *New York Times,* May 8, 1988, Section F, p. 5.

35. David Noble, *America by Design: Science, Technology, and the Rise of Corporate Capitalism* (New York: Oxford University Press, 1977), pp. 69–83.

36. "Full Text of the U.S. Supreme Court Opinions in *ASME* v. *Hydrolevel:* Standards Abuse and Liability," *I.E.E.E. Technology and Society Magazine,* Sept. 1982, pp. 3–12.

37. See the entire issue of *Perspectives on the Professions,* Vol. 3, No. 3 (Sept. 1983), published by the Center for the Study of Ethics in the Professions, Illinois Institute of Technology, Chicago, Illinois.

38. Thomas P. Hughes, *Networks of Power: Electrification in Western Society, 1880–1930* (Baltimore: Johns Hopkins University Press, 1983), pp. 106–139.

39. Francis Ventre found that the dominant influence on local building codes, for instance, was the vendors of building materials; see Francis T. Ventre, "Maintaining Technological Currency in the Local Building Code: Patterns of Communication and Influence," *Urban Data Service Reports,* Vol. 3, No. 4 (Washington, D.C.: International City Management Association, Apr. 1971).

40. Roy I. Wolfe, *Transportation and Politics* (Princeton, N.J.: Van Nostrand, 1963), pp. 37–38.

41. William Beard, *Government and Technology* (New York: Macmillan, 1934), pp. 40–41.

42. A rather optimistic view of this problem is given in Joseph G. Morone and Edward J. Woodhouse, *Averting Catastrophe: Strategies for Regulating Risky Technologies* (Berkeley: University of California Press, 1986). It is worth noting that this book was written before the Chernobyl incident and the discovery of the "ozone hole" in the Antarctic.

43. Sheldon Krimsky, Kostia Bergman, Nancy Connell, Seth Shulman, and Nachama Wilker, "Controlling Risk in Biotech," *Technology Review,* Vol. 92, No. 5 (July 1989), pp. 62–70.

44. See, for instance, Martin V. Melosi, *Pollution and Reform in American Cities, 1870–1930* (Austin: University of Texas Press, 1980).

45. An interesting earlier survey of such functions is William Beard, *Government and Technology* (New York: Macmillan, 1934); more recently, see Henry Lambright, *Governing*

Science and Technology (New York: Oxford University Press, 1976); Dorothy Nelkin (Ed.), *Controversy: The Politics of Technical Decisions,* 2nd ed. (Beverly Hills, Calif.: Sage, 1984); and Simon Ramo, "Regulation of Technological Activities: A New Approach," *Science,* Vol. 213 (Aug. 21, 1981), pp. 837–842.

46. Alfred E. Kahn, "Letters: Trucks Need More Inspection, Not Regulation," *New York Times,* Nov. 5, 1987.

47. Matthew L. Wald, "Critics Say NRC Lags on Inspections of Nuclear Plants," *Ann Arbor News,* Jan. 31, 1988, Section B, pp. 1, 3 (reprinted from the *New York Times*).

48. Even this agency made an important difference, however, as can be seen by comparing U.S. nuclear *power* with U.S. nuclear *weapons* production. The latter has involved far more serious contamination, accidents, and hazards; see John F. Ahearne, "Fixing the Nation's Nuclear-Weapons Reactors," *Technology Review,* Vol. 92, No. 5 (July 1989), pp. 24–29; and Matthew L. Wald, "Energy Chief Says Top Aides Lack Skills to Run U.S. Bomb Complex," *New York Times,* June 28, 1989, pp. 1, 41.

49. Joan Claybrook, and the staff of Public Citizen, *Retreat from Safety: Reagan's Attack on America's Health* (New York: Pantheon Books, 1984).

50. William Glaberson, "Is OSHA Falling Down on the Job?" *New York Times,* Aug. 2, 1987, Business Section, pp. 1, 6.

51. Mitchell L. Cohen and Robert V. Tauxe, "Drug-Resistant *Salmonella* in the United States: An Epidemiologic Perspective," *Science,* Vol. 234, (Nov. 21, 1986), pp. 964–969.

52. Bill Keller, "New Data Tie Human Illness to Antibiotics in Animal Feed," *New York Times,* Sept. 16, 1984, pp. 1, 19; and Irvin Molotsky, "Antibiotics in Animal Feed Linked to Human Ills," *New York Times,* Feb. 22, 1987, pp. 1, 20.

53. We will discuss this subject more fully in Chapter 17.

54. Paul Brodeur, "Annals of Chemistry: In the Face of Doubt," *New Yorker,* June 9, 1987, pp. 70–87. In January 1988, the evidence indicating that chlorofluorocarbons are involved in ozone depletion became stronger.

55. In March 1988, Dupont and other large CFC-producing firms had begun to cut back on CFC production; see Philip Shabecoff, "Industry Acts to Curb Peril in Ozone Loss," *New York Times,* Mar. 21, 1988, pp. 1, 11.

56. See Irving Janis, *Groupthink,* 2nd. ed. (Boston: Houghton Mifflin, 1982).

57. Eugene Bardach and Robert A. Kagan, *Going by the Book: The Problem of Regulatory Unreasonableness* (Philadelphia: Temple University Press, 1982).

58. Jordan D. Lewis, "Technology, Enterprise, and American Economic Growth," *Science,* Vol. 215 (Mar. 5, 1982), pp. 1204–1211.

59. And sometimes it is corruption. Some of the efforts of the Gorsuch management of the Environmental Protection Agency involved cooperation with industry that became a little too close. When the regulators lunched with the regulated, this was clearly an inappropriate interaction; see Jonathan Lasch, Katherine Gillman, and David Sheridan, *A Season of Spoils: The Story of the Reagan Administration's Attack on the Environment* (New York: Pantheon Books, 1984); and Melinda Beck, Mary Hager, William J. Cook, and Elaine Shannon, "Reagan's Toxic Turmoil," *Newsweek,* Feb. 21, 1983, pp. 22–25.

60. The sort of thing that comes to mind is the perhaps apocryphal story of the community library that was forced to close because it could not afford to build a ramp for wheelchair citizens, an action that was necessary as a condition for continued support from the federal government. The librarian protested that no citizen in the community used a wheel-

chair, and if one did, she would personally deliver books. But the agency was adamant, federal funds were withdrawn, and the library had to close. Only to a bureaucrat is rule enforcement of this kind a victory.

61. James Willard Hurst, *Law and Social Process in United States History,* The Thomas M. Cooley Lectures (Ann Arbor: University of Michigan Law School, 1960).

62. David Loth and Morris L. Ernst, *The Taming of Technology* (New York: Simon & Schuster, 1972).

63. Nicholas Wade, "When Judges Must Know More Than Law," *New York Times,* Dec. 27, 1987, Section 4, last page.

64. Sheila Jasanoff and Dorothy Nelkin, "Science, Technology, and the Limits of Judicial Competence," *Science,* Vol. 214 (Dec. 11, 1981), pp. 1211–1215.

65. See Howard Becker, *Outsiders: Studies in the Sociology of Deviance* (New York: Free Press, 1963), pp. 147–163.

66. Oscar E. Anderson, Jr., *The Health of a Nation: Harvey Washington Wiley and the Fight for Pure Food* (Chicago: University of Chicago Press, 1958).

67. See Harvey Wiley, *The History of a Crime Against the Food Law: The Amazing Story of the National Food and Drugs Law Intended to Protect the Health of the People Perverted to Protect Adulteration of Food and Drugs* (Washington, D.C.: Harvey Wiley, 1929).

68. Richard Berke, "Surgeon General Raises Estimate of Smoking Toll," *New York Times,* Jan. 11, 1989.

69. Louis Filler, *Crusaders for American Liberalism* (Yellow Springs, Ohio: Antioch Press, 1939).

70. See his autobiography: Upton Sinclair, *The Brass Check: A Study of American Journalism* (Pasadena, Calif.: Upton Sinclair, 1920).

71. Benjamin Parke DeWitt, *The Progressive Movement* (New York: Macmillan, 1915).

72. A good sense of these groups and their activities can be gained by reading a book written by one of the great reformers; see Florence Kelley, *Some Ethical Gains Through Legislation* (New York: Macmillan, 1905), pp. 209–228. Kelley was general secretary of the National Consumers' League and also served as a factory inspector for the state of Illinois.

73. See, for instance, Dorothy Nelkin, *Nuclear Power and Its Critics: The Cayuga Lake Controversy* (Ithaca, N.Y.: Cornell University Press, 1971), pp. 33–78.

74. David Dickson, "France Weighs Benefits, Risks of Nuclear Gamble," *Science,* Vol. 233 (Aug. 29, 1986), pp. 930–932. This secrecy has permitted two cover-ups: a whitewash of radioactive fallout dangers from Chernobyl and a major cooling system failure in 1984. Interestingly enough, explaining "irrational" fears of nuclear power seems to be a minor intellectual industry in France.

75. Frank Graham, Jr., *Since Silent Spring* (Greenwich, Conn.: Fawcett, 1970), pp. 96–153.

76. It is interesting that Everett Rogers, a sociologist who is an expert on innovation, notes that when *Silent Spring* appeared, he was studying agricultural innovation, and he accepted uncritically the general dismissal of the book by agricultural scientists. He also notes, however, that one of the farmers he interviewed had made interesting observations about the negative effects of pesticides, observations that Rogers dismissed at the time as irrational; see Everett Rogers, *Diffusion of Innovations* (New York: Free Press, 1984), p. 189. Rogers's experience illustrates the value of independent critics such as Carson, who did not accept the prevailing consensus. It also shows how dangerous it can be for sociologists to accept scientific opinion uncritically.

77. Tina Lam, "The Quiet Critic of Fermi," *Ann Arbor News,* July 6, 1986, Section G, pp. 1, 2.

78. David Kushma, "NRC Staff Proposes $300,000 Fermi Fine," *Detroit Free Press,* July 4, 1986, Section A, pp. 1, 4.

79. See, for instance, Margaret Sanger, *My Fight for Birth Control* (New York: Farrar & Rinehart, 1931).

80. It is worth noting that many drug deaths result from the use of barbiturates, which are physician-prescribed drugs, whereas marijuana, which is much safer and of considerable therapeutic value, is illegal. Whereas physicians have high social status, drug dealers are hunted down. The contradictions of drug legality are discussed in Stuart C. Hadden, "Drug Use," in Armand L. Mauss (Ed.), *Social Problems as Social Movements* (Philadelphia: Lippincott, 1975), pp. 237–280.

81. Ralph Nader, *Unsafe at Any Speed: The Designed-In Dangers of the American Automobile,* rev. ed. (New York: Bantam, 1973).

82. Thomas Whiteside, *The Investigation of Ralph Nader* (New York: Pocket Books, 1972).

83. Robert F. Buckhorn, *Nader: The People's Lawyer* (Englewood Cliffs, N.J.: Prentice-Hall, 1972), pp. 143–158.

84. Brian Dumaine, "Nuclear Scandal Shakes the TVA," *Fortune,* Oct. 27, 1986, pp. 40–48; and Bill Paul, "TVA's Troubles," *Wall Street Journal,* July 9, 1986, p. 1.

85. Telephone interview with Owen Thero, July 27, 1987.

86. Some of the following account comes from *NRC Regulation of TVA,* hearings before the subcommittee on Oversight and Investigations of the Committee on Energy and Commerce, U.S. House of Representatives (Washington, D.C.: U.S. Government Printing Office, 1987). I have also had extensive contact with Thero, who supplied many details missing from public accounts.

87. Indispensable background for understanding these events is supplied in articles by Randell Beck in the *Knoxville Journal,* Sept. 22 and 23, 1986: "TVA's Saltwater Network," and "Saltwater Crew Taking Liberty?"

88. Anonymous, "NASA Encourages 'Whistle Blowing'," *New York Times,* June 7, 1987.

89. Roger Boisjoly, speech at the University of Michigan, Nov. 6, 1987. The *Challenger* tragedy is treated more fully in Chapter 14.

90. In 1986 a U.S. House of Representatives panel revealed that a manual used in 1983 to train nuclear power plant employees of the General Public Utilities Nuclear Corporation contained tips on how to befuddle federal inspectors. This was the organization whose Three Mile Island power plant had been the scene of one of the most serious nuclear accidents up to that time, and it had been the only utility convicted of criminal violations; see Matthew A. Wald, "Tips at Atom Plant: Keep Inspectors in the Dark," *New York Times,* Sept. 1, 1985, Section L.

91. See, for instance, Ralph Nader, Peter J. Petkas, and Kate Blackwell (Eds.), *Whistle-Blowing: The Report of the Conference on Professional Responsibility* (New York: Grossman, 1972); Deena Weinstein, *Bureaucratic Opposition* (New York: Pergamon Press, 1979); John T. Edsall, "Two Aspects of Scientific Responsibility," *Science,* Vol. 212 (Apr. 3, 1981), pp. 11–14; Vivian Weil (Ed.), *Beyond Whistleblowing: Defining Engineers' Responsibilities* (Chicago: Illinois Institute of Technology, 1983); and Myron Peretz Glazer and Penina Migdal Glazer, *The Whistleblowers: Exposing Corruption in Government and Industry* (New York: Basic Books, 1989).

92. R. M. Anderson, R. Perrucci, D. E. Schendel, and L. E. Trachtman, *Divided Loyalties:*

Whistle-Blowing at BART (West Lafayette, Ind.: Purdue University Press, 1980), pp. 108–116.

93. Gordon Friedlander, "Fixing BART," *IEEE Spectrum,* Feb. 1975, pp. 43–45.

94. A. Ernest Fitzgerald, *The High Priests of Waste* (New York: Norton, 1973), pp. 224–282; see also Brit Hume, "Admiral Kidd Vs. Mister Rule," *New York Times Magazine,* Mar. 25, 1973, pp. 38–52.

95. Ben A. Franklin, "The T.V.A. Mothballs Its Nuclear Ambitions," *New York Times,* June 8, 1986, Section 4, p. 4.

96. Colleen O'Connor, Ginny Carroll, and Sylvester Munroe, "Troubled Times for the TVA," *Newsweek,* Jan. 17, 1986, pp. 23–24; and Ben A. Franklin, "Nuclear Program at T.V.A. Assailed," *New York Times,* Mar. 2, 1986, Section 1.

97. Weinstein, *Bureaucratic Opposition.*

98. N. R. Kleinfield, "The Whistle-Blowers' Morning After," *New York Times,* Nov. 9, 1986, Section 3, pp. 1, 10, 11.

99. Roland Wilkerson, "New Law Increases Whistle Blowing," *Detroit News,* Sept. 9, 1981, Section A, pp. 3, 9.

100. Neal Karlen with Ginny Carroll, "Nuclear-Powered Murder?" *Newsweek,* Nov. 4, 1985, p. 29.

101. At the national level, the federal Merit System Protection Board, created in 1978, was designed to protect whistle-blowers. But it has received little support from the White House and became what one critic called "a sting operation to smoke out critics of the government"; see the Associated Press story "Informers' Aid Agency Attacked," *Detroit News,* Nov. 3, 1980, Section A, p. 5. Under the Reagan administration, the board was accused of advising managers on ways to avoid trouble while taking action against their workers; see "She Blows Whistle on Own Office," *Detroit News,* May 18, 1982, Section A, p. 4.

102. A. Ernest Fitzgerald, *The Pentagonists: An Insider's View of Waste, Mismanagement, and Fraud in Defense Spending* (Boston: Houghton Mifflin, 1989), pp. 1–6. Fitzgerald was fortunate in that the Watergate tapes proved Nixon's collaboration in his harassment and firing; see "Why Nixon Settled a Suit out of Court," *Newsweek,* Aug. 24, 1981, p. 19.

103. Kleinfield, "Morning After," p. 10.

104. Samuel C. Florman, "Beyond Whistleblowing," *Technology Review,* July 1989, pp. 20, 76.

105. James T. Ziegenfuss, *Organizational Troubleshooters: Resolving Problems with Customers and Employees* (San Francisco: Jossey-Bass, 1978).

106. One of the more disturbing portraits of this "technological community" for agricultural chemicals is Robert Van Den Bosch, *The Pesticide Conspiracy* (Garden City, N.Y.: Doubleday, 1978).

C H A P T E R   1 6

Technology Assessment

We need technology assessment because we need protection against unforeseen problems.[1] Throughout our lives we are exhorted to "look before we leap," informed that "experience keeps a dear school," and advised "better safe than sorry." Yet often we ignore this well-meant advice and we get into trouble. We act rashly, "learn the hard way," or suffer from "20-20 hindsight." Although our individual mistakes can be serious enough, the same principle applies to society at large. And usually, the larger the group, the more serious the consequences. Thus it is important to consider, in a systematic way, the potential consequences of new technologies in society.[2]

The Importance
of Technology Assessment

Every technology is an intervention in our world. Each intervention produces effects, some potentially reversible, some not. Like a stone dropped into a pool, such interventions sent ripples of change through the social and physical structures that surround us. Some of these consequences are direct; such first-order effects are a technology's primary purpose. But there are also second-order consequences, side effects. Our attention is usually focused on the primary effects, and because we develop a technology to create them, we tend to monitor these effects carefully. The side effects however, precisely because they are *unintended* consequences, often escape our attention. We notice them in advance only if we are forced to do so. Thus a major motive for technology assessment is that it forces us to pay attention to side effects.

Ignoring side effects is one problem. Another is the way we learn. Human beings learn relatively well in simple trial-and-error situations. When we get immediate feedback from what we do, we tend to change our behavior quickly, as when we touch a hot stove. But with many modern technologies, this kind of trial-and-error learning is inadequate. With carcinogenic chemicals or radiation poisoning, for instance, we often must wait 20 years to experience the full extent of the damage. In situations like this, feedback is too late. We need some form of "feed-forward." We will discuss the dynamics of such predictive mechanisms in Chapter 17. But here it is important to note that the exercise of *foresight* is not automatic with us. Hence we need some kind of systematic procedure to force us to look forward and consider longer-range consequences.

Yet another problem is the scale of our current technological activities. The activities of primitive cultures made impacts on the environment, but these effects were small and local, even if sometimes enduring.[3] If an ancient group of people depleted its environment, it could always move on to new forests, new pastures, or new mountains. Now humanity is filling up the planet. We are involved in changing major aspects of life on earth. Are the institutions we have, is our ability to project and to understand, up to the task of identifying such impacts and making wise decisions about them? Do we have a global perspective appropriate to the global consequences of our activities?

In cybernetics there is a concept called *requisite variety.*[4] This simply means that a well-designed system must have a variety of responses that is equal to the variety of situations it faces. Requisite variety is, in other words, a kind of criterion of intellectual fitness. We can postulate a similar criterion for the performance of our own forward-looking institutions. Let's call this *requisite imagination,* the ability to imagine aspects of the future we are planning. As we will see in the next chapter, feedback about our actions often comes too late to help us avoid some severe negative consequences. This means that there has to be some kind of "feed-forward" mechanism to aid us in decision making.

But whereas the past is factual—it has already happened—the future is hypothetical, and there are several futures that could take place, depending on what we do now. Each of these futures—not only the ones that should be avoided, but also those that might offer a significantly better outcome than what we would expect—must be envisioned. A requisite imagination would give us access to the full spectrum of these futures.[5] This spectrum would in turn help us to identify dangers, envision opportunities, and anticipate possible choices. Institutions that embody a requisite imagination would generate a variety of forward-looking activities and would include such activities in the formation of policies and shaping of decisions.

We might examine how our institutions would stack up on an "index of fitness." The mental reach of our society must be adequate to the scope of the interventions we make, for if it is not, we will be like a blind person, unable to "see" something until we bump into it. And there are many things which ought to be seen before they are bumped into. One of the ways we can avoid such unpleasant futures is to learn ways to scrutinize technologies with an awareness of possible side effects and long-range impacts. Among the best ways to look to the future is technology assessment.

Technology assessment is an attempt to predict what the effects of a technology will be if it is implemented. Some people use the term *technological forecasting* when only the technical possibilities are of interest,[6] but technology assessment usually includes both the technical and human aspects of the technology because, as we have seen, they are closely interconnected.[7] By "technology assessment," people usually mean a report that examines the potential effects of, say, developing a new communications technology, using certain kinds of chemicals on the environment, or automating offices. Here technology assessment is a product, but first and foremost it is a process: It is a continuous inquiry into the nature of our technological activities and the enormous variety of consequences they might have.[8]

The History of Technology Assessment

As we have seen, William Fielding Ogburn and Sean Colum Gilfillan were very much interested in the assessment of technology's consequences, and they often reflected not only on particular technologies, but also on the process itself. But neither the term nor the importance of technology assessment was widely accepted until the 1960s. The term itself was coined in 1966, although the concept is certainly much older.[9] The idea was embodied in the creation of the congressional Office of Technology Assessment in 1976.[10]

Technology assessment developed as a means of coping with the rapid technological changes taking place, many of which people had begun to view with alarm. Exactly why the 1960s should have seen the flowering of this idea is not entirely clear. Rachel Carson's book *Silent Spring* (1962) may well have been a catalyst, but probably the strongest influence was the ferment so pervasive in American society during this time.

As we mentioned in Chapter 3, technology often seemed to be a dangerous and alien force during these years.[11] Accordingly, various attempts were made to bring it under control. Similar concerns also led to the development programs of "future studies" and to the Environmental Protection Act of 1969, which mandated the use of environmental impact statements.[12] Many people felt that technology was changing too rapidly; Alvin Toffler's book *Future Shock* expressed what many people felt.[13] Previously, people assumed that because something could be invented or developed, it ought to be; progress could not be held up! But such approval was no longer automatic, and without some way of determining which innovations were useful or safe, choices could not be made. Society needed a form of foresight, so methods were developed to explore future possibilities and consequences mentally before we in fact had to experience them. Thus one of the ways undesirable consequences might be avoided was technology assessment. An early technology assessment was instrumental in ending the project to develop a supersonic transport in the United States.[14] "Doing a technology assessment" became familiar to Americans in similar controversies over antiballistic missiles, pesticides, and fluoridation. Thus technology assessment became one procedure for coping with and anticipating technological impacts.[15]

When Should a Technology Assessment Be Done?

To many people today, a technology assessment is a simple, straightforward procedure. A committee is gathered, it meets, it researches the problem, it writes a report, and it delivers it to the people who asked for it, usually representatives of the government. But this simple description is deceptive and even dangerous. Let's see why.

The most unrealistic notion of technology assessments is that there should be a single document, produced at a single point in time, presumably years ahead of the development in question. On the contrary, any major development should have more than one assessment.[16] There are a variety of reasons why multiple technology assessments are useful. First of all, any expert group will have its own blind spots and biases.[17] The best way to cope with such biases is to convene other groups, each with different blind spots and biases, and ask them to examine the same problem. Similarly, examination is likely to be most beneficial if it is carried out several times. As the date for the device's actual use approaches, society may have narrowed its options for changes, but at the same time the technical details may have changed in important ways, and in any case more information will be available. This new information allows reassessments of the device's impact based on this new data.

In fact, some of the most valuable assessments may come after the early stages of a device's use.[18] Although these assessments may be "too late" in some respects, they are likely to offer much more realistic information about the nature of uses and users, good and bad side effects, and so forth. They may confirm that earlier assessments were correct, a useful piece of information in itself. If there are negative side effects, this may be a good reason to institute regulation or to discontinue use.[19] If there are positive side effects, they may be amplified by changing the device, the strategy for using it, or public perception of it in various ways.

As we have already seen in Chapter 9, Paul Ehrlich was very careful to assess the immediate effects of compound #606. The oral birth-control pill provides another important example of this need for continuing assessment. Although the pill in its early formulations was a very useful device, many of the medical side effects proved worrisome. Unfortunately, the medical community and the press were initially very reluctant to talk about these side effects.[20] Then reports of problems with cervical cancer and stroke surfaced, prompting a great deal of concern that the pill was dangerous. Investigation showed that the association with cancer was unjustified, but the cardiovascular problems were real enough. It was also discovered that the dosage of hormones in the pill could be reduced with no reduction in effectiveness. Thus feedback led to changing both the pill itself and the kinds of warnings included with it.[21] Even some three decades after introduction of the pill, new concerns are being raised.[22]

In surgery, new procedures are often greeted with much fanfare. As time goes on, however, their true effectiveness becomes more apparent. Sometimes the early promise holds up, and sometimes it proves false.[23] Accordingly, efforts have been made to establish national centers to evaluate surgical procedures and the technology that supports diagnosis and treatment. Federal support for these centers, however, has been fickle.[24] The National Center on Health Care Technology was allowed to disappear for lack of funding, and the federal government only recently created a partial substitute, the National Health Care Information Service Center. Even as the costs of medical-care technology rise, efforts to evaluate its effectiveness lag behind.[25] No doubt Ogburn would have been correct in seeing this as a "cultural lag!"

So a technology assessment should not be done once for all time. Rather, a continual upgrading of assessments of new technologies is a necessary activity. We need to shift from the judicial model in which we "weigh the evidence" once to an active model of continuing inquiry.[26]

How Technology
Assessment Works

Doing technology assessment is very much an art, like painting a picture, cultivating a garden, or designing a building. It cannot and should not be turned into a highly routinized activity, but rather it should be regarded as an essentially open system. Joseph Coates, a well-known practitioner of this art, defines it as "the use of human arts and sciences to achieve human objectives."[27] Although this definition is too broad, it raises an important point: Technology assessment is a form of inquiry, and although it might be possible to routinize it, such routinization would have undesirable consequences, for it would limit the tools available to technology assessments.

How should a technology assessment be done? First, a broad outline of the technology or intervention must be considered.[28] This broad outline would identify some of the technology's impacts and thus some of the potential stakeholders who would be affected when the technology is introduced or changed.[29] These stakeholders, or representatives of this group, should then be included in the technology assessment process, for they will have a major role in shaping the technology's fate and must live with the consequences.

The group thus gathered then engages in a series of exercises designed to help them identify and imagine future consequences. Some of these exercises are highly structured and seek estimates of, for instance, the number of users involved. Other techniques have more of a brainstorming quality and involve the construction of future scenarios. Among the commonly used techniques are:

1. *Mathematical modeling* (for example, of the economy), *simulation, and trend analysis*. Attempts are made to provide formal computer models that are specific enough to make quantitative predictions.[30]

2. *Technological forecasting*. Here the emphasis is on predicting the development of the technology and assessing its potential for adoption, including an analysis of the technology's market.[31]

3. *Cross-impact analysis*. Because technologies have interactive effects, realistic planning for the future must consider the multiplicity of impacts technologies may have on society. The state of society into which the technology under consideration will enter must be considered.[32]

4. *Scenario building*. First developed formally at the Rand Corporation by Hermann Kahn, scenarios are plausible sequences of action used to facilitate consideration of unforeseen possibilities.

5. *Gaming*. The use of gaming techniques is extensive in military affairs and, to a lesser extent, in business. Gaming forces planners to take account of the actions of others, who are likely to do surprising things, just as might occur in real life.[33]

6. *Delphi methods.* A Delphi procedure provides a method of pooling expert opinions, and in its more sophisticated forms it forces experts to reexamine their own opinions in the light of possibilities raised or assessments made by others. The key is not simply to create an "average" assessment, but rather to provide a spectrum of opinions and their rationales.[34]

It is important that the exercises provide scenarios that include surprises, for surprise stimulates thought. Imagination is difficult to stimulate and may be actively resisted. People often find it very difficult to imagine the consequences of policies they are considering, not only in futuring exercises, but in real life. As we saw in Chapter 8, people are often unwilling to project themselves into the future. People often fail to consider that the technology will evolve, and they may neglect important user groups and forget that there may be alternatives. Cross-impacts from other technologies can be neglected as well, which reduces the realism of the exercises. One value of technology assessment exercises is that they provide an environment in which imaginative foresight is encouraged and supported.

Often a technology assessment is viewed as an exercise in prediction. However, it might be better to consider it an exercise in stimulating requisite imagination. Simply getting relevant decision-makers to think about various outcomes, even if someone else has already predicted them, can be valuable. The formal exercises can be mind-stretching, forcing people to consider different applications, time frames, and uses from those that come most readily to mind. Studies by cognitive psychologists have shown that often the examples that come most readily to mind are not necessarily indicative of the full range of situations likely to occur.[35] In some ways the most important aspect of a technology assessment is the broadened perception of outcomes, whereby the imagination of participants is expanded so that they think about not only the consequences of what is being considered, but also what additional information or intervention might be needed. A technology assessment need not arrive at firm conclusions; it may be useful simply by stimulating further inquiry or by developing awareness that the future has in fact not been adequately thought through.

Thus the "output" of a competent technology assessment is likely to be both a document and a set of expanded minds, for even though the group doing the assessment may have to report to a larger body or even to the public, the key point is that they have been forced to do some thinking. Reports can be put on a shelf, as many fine committees have discovered. But when the stakeholders (i.e., those who will be affected by the decision) themselves have been forced to explore consequences, they are less likely to forget or ignore them. A technology assessment, then, is both a thinking process and a sensitization process. Two Dutch scholars, Ruud Smits and Jos Leyten, view technology assessment as having evolved from a "scientistic" phase, in which technology assessment was done in a neutral manner by scientific experts, into a more political model.[36] In the latter form, technology assessment is more likely to involve political actors and considerations. The reason for this shift is not that scientists develop inferior answers as compared to politicians. Rather, the key to useful assessments is having policy-makers as well as scientists think about a broad range of impacts.

There may be no obvious "rational" way to move from an assessment to implementation of a policy under scrutiny. During assessments, information is collected, exchanged, and analyzed. It still must be acted upon, and this involves not only information, but also values, political considerations, and the like. It is a misconception that scientists do research and politicians act on such research. In reality, research is only one input into the decision-making process; many other inputs enter in. Getting politicians and civil servants to think and to increase their understanding is probably the best we can hope for. If the public can be involved also, so much the better. Technology assessments increase the informational basis for decisions by forcing consideration of things typically left out of consideration. But what will be done with the information, if anything, is another matter.

The Role of "Interested" Experts in Technology Assessment

The usual view is that technology assessment is a process that is "rational" in a calculative way. We might more beneficially apply a generative rationality to technology assessment. What we need are not so much procedures that "settle" the issue one way or another, but rather an open form of inquiry that invites contributions from a variety of quarters and involves decision-makers in the process. Thus a technology assessment "opens up" inquiry rather than closing it off. Any group that might have useful information will be encouraged to generate it in the first place and then transmit it in a useful way. This means that we must carefully consider whom we should ask for an evaluation.

One problem with the single-shot idea of a technological assessment is the "committee of experts" idea. Although it is certainly valuable to get knowledgeable people involved in the assessment, some "experts" may not always perform very well. For instance, they may fail to imagine what the full range of effects is likely to be.[37] A full-scale assessment includes not only the future development of the device, but also the future behavior of society. Each of these aspects involves complex predictions, and not all experts are equally good at providing them. For many people, including the very bright, the future may be a threatening prospect; hence they desire to shape the future according to their conceptions of the present. But the real future—the one that is out there beyond the walls of academia—may not be so easily shaped and may surge on regardless of experts' predictions. Many of the people on expert committees have achieved splendidly in the past with current technologies; they may be very good at calculating present capabilities, but less able to imagine future applications and alternatives.

Furthermore, "objectivity" does not necessarily accompany expertise.[38] Real-world experts are very likely to have some vested interest in the technology they are evaluating.[39] Such intellecutal "conflicts of interest" are a general phenomenon, and it is important to take them into consideration, rather than trying to find a "perfectly neutral" body of experts, if such a thing exists.

Vested interests are evident, for instance, in expert evaluations of nuclear power. The reason for this kind of vested interest is simply that scientists and technologists are likely to derive their expertise by previous work on the technologies in question, often while they are employed by the sponsors of the technology. The nuclear scientist is often a consultant to the nuclear-power industry, the pharmaceutical chemist is often employed by a drug company, and so on.[40] More than one critic has pointed out that this means that much scientific advice is shaped by industry ties.[41]

After the Santa Barbara oil spill in 1969, community members discovered that scientists knowledgeable about petroleum technologies tended to get their support from the petroleum industry.[42] The rapid development of recombinant DNA technology turned many "dispassionate" scientists into industry promoters.[43] But commitment to the plans of government agencies can also influence opinions. The key problem here is that when people belong to a network with a common opinion, they naturally tend to agree with the industry consensus. And when insiders give evaluations that differ from this consensus, they may face severe sanctions.

Dr. Roy D. Woodruff was director of a $300-million weapons program at Lawrence Livermore Laboratory in California. In 1985 he criticized the arguments of Edward Teller, one of the major proponents of President Ronald Reagan's "Star Wars," the Strategic Defense Initiative. Woodruff claimed that Teller had been exaggerating the capabilities of the X-ray laser, Super Excaliber, an important part of "Star Wars." His remarks, "at first in secret memoranda and then to the press, angered colleagues and shocked some members of Congress." As a result, he has claimed, he was forced to resign his position. Later, after considerable protest, he was restored to his post.[44] The U.S. Secretary of Energy, John S. Herrington, let it be known that he was opposed to such public airing of scientists' disagreements, an ominous sign for future dissent because the Department of Energy funds laboratories like Livermore.[45] And other critics of powerful programs have not been so fortunate.[46]

The Role of Controversies in Technology Assessment

Allan Mazur has identified the difficulties involved in expert testimony in technological controversies. He notes that a technology's proponents are likely to overlook certain kinds of problems and hazards. Official committees, even bureaucracies set up to safeguard the public, are fallible. Thus it is important to have a free flow of discussion from independent agents, for

> there have by now been numerous instances when the process of social controversy has been more effective in identifying and explicating the risks and benefits of a technology than have any of the formal means which are supposed to do this. Critics attack with great vigor, stretching their imaginations for all manner of issues with which to score points against their target. Proponents counterattack, producing new analyses and funding new experiments in order to refute the critics. Each side probes and exposes weaknesses in the other side's arguments. As the controversy proceeds, there is a filtering of issues so that some with little substance become ignored while others proceed to the fore.[47]

Hence, rather than a single assessment by a single group, we benefit from a constant flow of information from multiple sources.[48] These expressions may include formal reports, articles in professional journals or newspapers, or even simple expressions of opinion. The free flow of information and opinions is important, for out of it may arise facts and issues that become the subject of still other, sustained investigations.

Controversies can thus have a valuable investigative function.[49] Of course, false issues, stereotypes, and misinformation may also result from public controversy, but these are hardly absent from expert reports. Some informal technology assessments may cause problems. In January 1989, Ralph Nader's organization Public Citizen announced that some home kits for detecting the radioactive gas radon gave "dangerously misleading results." This announcement induced many people to turn in their radon kits for a refund, and it had damaging economic consequences for the companies that sold the kits. Other experts, however, pointed out that even though the detection devices were not very good, they might well be good enough for most practical purposes, and these experts criticized Nader's group for its premature announcement.[50]

Controversies can also cause delay. In the introduction of new drugs, for instance, many people criticize the federal Food and Drug Administration (FDA) for being much too slow.[51] Commonly the FDA is considered to be excessively conservative. But sometimes the problem may be politics. In 1982 the FDA planned to issue a warning that giving aspirin to youngsters with flu or chicken pox might cause a potentially fatal disease called Reye's syndrome. After meetings with the aspirin industry, however, the Office of Management and the Budget decided that the evidence was not strong enough. Three years and 3,000 additional cases of Reye's syndrome later, the FDA was allowed to issue a warning.[52]

Controversies are likely to polarize opinion even among experts.[53] The different sides are likely to have on hand (or recruit) technical experts who will support their point of view. This kind of polarization took place, for instance, in the controversy over antiballistic missile systems. During this dispute, feelings in the scientific community ran fairly high. Professor Albert Wohlstetter of the University of Chicago and his opponents exchanged letters in the *New York Times*. Then Wohlstetter became upset over what he regarded as improper treatment of the data by his opponents, and he asked the Operations Research Society of America to look into the matter. They did, and they issued a substantial report largely supporting Wohlstetter.[54] The report led to rejoicing on one side and very hard feelings on the other.[55] One of the costs of controversy is precisely this kind of scientific polarization.

In Allan Mazur's estimation these costs of conflict are usually worth incurring.[56] The basic point is that we want as many sources of information and advice as possible. It may be difficult to anticipate in advance how or from where the critical facts may surface. The persistent scrutiny that controversy generates is not only helpful for getting needed information; it also helps educate those who scrutinize. It is important that people develop the habit of scrutiny, and public controversy is a means of achieving this goal. More people will pay attention when they know that their opinions matter. Controversy, then, may produce propaganda, misapprehensions, and polarization, but it can also produce awareness and interest.

Assessments Conducted by Business Firms

If we use a broad definition of technology assessment, we would find that much of it is done for private interests.[57] This broad definition requires some qualification, for technology assessments carried out by the "private sector" do not have the same purposes, and do not follow the same procedures, as those carried out for the government. Industries or firms conduct assessments for one primary purpose: to determine how well a product or a class of products will sell.[58] A secondary reason today is to assess such corporate risks as product liability that the firm might incur in marketing the items in question. What wider impacts the technology might have on society is usually ignored; firms exist to make money, not to save or protect society. Sometimes governmental technology assessment has been similarly limited to feasibility studies without regard for broader consequences.[59]

Business firms, then, routinely carry out technology assessments in the form of market forecasts out of necessity. Manufacturers often make plans for their products years ahead of their actual introduction to the market., which puts a premium on foresight. When plans miss the mark, the effects can be serious. When microwave ovens were becoming popular, many companies decided that they would produce cookware especially designed for the new technology. After tooling up and going into production, however, many companies found that their products simply were not selling very well. They had made two serious mistakes: They overestimated how many microwave ovens would be sold, and they did not anticipate that some ordinary dishes would work very well in the ovens. A final frustration was that too many firms entered the microwave cookware market, thereby leaving only a relatively meager market share to each one.[60]

An even bigger mistake was the Ford Motor Company's decision to produce the Edsel automobile, whose sales were disastrously lower than expected.[61] Altogether 110,810 Edsels were produced between 1957 and 1960, at a net loss to Ford estimated at $350 million. With a design gleaned from the results of opinion polls and "scientific" sources, the Edsel proved an impressive failure. Much of the blame for slow sales can be given to poor quality control, which contrasted badly with the car's publicity. But its failure was also due to the simple fact that it wasn't what people wanted to buy. The company's assessment had failed in one of its basic goals: It had misgauged how many people would use the device.

It is worth taking a few moments to reflect on the Edsel fiasco. We should first note that automobile companies tend to take their assessments very seriously. Further, the resources for corporate technology assessment are often quite impressive as compared to those in the public sector. The decision to produce a new line of cars today means the commitment of hundreds of millions, sometimes billions, of dollars. Ford's decision to produce the new Sable/Taurus line in 1986 cost $3.1 *billion.*[62] The company must retool entire plants, develop dealer networks, and attempt to manipulate consumer attitudes through advertising. Designers, engineers, manufacturing workers, and salespeople are all geared up to sell the new car. In spite of all this, sales are often wildly at variance with predictions. Part of the problem, of course, is the many kinds of cars from which buyers can choose, some of which are produced by other divisions of the same firm. Nonetheless, very large sums go into the private marketing reports, sums large enough to make university social scientists green with envy. That corporate studies produce dubious results is not reassuring because many of the same techniques are applied to test probable acceptance of other technologies, some of them much more radical than cars.[63]

It is well known, by the way, that market tests are not very useful on truly novel products because consumers often cannot anticipate their uses.[64] For instance, a company named Haloid tried to determine the potential market for a dry photocopying machine when most people were still using a wet photocopying process. The market survey suggested that the firm could not make a profit selling such machines. The firm, however, did not believe the market survey; it was so convinced the product was useful that it went ahead and marketed what it called the Xerox machine. The Haloid executives' "unyielding optimism" was justified and rewarded: Xerox machine sales climbed rapidly, and the brand name is now a household word.[65] On the other hand, market research firms can exaggerate a new product's potential.[66] The disparity in these observations may provide some clues about our ability to predict the future use of new technologies.

The reason for considering here these relatively narrow assessments carried out by business firms is that we can often learn a great deal about the process of prediction and assessment by noting the performance of the private sector. As we observe that they often make mistakes about relatively simple matters, we will learn humility in regard to our ability to make predictions about more complex ones. This humility is very helpful to us, for it creates a greater degree of vigilance—a very useful attitude whenever change is being considered.

The Basic Questions
Assessments Must Answer

There is no simple formula for carrying out technology assessments. Although individuals and organizations have developed skill at doing assessments, it is still an art, not a science. Where, for instance, do we start? We live in a world in which everything is connected to everything else. Any intervention is likely to have nearly endless ramifications, as Ogburn pointed out long ago. We must start somewhere, though, and we must agree to scrutinize some areas more closely than others. One way to begin is to start with certain basic questions. The answers to these questions may raise still other questions. There are four basic questions that technology assessment must answer:

1. *How will the technology evolve?* An assessment must determine what will happen with the technology itself. How much is it likely to change? Will it become cheaper? More efficient? Will competitive technologies spin off from it?

2. *What will the technology be used for?* Often the answer to this question is not at all clear at the outset, but it has a great deal to do with overall impacts.

3. *Who will the technology's users be?* The primary impact of a new technology will fall upon its users. Determining who they likely will be allows us to answer many other questions.

4. *Who will decide how the technology will be used?* Knowing who a technology's sponsor will be is valuable in anticipating its development; knowing who will resist it is equally important.

The answers to these questions lead to other questions: What technologies will the new one replace? How will it affect our work lives? What will its impact on the family be? To answer these additional questions, we must look at specific parts of society—the home, the workplace, transportation, and so on—in terms of how the new technology may change these institutions.

This brings up an important point: The ability to do good technology assessments requires a very good understanding of how society works. Ordinarily we would expect an assessment team to include social scientists, legal experts, engineers, and scientists. We would hope that professionals who do technology assessment would not only develop a good knowledge of history, but would also monitor the accuracy of their own previous assessments. And we would hope that some nonprofessionals would be involved in technology assessment, for experience shows they often think of questions that elude the professionals. It's probably also important to have a mix of ages among assessors, not only because we get different agendas with different age groups, but also because different age cohorts bring to the assessment different assumptions based on their experience (or lack of it). Similar considerations suggest involving both sexes and members of different ethnic groups.

Good technology assessments require wisdom, not just professional competence. Hazel Henderson, a thoughtful writer on this subject, has pointed out a pitfall of professional assessments:

> Now for the bad news. Although technology assessment is coming to be understood as a form of societal and governmental learning, one wonders whether there might not be a cheaper way than this for society to re-educate and recycle all of its overspecialized, miseducated, technological junkies, than by letting them all play in the expensive new conceptual sandbox of TA. Ordinary citizens invented TA; it might be much more effective and less costly to include more of them in TA processes than to over specialize still another group of analysts and train them in new forms of incapacity. I do see, for example, that there is a great deal of compulsive schematization going on, and rationalization; and there's a great deal of new jargon and new mystification.[67]

So there is danger both in too little and too much professionalism in technology assessment. We need some professionalism; otherwise, each new group of assessors would have to relearn the elementary mistakes of their predecessors. But if technology assessment becomes too professionalized, systematic errors could be perpetuated. Professionals have similar training, go to the same meetings, read the same publications, and tend to share the same biases.

Something of the complexity of real technology assessments can be grasped by considering assessments of technologies already in place.[68] Such retrospective technology assessments have been done, for instance, of the telephone, the railroads, and the submarine telegraph cable.[69] Interestingly, the initial assessors of the submarine cable were much more aware of its benefits than its drawbacks. The immediate benefits were obvious even before the cable was installed, but the good and bad side effects became evident only with time.[70] In the case of the telephone, an even more interesting analysis was also done: an after-the-fact assessment of before-the-fact assessments.[71] Unfortunately, this analysis could have been even more informative had it examined examples of predictions that were accurate versus those that were wildly off, or had it attempted to determine which groups were likely to make the best guesses about the future.

Now let's consider in more detail each of the basic questions assessments must answer.

How Will the Technology Evolve? Predicting the social effects of a technology is very complicated. Most technologies are changeable and are likely to evolve during the time they are in use. The issue of technology evolution during use is dramatically illustrated by the rapid development of computers made possible by smaller electronic circuits. Within four decades, the very meaning of "computer" and "circuit" has completely altered, and different systems have different impacts.[72] In 1950 a "computer" was a huge set of machines weighing several tons and using several dozen horsepower. Today the same calculations can be done on a system weighing a millionth as much and whose power requirements are a minuscule fraction of those of a "Univac" of the 1950s. Virtually no one in the 1950s foresaw the impact that miniaturization would eventually make.[73]

Similarly, in the first decade of the 1900s "airplane" inspired in few people the thought of a device the size and power of a Boeing 747. Before then some well-informed people doubted that aircraft could be made to work at all, and as late as the 1920s, there were intelligent people who doubted the importance of airpower in war.[74] Space exploration, anesthesia, and nuclear weapons all had intelligent detractors whose imaginations proved to be wanting.

But the opposite problem is also very real. Too often a new technology falls short of providing the cornucopia of uses it first appeared to offer. In the 1940s and 1950s the helicopter seemed to promise a totally new form of personal transportation. Ogburn and other sociologists glibly assumed that helicopters might one day be used for commuting into cities.[75] City skies filled with helicopters were envisioned by these enthusiasts (this vision is a nightmare to many of us). That helicopters might not improve sufficiently to permit ordinary people to fly them safely does not seem to have been considered. Obviously no consideration was given to the intractable traffic, pollution, and safety problems that widespread helicopter use in a city center would cause. Thus although helicopters have found a niche in modern society, routine use for transport and commuting is not part of it. The "helicopter age" did not materialize.

Today it is tempting to envision an "era of the ultralight aircraft." After all, ultralight aircraft are able to work using low-power (20-horsepower) engines, able to land on short landing strips, comparable to automobiles in cost, not restricted to areas with roads, and are about as easy to fly as automobiles are to drive.[76] I personally expect ultralight aircraft to have an important effect on life in desert and sparsely populated areas, and I predict ultralights will replace both fixed-wing aircraft and helicopters in many of their current uses. But I also suspect that there will not be an "ultralight era."[77]

Another blatant example of overoptimism occurred during the 1950s with respect to nuclear power. The "peaceful atom" was portrayed as an almost perfect power source: clean, efficient, and extremely cheap. It is painful to realize today, after Chernobyl, the difference between the earlier hopes and the current reality of nuclear power.[78] In part, of course, the reason nuclear power has proved so dangerous is that (in the United States at least) its sponsors have been incredibly careless in terms of safety. But even in a far more responsible society, it is doubtful that nuclear power as we now know it could be made safe on a large scale. This may change in the future, of course, but for the present it is important to absorb the lesson that technological evolution has not yet produced the safe, efficient energy provider that had been promised earlier.

Thus because a technology's evolution is unpredictable, its social effects must also be unpredictable, for social effects depend on what the technology is, and what the technology is, changes.

What Will the Technology Be ·Used for? We have already seen that a technology's uses are not always obvious in the beginning. It seems likely, therefore, that new uses for important technologies may sometimes prove to be more significant than those for which the technology was invented. Solar power was first developed to power satellites, for despite its expense, it is extremely compact. Only later was it seriously considered as a potential terrestrial power source. Similarly, few people imagined that a major use for computers would be word processing, or that radio would be important for mass communication. Clearly, social impacts are closely associated with the way in which a technology is used.

But the issue of use is important for another reason. Just as new technologies may allow people to do new things, other technologies may arise that can do them even better. And so, for instance, a technology like cassette tapes can have a rapid rise in popularity, only to be replaced in many of its niches by compact discs. Similarly, transistors, once a radical technology with very important impacts, were eventually replaced by integrated circuits.

New technologies not only offer alternatives, but synergistic applications as well. A new technology depends on other technologies, and other technologies, in turn, depend on it. For instance, space satellite technology depends on launch systems. Recently the United States found that the space shuttle's launch system had serious reliability problems.[79] Suddenly, the whole space program, satellites and all, was put in jeopardy due to problems with launchers. And the satellites also depend on other technologies as well, especially miniature electronic circuits.

This calls to mind a striking instance of the failure to anticipate the impact of related technologies. When Arthur C. Clarke first conceived of an artificial earth satellite, he did not anticipate that it would be unmanned. Clarke is a very imaginative futurist and a writer of science fiction. In an article published in *Wireless World* in 1945, Clarke proposed putting a communication relay into geosynchronous orbit, 22,300 miles above the earth.[80] Such satellites, Clarke predicted, would necessarily be very large in order to accommodate the large numbers of vacuum tubes, power sources, and crews to operate them.[81] He did not suspect that miniaturization would allow the use of very small satellites. For this reason, he did not patent his idea, and he did not anticipate that within two and a half decades it would become a reality. Now the "Clarke Orbit" is already becoming crowded.[82]

Who Will the Technology's Users Be? Who will use a technology is important because technologies tend to be designed and redesigned to fit their users. What users want becomes an important force in shaping the technology's form and substance. This was true with the development of bicycles, automobiles, and aircraft. In each case the initial models were largely used for sport.[83] As opportunities for larger groups of consumers to use them became available, they were redesigned with features that appealed to a broader range of users.[84] One of the reasons for the early importance of sport is that enthusiasts will put up with things that are not very practical, require a fair amount of mechanical knowledge, and even involve some danger.

Commercial sponsors, meanwhile, use sport to advance their claims for having the best vehicle. The sporting community provides both moral support for early users and a pool of expertise and problem-solving ability. Sport also provides strong incentives for improvements in performance. While the sporting community perfects the "curiosity" and takes the bumps and bruises, the transformation of the object into a commercially viable consumer product grows apace. Nonetheless, it is obvious that who the users are will have a strong impact on what "problems" need to be solved and what will come to be considered "solutions" for them.[85]

Thus the accuracy of a manufacturer's perceptions of users' needs may be important in shaping the technology's development. Henry Ford, for instance, was much more attuned to the needs of rural users than were his competitors.[86] When urban sales and product appearance became more important, Ford was unable to adjust rapidly enough, and General Motors became dominant in the American market.[87] The new, affluent urban users preferred appearance and comfort over ruggedness and were willing to pay the price of annual model changes.

Decisions about users also influence the direction of research. Edwin Land, founder of the Polaroid Corporation, was originally interested in developing polarized glass for the automobile industry. When he discovered that the automobile industry was not interested, he adapted his innovation for use in military applications and for sunglasses instead. Land was determined thereafter never again to put himself in a position in which there would be intermediaries between his firm and the user. Instead, Polaroid's products would be sold directly to consumers.[88] This attitude was one of the factors that led to Polaroid's decision to develop cameras capable of "instant" film development. Land's failure to interest the automobile industry demonstrates how a new technology may not be adopted by those for whom its developers intended it. Another group of users may prove to be far more important.

Who Will Decide How the Technology Will Be Developed? Another question assessment must answer involves its sponsors. The people who will manage the technology make a major difference in how rapidly it evolves and what its impacts will be. Similarly, the opponents of the technology can make a major difference, too. It is easy to see that powerful sponsors are more likely to be able to develop and deploy technologies rapidly. There are limits, of course. Even if solar-generated electrical power was embraced by the same utilities who have backed nuclear power, it is difficult to know how rapid its progress would be, although it would certainly develop more rapidly than it is currently. And it is obvious that in spite of powerful sponsorship, nuclear power must still overcome severe limitations.

But sponsors do make a difference. Charles Babbage's mechanical computers were makeable, and if the British Crown had decided that they merited it, they would have been completed, even the Analytical Engine. What impact this might have had on nineteenth-century technology can only be surmised.[89] Likewise, we might attempt to compare the development of passenger rail travel in the United States with that in other countries more dependent on rail travel. It is difficult to believe that "bullet trains," like the French Train Grande Vitesse or the Japanese Shinkansen, would have been widely used in the United States before the twenty-first century.[90] In the United States, the influence of powerful automotive sponsors has largely habituated short-distance travelers to the use of roads and long-distance travelers to the use of airlines. The railroads themselves have preferred to emphasize the transport of freight over that of people, and as a result they allowed railroad service to stagnate even as airlines and automobile companies were wooing passengers.[91]

Technologies can stagnate in societies in which there is little incentive for use or development; conversely, they can zoom forward in societies in which there is strong interest, as has happened recently with biotechnology.[92] We can only wonder what would have happened to the "microchip revolution" without the development of Silicon Valley, international competition, and so forth. It would be interesting to consider the evolution of telephone technology in the United States before and after the breakup of the A.T.&T. monopoly in 1984. In the early 1990s, the hottest new technology under development is higher-temperature superconductors. The applications of this technology are so important that competition is intense for the development of new superconducting substances.[93]

Some resistance may come from the inventors themselves, who consider the technology to be either unimportant or actually dangerous. John Napier (1550–1617) the inventor of logarithms, claimed to have invented a devastating weapons system, whose nature we can only guess at, to be used against a potential Spanish invasion. When the invasion failed to materalize, he suppressed the weapon because its immediate necessity had passed. An admirer of Napier claimed that the device could clear a plain with a four-mile circumference of every living thing above a foot in height. When, on his death bed, he was pressed for the device's plans, he is said to have claimed that

> for the ruin and overthrow of man, there were too many devices already framed, which if he could make to be fewer, he would with all his might endeavor to do; and that therefore seeing the malice and rancor rooted in the heart of mankind will not suffer them to diminish, by a new conceit of his the number of them should never be increased.[94]

Similarly, Leonardo da Vinci refused to divulge the secret of his submarine because he feared quite correctly that it would be used to destroy ships and their crews.[95] Today such scruples on the part of inventors are rare.[96] If inventors with financial support do not develop viable inventions, the reason tends to stem from fear of lawsuits by accidentally injured users, rather than from such basic ethical concerns.

The Complexities
of Technology Assessments

The Complex Effects of Technology. When we talk about the "impacts" of technology, we are using a metaphor. This expression is partly a carryover from the Ogburn school, which contended that society reacted to technology and not the reverse.[97] The metaphor has a certain persuasive value: It alerts us to pay attention to the changes technology may cause. But it may also blind us because it directs our attention to immediate effects. It implies that a technology is implemented within a social structure, makes certain changes, and is thereafter a passive factor. The advent of snowmobiles in Lapland certainly seems to have had this quality. Use of snowmobiles violently altered the social structure, the terrain, and the fauna in a few years.[98]

In other cases the "impact" metaphor can be misleading. For one thing, most technologies must be adopted by a society. In some cases this may seem automatic, but there is in fact always a choice. The choice is greater in leisure technologies than in those related to industrial and agricultural productivity. But even in international trade, in which competitive advantages are important, certain advances are still refused or delayed.

And influences often exert themselves over a fairly long period of time. Technologies are taken up continually as part of an ongoing stream of activity by potential sponsors. Thus adoption may be widespread or rare, use may be intensive or occasional, and acceptance may be complete or partial. Most technologies have alternatives, and local conditions may make one technology superior to another. In France, for instance, lack of natural gas and coal led to a more intensive use of nuclear energy than in the United States.[99] Also in France, the deployment of the "Minitel" home terminal took place as a national experiment, rather than as a private venture.[100] As a result, extensive use and experience with home terminals exists in France, whereas this technology has stagnated in the United States.

When a technology is complicated it may, so to speak, "spawn" its own technical community, which may in turn become a sponsor of the technology and shape it in important ways, including further technical refinement. The technological community may also develop its own ways of gathering, using, and conceptualizing information on the technology's performance.[101] As the technical community grows, offshoots develop, sometimes forging new uses for the current technology, sometimes developing competing ones.

While all this is going on, jobs are being destroyed and created, home life is being shaped, cities are being restructured, and minds are being changed. The nature of the American farm, for instance, has been altered by a long-term trend toward mechanization. First steel plows, then mechanical harvesters, then gasoline-powered tractors, and finally giant high-tech machines have driven millions of families off farms and brought the advent of hybrids, fertilizers, pesticides, and agribusiness empires.[102] Machines and chemicals make it cheaper to farm on a large scale, at least in a society in which produce is trucked from one end of the country to another to be sold through supermarket chains. Universities develop departments to train farmers to think in terms of machines, chemicals, and the bottom line, and they even develop devices to harvest the specially bred tomatoes needed if machines are to pick them without breakage. When devices displace workers, protests ensue for a time, but the system cranks on to produce more and better machines.[103] Meanwhile, the landscape is radically altered, and the food we eat is increasingly processed and filled with potentially harmful residues of a variety of chemicals.

Advances in electronics led to the development of television. A technological wonder, television developed into a consumer product that invaded our homes. An entire industry sprang up to support television broadcasting and programming. The role of newspapers shifted as television news displaced some of the functions of newspapers. Families spend time watching television together—interested, amused, amazed, and virtually glued to the screen.[104] Studies show that children spend a third of their waking life watching TV. Meanwhile, national literacy scores have dropped, as have student achievement scores. Television has become important in politics and advertising, and it has allowed the rapid dissemination of VCRs and tapes. Television has encroached on movie theater audiences, just as films have encroached on audiences of live theater.[105] With the popularity of television came refinements: color, stereophonic sound, recording, and giant screens.[106] There is no single impact of television; rather, television has entered the mainstream of American life, at once shaping and being shaped, its ripples spreading throughout society.

The Complexity of Making Choices. But how are choices made among alternative paths of development? And who does the choosing? Some indications of how complicated choice can become can be gained from considering a seemingly simple example. In 1974 the Denver Police Department wanted to adopt a new hollow-point bullet, which was expected to have greater "stopping power." When this choice was challenged by citizen groups, assistance was sought from ballistics experts, who knew a lot about bullets but not very much about how to make ethical judgments. The dilemma was finally resolved by having the ballistics experts rate the physical characteristics of 80 bullets and then allowing decision-makers to express their judgments concerning the desirable mixes of physical and "social" characteristics (stopping power, injury to target, bystander threat). The physical and social characteristics were then fed into a relatively simple equation, and individuals could see how their judgments affected the choice of bullet. As a result of this exploration, a hollow-point bullet was selected for use, but it was a different one from the bullet originally chosen for adoption.[107]

This example neatly demonstrates some of the complexities of choice. On the one hand, we need information about what the actual choices will involve; this is an issue of fact. On the other hand, we need information about what things people value. In this case, the physical facts were straighforward and already known, for a bullet is not a particularly complicated device; still, there were 80 different kinds from which to choose. Value judgments in this case involved an equally straightforward interpretation concerning how much perpetrator and bystander injury people were willing to endure in order to get a certain amount of stopping power. In this case, values varied from one person to another, but even so, one bullet seemed to offer the best mix of stopping power (a benefit) and injury to perpetrators and bystanders (a cost). This is a real-life example involving an important and emotion-laden decision.

Thus using quantitative methods for assessing the "goodness" or "badness" of impacts can be useful, but only in certain, limited cases.[108] The common denominator of situations in which quantitative cost/benefit analyses are useful is a well-understood goal for which there exist several different ways of achieving it. Such situations are ideal for what used to be called "systems analysis," and they are common for technologies with an obvious input or output measure and whenever important side effects can occur only in limited ways. Military applications and energy-conservation problems lend themselves well to this kind of analysis. For instance, we can compare the energy-intensiveness of using recyclable glass bottles versus the energy use related to throw-away bottles.[109] When broader social impacts are taken into account, however, quantitative methods are much less useful and often appear artificial and unconvincing. This is not to say it can't be done, but rather that it is a much more difficult enterprise.

The Balance Between Costs and Benefits.

Broad-gauged technology assessments are necessarily complicated in that, instead of a narrow range of impacts, as in the military and energy examples just mentioned, a wide range is considered. They involve extrapolations, not direct data, about physical performance. Predicting social reactions is even more conjectural and involves sociological rather than physical speculations. For elaborate technology assessments, it may be impossible to arrive at simple quantitative measures and thereby balance, in any meaningful way, costs against benefits. It is the process of exploration that is important, for it gets those involved in the planning process to identify and think about the variety of consequences that may be involved.

For instance, in substituting mass transit systems for freeway construction, large numbers of difficult-to-predict variables enter in. In constructing the Bay Area Rapid Transit system in California, for instance, traffic predictions turned out to be excessively high, causing underutilization of the system.[110] Values also become more difficult to define when we must deal with chains of events rather than first-order consequences alone. Extremely important events may occur only at the end of a long chain of events. But considering such "inconceivable events" is precisely the reason for doing technology assessments, for sometimes the "inconceivable" can become the actual.[111] The possibility that developing nuclear power might lead to the possession of nuclear bombs by terrorists is one example.

It is really quite difficult to understand what kind of changes using a technology will cause. Consider, for instance, the development of X rays. Discovered by Wilhelm Roentgen in 1895, X rays excited intense interest first among physicists, then in the medical community.[112] Although X rays allow seeing "through" inanimate objects, their greatest asset was their ability to reveal the structure and functioning of the body. We have already mentioned their use in killing diseased tissue, another important asset. But X rays can be dangerous. As the decades rolled by, their ability to cause cancer became more evident. Dosages began to be lowered, and some uses— such as the "see through your foot" machine that formerly existed in some shoe stores—were eliminated altogether. But the dangerous side effects of X rays were totally unanticipated.

X rays also have unexpected positive benefits. Around X ray machines developed a corps of specialists—physicians called radiologists.[113] This group became expert at looking at films that can reveal changes in the body's internal structures. Out of radiology developed a subgroup that specialized in working with children. These *pediatric radiologists* were among the first to notice that many children had unexplained broken bones.[114] These same children often showed other suspicious symptoms suggesting abuse by their caretakers.

As a result of collecting these observations, pediatric radiologists were the first to discuss widespread evidence of child abuse.[115] The X-ray machine thus had a positive impact that would have been virtually impossible to predict. Still, the ability to see broken bones was not the only factor in recognition of child abuse, for physicians had to be willing to see it and society had to be ready to cope with it. Furthermore, it was not the radiologists but a pediatrician, C. Henry Kempe, who transformed child abuse from a medical issue to a social problem.[116] Even though it is likely that child abuse would have become a medical problem even if X rays were never used in medicine, X rays nonetheless contributed to detecting and "discovering" child abuse.[117]

Conclusion

Our security rests on having foresight, on having the requisite imagination to see the potential long-range effects of the things we do. Only in the last two decades, however, have we begun intentionally to cultivate this foresight about technology.[118] We are swimming in a new ocean, and we are not very good at it yet. But we will have to become good at it, if we are to avoid the kind of negative impacts threatened by many of our technologies, particularly those involving noxious chemicals and radioactivity. And we will have to become good at it if we also intend to cultivate systematically certain kinds of desirable trends. We should cultivate, for instance, the development of technologies that improve our health and sense of well-being, those that encourage the development of human skills, and those that have less negative effects on our environment.

It is essential that we encourage this ability to foresee not only among the technological elite, but also among ordinary citizens.[119] We cannot expect people to vote intelligently unless they understand these issues, nor can we expect them to cope adequately with such things as fluoridation, automation, and pesticides in their own lives. Exercises in foresight belong in the schools, in offices, and in factories. They belong in colleges and universities, as well as in institutes dedicated to special causes. We are a powerful society, but we need to become a responsible one as well, and we cannot be responsible unless we learn to foresee the effects of the things we do.

Notes

1. A very useful short overview of technology assessment is Joseph F. Coates, "Technology Assessment," in *McGraw-Hill Yearbook of Science and Technology* (New York: McGraw-Hill, 1974). I am grateful for this and other materials very supplied to me by Joseph and Vary Coates for use in this chapter. Although their influence on this chapter is manifest, we do not always agree, and the final product represents only my own views.

2. I cannot recommend highly enough Jib Fowles (Ed.), *Handbook of Futures Research* (Westport, Conn.: Greenwood Press, 1978). This is a valuable introduction to this subject.

3. William L. Thomas, Jr. (Ed.), *Man's Role in Changing the Face of the Earth* (Chicago: University of Chicago Press, 1970), Vol. I.

4. Ross Ashby, *Introduction to Cybernetics* (New York: Wiley, 1966), p. 206.

5. In a study of trends among futurists, Harold Shane found that their styles of thought became more sophisticated over time. First, futurists tended toward simple linear extrapolation; then, they began to consider a "fan" of multiple futures; finally, they began to consider cross-impact analyses; see Harold G. Shane, *The Educational Significance of the Future* (Bloomington, Ind.: Phi Beta Kappa Society, 1973).

6. See, for instance, the fine essay by James R. Bright, "Technology Forecasting Literature: Emergence and Impact on Technological Innovation," in P. Kelly and M. Kranzberg (Eds.), *Technological Innovation: A Critical Review of Current Knowledge* (San Francisco: San Francisco Press, 1976), pp. 299–334. James R. Bright (Ed.), *Technological Forecasting for Industry and Government* (Englewood Cliffs, N.J.: Prentice-Hall, 1968), is still a valuable introduction to this field. See also Erich Jantsch, *Technological Forecasting in Perspective* (Paris: O.E.C.D., 1967).

7. To make this point clearer, consider how useless a forecast concerning nuclear power would be if it did not take into account the social effects of the kinds of errors that were made at Three Mile Island and Chernobyl.

8. Joseph Coates remarks that the process may be more important than the formal assessment. Put another way, getting planners to think about impacts may be more important than the specific conclusions at which they arrive.

9. Vary T. Coates, "Technology and Public Policy: The Process of Technology Assessment in the Federal Government, Summary," July 1972, in *Readings in Technology Assessment* (Washington, D.C.: George Washington University Program of Policy Studies in Science and Technology, 1975), pp. 1–47. With respect to earlier technology assessments, note for instance Ogburn's work on the future impacts of aviation and Gilfillan's work on technological forecasting in Chapter 3 of this textbook.

10. The best perspective on the importance and development of technology assessment in the federal context is a report entitled *Technical Information for Congress,* 3rd ed., written largely by Franklin P. Huddle, and prepared for the U.S. House Committee on Science and Technology (Washington, D.C.: U.S. Government Printing Office, 1979). This long and informative report not only provides background for the development of the Office of Technology Assessment; it also gives capsule histories of a number of the controversies in which technology assessment became important. It is highly recommended to anyone approaching this subject for the first time.

11. See Melvin Kranzberg, "Historical Aspects of Technology Assessment," 1969, in *Readings in Technology Assessment,* pp. 1–21. Kranzberg comments on many of the "antitechnology" intellectuals of the time.

12. See Serge Taylor, *Making Bureaucracies Think: The Environmental Impact Statement Strategy of Administrative Reform* (Stanford, Calif.: Stanford University Press, 1984).

13. Alvin Toffler, *Future Shock* (New York: Random House, 1970).

14. Mel Horwich, *Clipped Wings: The American SST Conflict* (Cambridge, Mass.: MIT Press, 1982).

15. I first became acquainted with such procedures while working as a consultant for the Rand Corporation in 1969. At that time, "systems analysis," developed for use on future

weapons systems, included a fair amount of what would now be called technology assessment. Rand, Mitre, and other "think tanks" then extended these techniques into the civilian sphere; see Paul Dickson, *Think Tanks* (New York: Atheneum, 1971). It is interesting to note that cost-benefit analysis, so prominent in many earlier analyses, is no longer as popular as it once was. This suggests a movement toward broader-gauged assessments.

16. Joseph Coates argues that an assessment should be a three-stage project. First, there should be an attempt to understand the technology's scope to determine the overall territory to be considered. Then, a draft of the assessment process itself is carried out, in whatever depth and complexity the budget will allow. Finally, a draft of the report should be sent out for critical review before the final version is written.

17. For instance, consider the physicians who for years irradiated children who had the condition called "enlarged thalamus," causing many of them to die subsequently of cancer; see Allan Mazur, *The Dynamics of Technical Controversy* (Washington, D.C.: Communications Press, 1981), pp. 1–7. Anyone doubting the need for technology assessment would be advised to read this account.

18. Edward W. Lawless, *Technology and Social Shock* (New Brunswick, N.J.: Rutgers University Press, 1977), p. 490; in it the author observes that in 40 percent of the 45 cases of technological "alarms" he examined, an early warning that was available was ignored by the sponsors or others responsible. Furthermore, in half the cases, the threat was allowed to increase even after the first signs of danger.

19. People in the United States were fortunate that such a midcourse assessment took place with Thalidomide/Kevadon, allowing us to avoid some of the serious medical consequences people of European countries suffered; see Huddle, *Technical Information for Congress,* pp. 397–447.

20. Edward W. Lawless, "The Oral Contraceptive Safety Hearings," in his *Technology and Social Shock,* pp. 28–45.

21. On the broader effects of the pill, it would be interesting to know whether anyone anticipated that it would affect premarital sexual intercourse more than marital intercourse.

22. Gina Kolata, "Cancer Experts See a Need for Caution on Use of Birth Pill," *New York Times,* Jan. 7, 1988, pp. 1, 11; and Gina Kolata, "Ambivalence over Pill Grows with Risk Data," *New York Times,* Jan. 8, 1989, p. 14.

23. J. P. Bunker, D. Hinkley, and W. V. McDermott, "Surgical Innovation and Its Evaluation," *Science,* Vol. 200 (May 26, 1978), pp. 937–941.

24. See, for example, John P. Bunker, "Sounding Board: Hard Times for the National Centers," *New England Journal of Medicine,* Vol. 303 (Sept. 4, 1980), pp. 580–582; and John Bunker, Janet Fowles, and Ralph Schaffarzick, "Evaluation of Medical-Technical Strategies: Proposal for an Institute for Health Care Evaluation," *New England Journal of Medicine,* Vol. 306 (Mar. 11 and 18, 1982), pp. 687–692. I am grateful to Dr. Bunker for calling my attention to these references.

25. John Bunker, personal communication, Aug. 31, 1988; and Martin Tolchin, "U.S. to Establish a Data Bank on Medical Care It Finances," *New York Times,* Oct. 29, 1988.

26. Joseph Coates disagrees with this idea of continual assessment. He points out that if assessment is to be done seriously, it is likely to be done by paid professionals, at best once or twice, in each case as a discrete "project" rather than as an ongoing process. Vary Coates notes that public interest is fickle, sometimes waxing enormously and at other times fading away. Compare these views with those in Armand Mauss, *Social Problems as Social Movements* (Philadelphia: Lippincott, 1975), pp. 61–70.

27. Joseph Coates, personal communication, Aug. 25, 1988.

28. Nothing in the treatment of technology assessment here would limit it to mechanical or biological technology; the same analysis could be applied to a proposed social intervention.

29. According to Joseph Coates,
> What has worked in almost every case, when one is starting a technology assessment, is to define the system that one is dealing with or impacting on, and use that system as fully elaborated as one can, as the spine around which to hang concepts, promote exploration, engage in search, look for implications. By defining that system, one is carried automatically into what are the necessary collateral or supporting elements, who the actors are, who the stakeholders are, what the system impinges upon, in terms of energy, materials, transportation, and so forth.

(Personal communication, Aug. 25, 1988.)

30. See, for example, Joseph Coates, "The Role of Formal Models in Technology Assessment," *Technological Forecasting and Social Change,* Vol. 9 (1976), pp. 139–190.

31. See, for instance, Gordon Wills, with Richard Wilson, Neil Manning, and Roger Hildebrandt, *Technological Forecasting: The Art and Its Managerial Implications* (Harmondsworth, England: Penguin, 1972).

32. John G. Stover and Theodore Gordon, "Cross-Impact Analysis," in Fowles, *Handbook of Futures Research.*

33. Richard D. Duke, "Simulation Gaming," in Fowles, *Handbook of Futures Research.*

34. Harold A. Linstone and Murray Turoff (Eds.), *The Delphi Method: Techniques and Applications* (Reading, Mass.: Addison-Wesley, 1975).

35. Richard Nisbett and Lee Ross, *Human Inference: Strategies and Shortcomings of Social Judgment* (Englewood Cliffs, N.J.: Prentice-Hall, 1980).

36. Ruud Smits and Jos Leyten, "Key Issues in the Institutionalization of Technology Assessment," *Futures,* Feb. 1988.

37. On the importance of amateurs in technological assessments, see Colin Norman, "Isotopes the Nuclear Industry Overlooked," *Science,* Vol. 215 (Jan. 22, 1982), p. 377.

38. See Dorothy Nelkin, "The Political Impact of Technical Expertise," *Social Studies of Science,* Vol. 5, No. 1 (1975), pp. 35–54; Yaron Ezrahi, "Utopian and Pragmatic Rationalism: The Political Context of Scientific Advice," *Minerva,* Vol. 18, No. 1 (Spring 1980), pp. 111–131; and Beth Savan, *Science Under Siege: The Myth of Objectivity in Scientific Research* (Montreal: CBC Enterprises, 1988), pp. 73–88.

39. This vested interest is sometimes a financial interest, but not always. The important point is that this expertise comes from people who are used to thinking about the technology in a certain way. Financial gains are not necessary, and blatant bribes might in fact be indignantly refused by many whom outsiders would consider biased.

40. Allen Schnaiberg, *The Environment: From Surplus to Scarcity* (New York: Oxford University Press, 1980), pp. 277–315.

41. David Robbins and Ron Jonston, "The Role of Cognitive and Occupational Differentiation in Scientific Controversies," *Social Studies of Science,* Vol. 6, Nos. 3 & 4 (Sept. 1976), pp. 349–368; and Philip Boffey, *The Brain Bank of America* (New York: McGraw-Hill, 1975), pp. 53–88.

42. Harvey Molotch, "Oil in Santa Barbara and Power in America," *Sociological Inquiry,* Vol. 40 (Winter 1970), pp. 131–144.

43. Susan Wright, "Recombinant DNA and Its Social Transformation, 1972–1982," *Osiris,* 2nd Series, Vol. 2 (1986), pp. 303–360.

44. William J. Broad, "In from the Cold at a Top Nuclear Lab," *New York Times,* Dec. 27, 1987, Section IV, p. 14; and Robert Scheer, "The Man Who Blew the Whistle on Star Wars," *Los Angeles Times Magazine,* July 17, 1988, pp. 7–13, 29–32.

45. Dan Morain, "Energy Secretary Warns Weapons Scientists Not to Disagree in Public," *Los Angeles Times,* July 23, 1988, Part I, p. 24.

46. Rosemary Chalk, "Making the World Safe for Whistle-Blowers," *Technology Review,* Jan. 1988, pp. 48–57.

47. Mazur, *The Dynamics of Technical Controversy,* pp. 126–127. Copyright 1981 by Allan Mazur. Reprinted with permission.

48. This is not to say that controversy is a substitute for a more formal investigation; rather, it is a very important supplement to it.

49. See Arie Rip, "Controversies as Informal Technology Assessment," *Knowledge: Creation, Diffusion, Utilization,* Vol. 8, No. 2 (Dec. 1986), pp. 349–371.

50. Michael deCourcy Hinds, "Experts Call Warning of Radon Test Failures a False Alarm," *New York Times,* Apr. 8, 1989, p. 11.

51. Keith Schneider, "Safety Tests Are Called Woefully Inadequate for Drugs, Chemicals," *St. Louis Post-Dispatch,* Apr. 11, 1985, p. 5; and Robert Reinhold, "Pills and the Process of Government," *New York Times Magazine,* Nov. 9, 1980.

52. Steven Waldman, "Watching the Watchdogs," *Newsweek,* Feb. 20, 1989, p. 34.

53. Dorothy Nelkin, *Nuclear Power and Its Critics: The Cayuga Lake Controversy* (Ithaca, N.Y.: Cornell University Press, 1971); Richard C. Petersen and Gerald E. Markle, "Controversies in Science and Technology," in D. Chubin and E. Chu (Eds.), *Science off the Pedestal: Social Perspectives on Science and Technology* (Belmont, Calif.: Wadsworth, 1989), pp. 5–18.

54. Operations Research Society of America ad hoc Committee on Professional Standards, "Guidelines for the Practice of Operations Research," *Operations Research,* Vol. 19, No. 5 (Sept. 1971), pp. 1123–1258. As a student of Wohlstetter at the time, I remember my amazement that this professional body would interest themselves in this matter. This report is noteworthy as one of the rare occasions when an official scientific body has spelled out fairly clearly what it considers to be the norms of science.

55. Paul Doty, "Can Investigations Improve Scientific Advice? The Case of the ABM," *Minerva,* Vol. 10, No. 2 (Apr. 1972), pp. 280–294.

56. Mazur, *Dynamics of Technical Controversy.*

57. Technology assessment can also be done for the much more narrowly defined interests of government bureaucracies or the military; see Michael H. Gorn, *Harnessing the Genie: Science and Technology Forecasting for the Air Force 1944–1986* (Washington, D.C.: Office of Air Force History, 1988).

58. Vary T. Coates and David M. O'Brien, "Technology Assessment and the Private Sector," in D. M. O'Brien and D. A. Marchand (Eds.), *The Politics of Technology Assessment* (Lexington, Mass.: Heath, 1982), pp. 21–32. Coates and O'Brien take some pains to deny the label "technology assessment" to most of the activities of this kind carried out by the private sector. Their argument is cogent, but I am not entirely persuaded for the reasons indicated in the text.

59. This has been one of the criticisms of the federal Office of Technology Assessment; see

David Dickson, *The New Politics of Science* (New York: Pantheon, 1984), pp. 255–258. Daryl Chubin, a senior analyst at the OTA, contends that whereas Dickson's criticism may have been well founded a decade ago, today it is unjustified (personal communication Sept. 1988).

60. Claudia H. Deutsch, "What Microwave Mania Missed," *New York Times,* Jan. 3, 1987, Section IV.

61. This story is chronicled in great detail by John Brooks, in *Business Adventures* (New York: Bantam, 1970), pp. 23–69.

62. Robert England, "As Ford Constellation Rises, Taurus Is the Brightest Star," *Insight,* July 13, 1987, pp. 15–17.

63. Claudia H. Deutsch, "What Do People Want, Anyway?" *New York Times,* Nov. 8, 1987, Business Section.

64. Edward M. Tauber, "How Market Research Discourages Major Innovation," *Business Horizons,* June 1979, pp. 22–26.

65. See John H. Dessauer, *My Years with Xerox: The Billions Nobody Wanted* (New York: Manor Books, 1971), p. 89.

66. Herb Brody, "Sorry, Wrong Number," *High Technology Business,* Sept. 1988, pp. 24–28.

67. Remarks by Hazel Henderson, as quoted in Mark A. Boroush, Kan Chen, and Alexander N. Christakis (Eds.), *Technology Assessment: Creative Futures* (New York: North Holland, 1980), p. 32. Reprinted by permission of the publisher. Copyright 1980 by Elsevier Science Publishing Co., Inc.

68. Joel Tarr (Ed.), *Retrospective Technology Assessment–1976* (San Francisco: San Francisco Press, 1977). See also the fine review of this subject in Howard Segal, "Assessing Retrospective Technology Assessment: A Review of the Literature," *Technology in Society,* Vol. 4 (1982), pp. 231–246.

69. Bruce Sinclair (Ed.), *The Railroad and the Space Program: An Exploration in Historical Analogy* (Cambridge, Mass.: MIT Press, 1965); Ithiel de Sola Pool (Ed.), *The Social Impact of the Telephone* (Cambridge, Mass.: MIT Press, 1977); and Vary T. Coates and Bernard Finn, *A Retrospective Technology Assessment: Submarine Telegraphy—The Transatlantic Cable of 1866* (San Francisco: San Francisco Press, 1979). Of these, the last is the most insightful concerning the problems with technology assessment; see especially Appendix I by Robert Anthony, pp. 198–226.

70. Robert Anthony, in Coates and Finn, *Submarine Telegraphy,* p. 198. Anyone who peruses this volume and thinks about current technology assessment procedures should become uneasy at all the impacts the Victorians did not foresee.

71. Ithiel de Sola Pool, *Forecasting the Telephone: A Retrospective Technology Assessment of the Telephone* (Norwood, N.J.: Ablex, 1983).

72. A flaw in Sherry Turkle's book *The Second Self: Computers and the Human Spirit* (New York: Simon & Schuster, 1984) is its insensitivity to the actual contours of the technology. Different computers (for example, mainframes versus microcomputers) have different social impacts. This shortcoming detracts from this book's otherwise useful insights.

73. See Paul Ceruzzi, "An Unforeseen Revolution: Computers and Expectations, 1935–1985," in Joseph J. Corn (Ed.), *Imagining Tomorrow: History, Technology, and the American Future* (Cambridge, Mass.: MIT Press, 1986), pp. 188–201.

74. Marion Whitworth Acworth ("Neon"), *The Great Delusion: A Study of Aircraft in Peace and War* (London: Ernest Benn, 1937).

75. William F. Ogburn, *The Social Effects of Aviation* (Boston: Houghton Mifflin, 1946), p. 555; Francis R. Allen, "Aviation," in F. R. Allen, Hornell Hart, Delbert C. Miller, William F. Ogburn, and Meyer F. Nimkoff (Eds.), *Technology and Social Change* (New York: Appleton-Century-Crofts, 1957), p. 204; and Jacob Shapiro, *The Helicopter* (New York: Macmillan, 1958), pp. 222–269.

76. Adapted from Luis Marden, "The Bird Men," *National Geographic,* Aug. 1983, pp. 198–217.

77. This impression was reinforced by a conversation with Paul MacCready, who is very knowledgeable about lightweight aircraft. The conditions under which ultralights can be operated are too restrictive, he argues, to make them a useful replacement for automobiles. I wonder, though, what effects future technical improvements might have on operating conditions. Automobiles once suffered under similar handicaps, which they overcame.

78. Christopher Flavin, *Reassessing Nuclear Power: The Fallout from Chernobyl,* Worldwatch Paper 75 (Washington, D.C.: Worldwatch Institute, 1987).

79. Gregg Easterbrook, "Big Dumb Rockets," *Newsweek,* Aug. 17, 1987, pp. 46–52. Even before the explosion of the *Challenger,* there were other serious problems with the shuttle; see John M. Logsdon, "The Space Shuttle Program: A Policy Failure?" *Science,* Vol. 232 (May 30, 1986), pp. 1099–1105; and William J. Broad, "In the Harsh Light of Reality, The Shuttle Is Being Re-Evaluated," *New York Times,* May 14, 1985, pp. 19–20.

80. Arthur C. Clarke, "Extra-Terrestrial Relays," *Wireless World,* Oct. 1945. This article is reproduced in John R. Pierce, *The Beginnings of Satellite Communications* (San Francisco: San Francisco Press, 1968), pp. 37–43. It is interesting that Clarke contemplated using solar power for the satellites, although he envisioned the need for much more of it than turned out to be required! Most communication satellites now are designed for less than 10 kilowatts (Kw); Clarke wrote of using 10,000 Kw.

81. Dave Dooling, "Voices from the Sky," in Kenneth Gatland (Ed.), *Space Technology* (New York: Harmony Books, 1981), p. 86.

82. Joseph N. Pelton, "The Proliferation of Communication Satellites: Gold Rush in the Clarke Orbit," in James Katz (Ed.), *People in Space* (New Brunswick, N.J.: Transaction Books, 1985), pp. 98–109.

83. This statement requires some modification. The aircraft was obviously going to be used by the military; thus Wilbur and Orville Wright made efforts to sell airplanes to the armed forces of Great Britain, the United States, and France. Ironically, the Wrights were pacifists and hoped the airplane would be a deterrent to war.

84. With respect to bicycles, see Trevor J. Pinch and Weibe J. Bijker, "The Social Construction of Facts and Artifacts: Or How the Sociology of Science and the Sociology of Technology Might Benefit Each Other," in W. E. Bijker, T. P. Hughes, and T. Pinch (Eds.), *The Social Construction of Technological Systems* (Cambridge, Mass.: MIT Press, 1987), pp. 17–50. This article first called my attention to the importance of users. My sources regarding automobiles bear less directly on this point. See James J. Flink, *The Car Culture* (Cambridge, Mass.: MIT Press, 1978), pp. 1–17; Jacques Rousseau, *Histoire Mondial de l'Automobile* (Paris: Hachette, 1958), pp. 18–34; and Roger Burlingame, *Engines of Democracy* (New York: Scribner, 1940), pp. 385–387.

85. Pinch and Bijker, "Social Construction," pp. 17–50. The role of users in determining performance criteria can be overemphasized, for sponsors may have a great deal to say about this as well.

86. Raymond Wik, *Henry Ford and Rural America* (Ann Arbor: University of Michigan Press, 1973).

87. Allen Nevins and Ernest Hill, *Ford: Expansion and Challenge* (New York: Scribner, 1957), pp. 409–478.

88. Mark Olshaker, *The Instant Image: Edwin Land and the Polaroid Experience* (New York: Stein & Day, 1978), pp. 36–37.

89. It is interesting to note what investment during the nineteenth century could do for a favored technology, such as submarine cables; see Coates and Finn, *Retrospective Technology Assessment*. If the same kind of resources had been applied to European efforts to develop computers, then the history of Europe might have been very different.

90. Nonetheless, I was told by Bernard Blood of the U.S. Department of Transportation that some key technologies in foreign high-speed trains had actually been developed in the United States. Personal communication, Feb. 8, 1990. See also Dwight B. Davis, "High-Speed Trains: New Life for the Iron Horse?" *High Technology*, Sept. 1984, pp. 28–39.

91. Robert Fellmeth, *The Interstate Commerce Omission: The Public Interest and the ICC* (New York: Grossman, 1970), pp. 285–310.

92. Wright, "Recombinant DNA Technology," pp. 303–360.

93. T. A. Heppenheimer, "Superconducting: The New Billion-Dollar Business," *High Technology*, July 1987, pp. 12–18.

94. John U. Nef, *War and Human Progress* (New York: Norton, 1963), p. 122.

95. Nef, *War and Human Progress*, p. 118.

96. Norbert Weiner, the developer of cybernetics, belongs in the company of Napier and da Vinci for refusing to work on numerical control technology that might degrade the jobs that workers do; see David Noble, *Forces of Production: A Social History of Industrial Automation* (New York: Knopf, 1984), pp. 71–76.

97. Much of the impetus for looking at the matter this way comes from anthropological studies, in which technologies were much more fixed and the changes in affected societies were much more graphic; see, for instance, Edward H. Spicer (Ed.), *Human Problems in Technological Change* (New York: Wiley, 1965) and H. Russell Bernard and Pertti Pelto (Eds.), *Technology and Social Change* (New York: Macmillan, 1972). I do not wish to imply, however, that the anthropological studies are less sophisticated than the sociological ones; if anything, the reverse is true.

98. Pertti J. Pelto, *The Snowmobile Revolution: Technology and Social Change in the Arctic* (Menlo Park, Calif.: Cummings, 1973). Anyone who equates "progress" with the mere implementation of advanced technologies should read this study.

99. David Dickson, "France Weighs Benefits, Risks of Nuclear Gamble," *Science*, Vol. 233 (Aug. 29, 1966), pp. 930–932.

100. Jeffrey A. Hart, "The Teletel/Minitel System in France," *Telematics and Informatics*, Vol. 5, No. 1 (1988), pp. 21–27.

101. Geof Bowker, "A Well-Ordered Reality: Aspects of the Development of Schlumberger, 1920–39," *Social Studies of Science*, Vol. 17, No. 4 (Nov. 1987), pp. 611–656.

102. Maisie Conrat and Richard Conrat, *The American Farm: A Photographic History* (Boston: Houghton Mifflin, 1977).

103. Marjorie Sun, "Weighing the Social Costs of Innovation," *Science*, Vol. 223 (Mar. 20, 1984), pp. 1368–1369; and Philip L. Martin and Alan L. Olmstead, "The Agricultural Mechanization Controversy," *Science*, Vol. 227 (Feb. 8, 1985), pp. 601–606.

104. See, for instance, Gary A. Steiner, *The People Look at Television: A Study of Audience Attitudes* (New York: Knopf, 1963); and Frank Mankiewicz and Joel Swerdlow, *Remote*

Control: Television and the Manipulation of American Life (New York: Ballantine Books, 1978).

105. Just as, one might note, Gilfillan said it would; see S. C. Gilfillan, "The Future Home Theatre," *The Independent,* Vol. 73 (1912), pp. 886–891.

106. See Harry F. Waters with Janet Huck, "The Future of Television," *Newsweek,* Oct. 17, 1988, pp. 84–93.

107. Kenneth R. Hammond and Leonard Adelman, "Science, Values, and Human Judgment," *Science,* Vol. 194 (Oct. 22, 1976), pp. 389–396.

108. This observation is made in Lynn White, Jr., "Technology Assessment from the Stance of a Medieval Historian," *American Historical Review,* Vol. 79, No. 1 (1974), pp. 1–13.

109. See, for instance, Bruce Hannon, "Bottles, Cans, and Energy Use," in B. Commoner, H. Boksenbaum, and M. Corr (Eds.), *Energy and Human Welfare: A Critical Analysis,* Vol. III *Human Welfare: The End Use for Power* (New York: Macmillan, 1975), pp. 105–118.

110. Peter Hall, "San Francisco's BART System," in his *Great Planning Disasters* (Berkeley: University of California Press, 1982), pp. 109–137.

111. See Gustaf Ostberg, "Evaluation of a Design for Inconceivable Event Outcome," *Materials and Design,* Vol. 5, No. 1 (Apr./May 1984), pp. 88–93.

112. Bern Dibner, *The New Rays of Professor Roentgen* (Norwalk, Conn.: Burndy Library, 1963).

113. X rays are also routinely used by dentists and chiropractors.

114. John Caffey, "The First Sixty Years of Pediatric Radiology in the United States—1896 to 1956," *American Journal of Roentgenology, Radium Therapy and Nuclear Medicine,* Vol. 76, No. 3 (Sept. 1956), pp. 437–454.

115. Frederic N. Silverman, "Unrecognized Trauma in Infants, The Battered Child Syndrome, and the Syndrome of Ambroise Tardieu," *Radiology,* Vol. 104 (Aug. 1972), pp. 337–353. I am grateful to Dr. Silverman for bringing this article to my attention.

116. See Nigel Parton, *The Politics of Child Abuse* (London: Macmillan, 1985), p. 53.

117. Certainly social workers such as Elizabeth Elmer did not need X rays to become convinced that there was a problem (personal communication with Elizabeth Elmer, Oct. 26, 1986). I am currently working on an article about the medical discovery of child abuse based on interviews with some of the pioneers involved in its recognition. See also Ron Westrum, "Social Intelligence About Hidden Events: Its Implications for Scientific Research and Public Policy," *Knowledge: Creation, Diffusion, Utilization,* Vol. 3, No. 3 (Mar. 1982), pp. 381–400, also reprinted in D. Chubin and E. Chu (Eds.), *Science off the Pedestal: Social Perspectives on Science and Technology* (Belmont, Calif.: Wadsworth, 1988), pp. 19–30.

118. Again, a qualification: Obviously, researchers such as Ogburn and Gilfillan were in favor of developing foresight. Such was Ogburn's intention in developing research, for instance, into "social trends." But this interest did not lead to the development of larger futuristic or policy-deliberation enterprises of the kind now in operation.

119. See Duncan MacRae, Jr., "Building Policy-Related Technical Communities," *Knowledge: Creation, Diffusion, Utilization,* Vol. 8, No. 3 (Mar. 1987), pp. 431–462.

Foresight & Social Intelligence

In the previous chapter we discussed how the effects of technologies are assessed. In this chapter we seek answers to some related questions: What institutions do the assessing? Is society ready to cope with the interventions its technologies represent? Are the intellectual institutions of our society adequate to cope with our technology? This matter of institutional fitness is crucial, for our major guarantee of survival is the ability to anticipate and monitor the effects of our actions.[1] Yet, as we saw in the previous chapter, the requisite imagination for coping with change may be absent. In any case, knowledge (like technology) does not float freely in space; it is created, maintained, and propagated by institutions. These institutions, taken as a whole, can be considered a vast learning system. The obvious question is: Is this learning system good enough to do the things we need it to do?

The Nature of Our Learning System

The Importance of Immediate Feedback. The notion of a "learning system," is fraught with ambiguity. Learning can be done on various levels and with different degrees of efficiency. We learn best from experience, but such learning can also lead to an ability to cope with situations more generally. People are better at learning from experience than they are at anticipating, and they are better at learning when feedback is immediate. Thus human beings do well when the feedback loop is short. As the feedback loop lengthens—as the delay in feedback increases—human performance deteriorates rapidly, a basic fact of life brought home to me when as a college student I worked at the Mental Health Research Institute at the University of Michigan.

At that time the institute was conducting some experiments on the effects of tranquilizers on simulated driving. The purpose of these experiments was to determine how dangerous it might be for real drivers to drive after taking these tranquilizers. The apparatus we had for measuring driving ability was crude—this was before the microchip revolution—and it probably did not simulate reality very well. The machine, which was very similar to a contemporary arcade game, consisted of a toy car about six inches long, a moving roadway, and a rod that linked the car to a "steering wheel" the subject turned. The machine moved the roadway, a sort of giant rubber strip, by rollers at each end. The effect produced was that the "road" was an irregularly undulating ribbon that moved toward the subject at variable speed. The subject's task was to keep the car on the road.

Subjects were usually able to keep the car on the road at moderate speeds. However, if the speed was increased beyond a certain point, subjects tended to oversteer. The subjects' steering corrections frequently became wild and the car went out of control, a phenomenon called "hunting" in cybernetics. The reason for "hunting" was that the delay between seeing what the car was doing and correcting the deviation was just enough so that the correction was out of phase with the car's action. First the correction did not seem to be strong enough, then suddenly it became too strong. Eventually the car hit the side of the driving machine. Steering deviations were thus amplified, as they sometimes are with real drivers. The accuracy of feedback-directed motions, then, reflects how fast the system is moving versus how fast its operator can process information and respond. If the system moves too rapidly or if the delays are excessive, corrections may come too late to avert disaster or may actually increase its likelihood.

Even a small delay in feedback can be disorganizing. The Soviets discovered this when they put a wheeled robot explorer, *Lunokhod,* on the moon and controlled it by radio from Earth. Because the moon is about 240,000 miles away and the speed of radio waves is roughly 186,000 miles per second, it took the signal about a one and a quarter seconds to go each way. Thus a delay of two and a half seconds was introduced between ordering an action and seeing its effects, and this delay caused serious problems:

> Lunokhod 1 was driven by a 5-man team—commander, driver, navigator, systems engineer and radio operator—working in the Deep Space Communications Centre, believed to be near Moscow. The need for coordinating their efforts in the early days was said to put great psychological strain upon them; the technique was reported to be so difficult that it beat many highly experienced drivers and pilots. This was mainly because of the time-delays resulting from sending commands, and waiting to observe the response over a distance of 386,240 km. They had to remember that the robot was already several metres ahead of the slow-scan TV picture they were watching; and that if there were rocks or boulders to be avoided, it would have moved on still further before the signals to take avoiding action would reach it.[2]

Had the robot been on Mars, the problem would have been much worse, for it takes 13 *minutes* for electromagnetic signals to travel from Earth to Mars and back!

To get some idea of the stresses the *Lunokhod* driver experienced, imagine steering a car at 10 mph with this same delay. At this speed, you are traveling about 15 feet per second. If you turn the wheel sharply, you will have gone 37 feet before you see anything happen; then you will see the car begin to turn. At 60 mph, you would have gone 220 feet before you see anything happen. It is not difficult to imagine the stress such a situation would produce. Because similar delayed reactions occur in industrial process applications, we can appreciate how stressful a "crisis" situation can be in a chemical plant, or even more so, in a nuclear reactor.

One of the reasons for the explosion of the Soviets' reactor at Chernobyl was exactly this kind of problem. The reactor was being given a safety test when the power dropped beyond acceptable levels. After having slowed down the coolant system, the operators pulled out all but six of the moderator rods to increase the speed of the nuclear chain reaction. In five seconds the reactor exploded, causing the most serious nuclear accident ever made public.[3] The accident suggests the existence of a growing mismatch between human abilities and the technological requirements of our constructed environment.[4]

Let's try to put these thoughts into a diagram (see Figure 17-1). In this situation we have an action system (in reality, usually a set of systems) that produces some physical or social effect. Between the effect and the action is some kind of learning or sensory system that weighs the action in terms of its effect.

At first, we might find this feedback system reassuring: The effects of our technological actions are weighed by the learning system and subsequent actions are adjusted accordingly. But there is a catch. Consider Figure 17-2.

Figure 17-2 is identical to Figure 17-1 except that we have added labels for the communication pathways between each of the parts. Each pathway's designation begins with a D because each represents a delay. D_1 represents the time it takes for actions to have effects; D_2 represents the time required for effects to be recognized and understood; and D_3 represents the time required to formulate some response to influence subsequent actions. This seems simple and straightforward, but the

Figure 17-1 A Feedback Loop

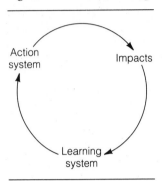

Figure 17-2 A Feedback Loop, Including Communication Delays D_1-D_3

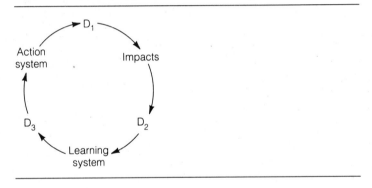

consequences of delays can be very serious and may even be deadly. Let's consider each of the delays in a bit more detail.

D_1 is an *impact lag,* which in turn produces a "delayed effect," or in more serious cases, a "time bomb." Some experts predict that the impact lag of current chlorofluorocarbon use (a problem we'll discuss below) may be as long as 50 years.

D_2 is a *recognition delay;* it includes not only perception of the impact, but also the understanding of what produced it. The acid rain problem has a long recognition delay. For many years its cause has remained a mystery, and now that we have a way of determining its geographical origin, we are closer to identifying the exact cause.

D_3 is a *response delay;* the learning system has identified the problem, but those responsible for action must be convinced of the need to do something. One example of a response delay is the lag between the recognition of the medical dangers of Thalidomide and full stoppage of its use.

Each of these delays is serious because each compounds the problems produced by actions that are detrimental. The longer the delay between the action and its stoppage, the more serious a problem becomes. This "decision cycle" must be shortened in whatever ways are possible; we are more likely to be able to reduce recognition and response delays because impact lags are likely to be governed by natural laws we cannot change.

Learning and Technological Impacts. This kind of lengthy feedback loop is inherent in such problems as the air pollution of Los Angeles. Even as the problem of automobile emissions in Los Angeles becomes more severe, the residents often fail to relate their own behavior to the overall situation. This attitude makes taking social action to solve the problem more difficult. Gary Evans, a psychologist, was surprised to find out that Los Angelinos were insensitive to pollution problems, particularly in relation to their health: "The longer people live here, the less likely they are to acknowledge the presence of pollution," he concluded. L. A. residents routinely ignore air-quality health advisories published in daily newspapers. "They keep jogging," Evans says. "Environmental problems don't give immediate feedback like touching a hot stove. When you drive a car, you don't experience long-term ecological repercussions."[5]

In large-scale social systems, we can see how these kinds of feedback delays can be important. For instance, between the time a social policy is chosen and implemented, there is always an inevitable delay as organizations are set up, staffed, vetted, and put into action. Similarly, evaluation of social policies that are in effect usually comes only after a period of years, and the conclusions are often ambiguous. But even when the cause of a problem is apparent in the minds of observers, getting attention and social effort directed to it is often difficult, as we have seen with struggles to improve scientific and technological literacy in the United States.

"Learning" in this sense is thus a reaction to some previous experience. This "glance in the rearview mirror," however, is not likely to be adequate in coping with new technologies. What we need is a system that not only learns, but one that *anticipates*—that is capable of considering the future. We might call such a system a "thinking system." A useful thinking system not only has the ability to use feedback, but also the ability to "feed forward" to see what the effects of its actions are likely to be. Obviously, technology assessment is a form of "feed forward." This ability to anticipate, to project into the future, is necessarily part of an adequate thinking system.

The impacts of technological systems affect the entire world around us, and now they are beginning to extend into space. Human beings have done much to alter the face of the earth, as a result of settlement, exploitation of natural resources, war, and disposal of wastes. When the human population was small, these impacts had little lasting effect. Now that humans are more numerous and much more powerful than ever, the impacts are serious.[6] We live on the surface of a giant globe roughly 8,000 miles in diameter. From a geological point of view the damage we do is relatively small and transient. But the biosphere, the arena in which we live, can be very adversely affected by what we do, by the changes we make, by the loss of species we exterminate. We cannot afford to neglect the impact of the changes we make in the biosphere, for it is our only home.

We have changed the biosphere in a variety of ways.[7] First, humans have killed off a great many species of animals and plants, either through direct extermination or by destruction of their habitats.[8] This direct and indirect extermination has reduced the variety of species available to us for food, medicine, clothing, and many other uses.[9]

Even more importantly, this extinction of species has moved us down an irreversible path to environmental instability. Second, we have carved up the landscape through mining, deforestation, and redirection of rivers.[10] Third, we have dumped toxic chemicals and just plain junk all over the place, and are likely to do even greater damage in the future as we dispose of radioactive wastes. Our machinations have begun to reduce the ability of the atmosphere to sustain and protect us.[11] Finally, the human population is exploding, shrinking the space and resources remaining for future generations.

In agriculture we have made other impacts. We have strikingly reduced the number of species of plants used in agriculture in practicing *monoculture,* the cultivation of single species. A single species is easier to manage in the short run—to plant, maintain, and harvest, especially with methods that are highly mechanized and rely heavily on chemicals—but inherent in monoculture are serious long-term vulnerabilities to climatic change, disease, and animal pests.[12] We have larger farms that are growing fewer varieties with more machines and more chemicals. Now, as we begin genetic experiments, our ability to influence global plant and animal communities is even greater.[13]

In the face of all these changes, how adequate are our institutions for coping with the future? Are we ready to assess, and to act upon, what appears to be likely to happen? Are we able, in short, to manage the kinds of changes we are now causing? Our ability to process and act upon the information we have may not in fact be good enough.

Anticipating Impacts on Society

The Failure of Foresight. We can get some indication of the role of feedback in social processes by considering the process by which an architect designs a building. Traditionally, architects have been influenced more by the visual and aesthetic tradition of architecture itself than by feedback from other previous architectural projects or the people who use the buildings. Architects' sketches and illustrations of buildings in architectural magazines typically omit people; buildings have a "cleaner" appearance without people.[14] In theory a building is designed for its users, but the status system in architecture values a good-looking building more than a comfortable or useful one. In many cases, architects do a very poor job of anticipating what the consequences of their designs will be. Only in recent years, with the feedback provided by postoccupancy evaluations, have architects begun to look at users' reactions to their creations.[15] They have also gotten users (or dwellers) more involved in the planning process, which has provided not only a better fit with users' needs, but also a wider range of ideas.

We find similar problems in the realm of social planning. The record of intellectuals in anticipating and coping with social problems is generally horrendous. Social changes are sometimes envisioned by people whose experience in actual social planning is modest or nonexistent. Although we want intellectuals to dream wildly, the execution of their dreams often carries considerable dangers if it is not tempered by experience. Intellectuals often lead lives that are closely circumscribed: friends, universities, public libraries, coffeehouses, correspondence. Without realistic experience in the politics and setbacks of planning, however, a social architect's plans may be badly conceived.[16] Although this may make little difference to the popularity of his or her ideas or the sale of his or her books, it makes a huge difference when plans are carried out. Small errors of conception may have grave consequences when translated into social structures. And very often the delay between the conception and execution of an intellectual's plans for societal reform is several decades.[17]

In some cases—Marx is an example—the intellectual dies before his or her plans are given a large-scale test. This leads to two sources of problems. First and most obviously, the intellectual is not around to see that the plans are implemented accurately. Those who come later may make major changes that undo the good effects, simply because of the natural evolution of organizations and because of power conflicts.[18] If violence is used to bring about the changes, this introduces further distortions.[19]

The second, often equally important source of problems is the distance in time and space between the social architect and the execution of the plans. Marx and Engels could write in *The Communist Manifesto* that politics would vanish with the absence of classes, but this contention turned out to be a cruel illusion, and the resulting trial-and-error process involved in "developing Communism" over time has involved grievous social costs.[20] Workers of the world had other things to lose besides the chains about which Marx and Engels wrote.

Similar problems also appear with technological planning. In the assessment of the physical and social effects of a new device or process, failures of foresight can be extremely serious.[21] By the time the thinking system has identified and evaluated the impacts, damage may be irreversible. If we consider the course of events with nuclear power, for instance, there seems a gross mismatch between early promises and current achievements.[22] Once technologies are in motion, they are difficult to stop.[23]

It is evident that there is a mismatch between our ability to act and our ability to anticipate. The thinking system in which these acts of planning are embedded is not efficient, and thus the planning process often cannot learn from its mistakes, much less anticipate the future. Instead, practices are institutionalized and encapsulated, and planners are separated from the observers who are in a position to see what is wrong. We have blind planners and mute users. The system will continue to make the same mistakes because it cannot learn.[24] How, then, can we engage in the continual education of the planners and improve the influence of the users?

Incorporating Foresight into Thinking Systems. Research on and theorizing about thinking systems is only just beginning. Probably the most powerful line of inquiry involves the related ideas of "organizational learning" and "reflective practice" advanced by Chris Argyris and Donald Schon.[25] These ideas are closely related to "organizational memory"; this idea, however, has a variety of meanings, including some in which the memory is purely passive, which is not what is intended here. Earlier, C. West Churchman suggested the idea of "inquiring systems."[26] Others have talked about "governmental learning."[27] There has also been extensive work on the "learning curve" by which technological lessons are institutionalized in an industry.[28] One of the great strengths of Japanese industry has been its ability to use the learning curve by paying careful attention to small details and thereby continually improving its performance.

On a larger scale, society needs thinking systems to anticipate and cope with the often interrelated impacts of technologies. Surprisingly little attention has been devoted, however, to setting up such thinking systems. Although there are many university departments that treat the initial planning and design of things—architecture and urban planning departments, for instance—there is no department that deals with the important process of using feedback from experience to correct or redesign a system. In fact, a reluctance to examine the likely impacts of laws and technological policies is common.[29] In the passage of laws, moral outrage is likely to play a larger role than long-range anticipation. This is no excuse, however, for the subsequent failure to examine the consequences of laws once passions have cooled.

But the passage of laws is only one way in which social action takes place on a large scale. To illustrate the difficulties associated with such large-scale action, let's consider an example of potentially severe, global damage associated with technological actions—the case of the chemicals called chlorofluorocarbons (CFCs).

The Need for Foresight—A Case History.

Chlorofluorocarbons (CFCs) are a group of organic compounds that contain at least one chlorine or flourine atom. One of them, dichlorodifluoromethane, has been very widely used as a refrigerant since the 1930s, and since the 1950s it has been mixed with trichlorofluoromethane for use as an aerosol-spray propellant. Not until 1974 did it become apparent that aerosols might be dangerous to the atmosphere. At that time, two professors of chemistry at the University of California, Irvine, Sherwood Rowland and Mario Molino, announced that the release of chlorofluorocarbons into the atmosphere might harm the ozone layer.

The ozone layer, part of the upper stratosphere about 20 miles above the earth, protects us from ultraviolet radiation, the same radiation that causes sunburn—and skin cancer. Chlorofluorocarbons, the chemists suggested, deteriorate in the stratosphere, and the released chlorine ions could destroy ozone. Note here a considerable lag—nearly 40 years—between implementation of the technology and the recognition that it was dangerous.

Note also the chain of events that occurred: An industry begins production of a chemical. Thanks to technical ingenuity, all sorts of uses are found for the chemical, and worldwide production takes place in large quantities. Then, through a stroke of imaginative genius—a prime example of "feed forward"—the consequences of this widespread use are suspected, and a hypothesis is published. However, a suspicion is not a certainty; certainty requires proof.[30]

We will not follow in detail the scientific debate that followed in the wake of Rowland and Molino's publication.[31] Investigations by government bodies were called for and took place; the CFC industry protested that the results of these studies were not clear; consumers began to reduce their use of aerosol sprays; new studies and experiments were planned and carried out; more protests were made about CFC regulation by the industry. The Environmental Protection Agency banned aerosol sprays containing CFCs in 1979. But this last development did little to protect the atmosphere from other large sources of CFCs: refrigerants and use of CFCs in manufacturing plastics products. About a million tons of CFCs were still being produced worldwide each year. U.S. appeals for cooperation by other countries were for the most part ignored.

A key issue in this debate was uncertainty. In this matter the upper atmosphere represents a vast laboratory, and figuring out its dynamics is both intellectually challenging and terribly frustrating. Computer models to predict what will take place in the upper atmosphere are only as good as the assumptions that go into constructing them. Different models produce different predictions. The lack of scientific certainty was used by the CFC industry as a rationale for their contention that regulation should wait until the evidence was conclusive. But of course by this point, it might well be too late. Calculative rationality can be lethal in this case because during the ensuing delays, CFCs continue to accumulate in the atmosphere.

CFCs, like many other risky chemicals, have powerful sponsors. Freons and other CFCs are manufactured by chemical giants like DuPont and Kaiser Aluminum, and the industry has not only its own scientists but also a powerful political lobby.[32] Environmental groups also have scientists and some grass-roots political power, but their energies are spread among a variety of issues.[33] In the middle are such government agencies as the Environmental Protection Agency, whose mandate is public protection but whose decisions are often influenced by political currents.[34] Often the net result of these conflicting political forces is inaction and a significant response delay. Sherwood Rowland, whose curiosity had led to the initial recognition of the problem, commented to journalist Paul Brodeur:

> After all, what's the use of having developed a science well enough to make predictions, if in the end all we're going to do is stand around and wait for them to come true? But, from what I've seen over the past ten years, nothing will be done about this problem until there is further evidence that a significant loss of ozone has occurred. Unfortunately, this means that if there is a disaster in the making in the stratosphere we are probably not going to avoid it.[35]

In May 1985 came a stunning revelation. British scientists in Antarctica had been observing a decline in the ozone cover above Halley Bay each October for a seven-year period. In spite of the importance of the data, the British held back their findings until they began to observe similar losses of ozone over another station at the Argentine Islands, about a thousand miles to the northwest. An article they then published in *Nature,* a leading scientific journal, prompted Americans to reexamine data gathered by the pollution-monitoring satellite Nimbus 7. One of the sensors on Nimbus 7 was a "Total Ozone Mapping Spectrometer" (TOMS) that generated about 190,000 data points a day. However, the computer that analyzed this data had been programmed to ignore certain low ozone values because these low values were thought to represent malfunctions of the equipment.[36] Thus the key observations were missed because a computer program threw them out. NASA scientists subsequently reprogrammed the computers and confirmed the British measurements. Now, as a result of measurements from satellites and other sources, we suspect that there is in fact a worldwide decline in atmospheric ozone.

This example highlights the enormous weakness of our social intelligence system when faced with the kind of delays we have discussed above.[37] We may face similar delays with buried toxic wastes and various kinds of radiation. We cannot always rely on foresight alone, for our imagination may be deficient. Monitoring events after the fact is chancy, too, for such a strategy depends on information that may exist only at the whim of scientific policy decision-makers—as was the case for the TOMS sensor on the Nimbus 7.[38] Even when partial understanding is achieved, it may be challenged by interested technology sponsors, and action may be delayed. The cost of this delay in the matter of CFCs may yet be severe damage to human bodies resulting from skin cancer and to the natural environment because of increased ultraviolet radiation.

Groups and Institutions Responsible for Foresight

The CFC example takes us back to our basic question: Are our institutions able to cope with the kinds of effects our technology can have? And it leads us to a related question: Which institutions in our society are responsible for foresight?

There are a variety of answers to this question. For instance, even though government is certainly supposed to plan wisely for the future, a survey of governmental institutions quickly shows that this responsibility is often shirked. Politicians must strive to get reelected, and if taking care of the future involves unpopular decisions, they may not make them. Middle-level bureaucrats are in the same job longer than most politicians, but their bosses, the top-level bureaucrats, frequently are replaced as often as politicians. In such a system, taking care of, say, natural resources may "fall through the cracks."

One presidential administration may be very concerned about any particular issue, but the next may not be. The same thing is true of technology. If a technology is dangerous, one presidential administration may resist it, but the next may have different views and let it through. It is notable that the first major "think tanks," such as the Rand Corporation, were for the military; the office of Technology Assessment came much later. Put differently, we tried to figure out what to do *with* technology before we tried to figure out what to do *about* it.

Consider, for instance, how the educational system in the United States has been allowed to decay. A vital resource for creating and coping with technology, and important for the national security and well-being, the U.S. educational system at the lower levels is one of the least successful among those of the developed countries.[39] Yet with increasingly complicated technologies, we nonetheless have an expansion of what philosopher Langdon Winner calls "relative ignorance." Technological literacy seems to be decreasing in the United States. Although U.S. Secretaries of Education have complained about the quality of the education we give our students, very little has been done to change it.

In spite of many reports pointing to a very serious problem, and a trade balance that gets steadily worse, there seems to be little more than lip service with respect to improving the schools. This suggests that people must be able to see connections between tax dollars spent on education now and potential payoffs in national productivity in the future. The failure to mobilize resources to bring the educational system up to international standards does not bode well for the future prospects of the United States. Similar problems exist with respect to retraining workers; in spite of an evident need, nothing much gets done.[40] Many technological problems are like education: They are difficult, they require slow building and long-term efforts, and they are similarly neglected.[41]

Government in the United States was designed to be responsive to citizens within a framework of checks and balances set forth in the Constitution. But even a responsive government is not necessarily a wise one, and it may have little inclination toward long-range planning because public attention is relatively fickle. In other countries, such as France and Japan, more national planning is expected, and institutions for this purpose are built into the framework of government.[42]

Government agencies in the United States plan ahead, of course, although not as often as we might like them to do. The original plan for an Office of Technology Assessment for the Congress, for instance, envisioned an agency that would act as an "early warning system" and set at least part of its own agenda. The current OTA does carry out some broad-gauged technology assessments, but much of its work consists of the narrower task of feasibility studies.[43] Formally, the OTA responds only to requests for information from congressional committees. It is limited by statute to 150 people, and so it contracts out much of its research. But informally the relationship between OTA personnel and congressional staffers is such that many requests are in fact initiated from within OTA.[44] The quality of work carried out by OTA is widely respected, not only in the United States, but also in foreign countries, many of which would like to have "their own OTAs."[45] Still, the resources of OTA, and therefore its scope of action, are very much smaller than we might like. Most large pharmaceutical companies have a marketing staff that is larger than the entire Office of Technology Assessment staff.[46]

Universities are expected by many to think about the future, but just which department is responsible? The greatest contribution of universities is to train people in disciplines that contribute to future studies in such fields as the physical, natural, and social sciences. Very few (if any) universities in the United States have "Departments of Future Studies," although many have futures institutes and futures courses.[47] "Future studies" has a very low prestige generally, especially compared to the sciences, because it is interdisciplinary,[48] and even in the sciences, those specializations that evaluate technological impacts (such as ecology) are much less prestigious than those involved in "basic" science.[49] High-prestige professors are much more likely to consult for the military than they are for the environmental movement.

Nonetheless, a large number of individual professors are interested in future studies and think of themselves as futurists. As we have seen in the case of CFCs, occasionally they can have an important impact. But this is a matter of individual initiative, not regular institutional activity.[50] No doubt many agree with political scientist Yehezkel Dror that more useful work on the future goes on at "Rand Corporation–type think tanks" than at universities.[51]

Scientific committees and commissions do a great deal of futures work. Some of them are commissioned by government and some by private sources, but these groups are often well staffed and well published. For instance, a very interesting Commission on the Year 2000 was put together by the American Academy of Arts and Sciences. In 1967 the deliberations of this commission were published in *Daedalus*,[52] a periodical available from at least some highbrow bookstores and magazine stands.

This exploration was valuable in a number of ways. In the first place, it served to inform the participants, and the list of participants included many people who were later to be influential in shaping government policy; Zbigniew Brzezinski (later Secretary of Defense), Fred Charles Iklé (later Undersecretary of State), and Senator Daniel P. Moynihan. Second, it informed readers, no doubt stimulating some of them to become critics or to elaborate further the ideas presented by the commission. Third, it created a precedent for similar efforts. Such commissions, however, are temporary, even though constant attention to many issues is required; some of this attention is supplied by private groups. Reports by private research institutes—the "think tanks"—are often commissioned by foundations, and they are important.

The formation of private groups dedicated to futures studies has been a significant development, although its impact is difficult to judge. According to one source, well over 200 institutes with futuring activities existed in 1979.[53] Many of these are "think tanks" with several activities, of which futuring is only one. The largest of these consulting organizations, such as the Science Applications International Company, the Stanford Research Institute, and the Rand Corporation, do most of their business with the military and "big science." A number of other institutes and organizations are specifically dedicated to futuring. The largest of these is the World Future Society, which, since its founding in 1966, has provided a significant forum for discussion about the future. Through its annual meetings, informative publications such as the annual *Future Survey*, book service, and bimonthly magazine, the World Future Society offers a series of opportunities and resources for thinking about the future.[54] Although it has only 30,000 members, many of them belong to the intellectual elite, and it has a larger impact on educated opinion than its numbers might suggest.

Another significant development for understanding the future was the formation of the Club of Rome in 1968. The Club of Rome is a wonderful example of an idea that "recruited" people to carry it out. An industrialist (Aurelio Peccei) and a scientist (Alexander King) discovered that they were both concerned with international developments greater in scope than any single scientific discipline could cover. They decided to bring together a small group of eminent people to ponder these issues and to commission studies. This international group of intellectuals, with representatives from many countries, considered its mission to be the examination of how such trends as population growth, economic growth, technological expansion, and escalation of armaments affected the future of life on Earth.[55] Although the group remained small, its influence was large.

One of the Club of Rome's most visible activities was to sponsor the development of a computer model of the world economic system by a team at the Massachusetts Institute of Technology under the direction of Jay Forrester. Whatever the opinion of the specific performance of the model, the exercise was significant. The results of the computer modeling were compiled in a controversial book called *Limits to Growth,* first published in 1972.[56] The value of this work in stimulating interest and debate on the dynamics of the world population and economy should not be underestimated: Some five million copies of the book were sold in some 30 languages. The book led to a variety of discussions and much soul-searching in the countries in which it was widely circulated. It also attracted substantial criticism, some of it simply dismissive. But at least one group at the University of Sussex took the time to write a detailed rebuttal.[57] Far from discrediting computer modeling, the debate stimulated new attempts to make modeling more accurate.[58] Computer modeling of the future is now a standard "foresight" technique.

Awareness of environmental dangers has begun to become stronger. A variety of crises—the greenhouse effect, the ozone problem, Chernobyl, and the destruction of the Amazon rain forests—have begun to focus worldwide attention on environmental problems. Although computer predictions have done little to convince people, the recognition that problems are occurring *now* is beginning to stir countries to action. Very helpful to this growing awareness is the existence of the Worldwatch Institute, founded by Lester Brown in 1975 and based in Washington, D.C.

Worldwatch grew out of Brown's personal experience in coping with agricultural problems during the Kennedy administration. Brown had received college training in agriculture and joined the U.S. Department of Agriculture in 1959. By 1965 he was working in India, where his rapid assessment of a dangerous food shortage led to successful international efforts to avert a famine. He rose rapidly in the U.S.D.A. to become the youngest divisional manager in the United States. His experiences in India made clear to him the need for a more systematic effort. In 1972, he joined a public-interest research group that focused on global issues.[59] With a $500,000 grant from the Rockefeller Brothers Fund, Brown and his associates developed a small but influential intellectual enterprise.[60]

The Worldwatch Institute provides an accessible, easy-to-consult source of information about various aspects of the global environment. It monitors worldwide trends in such subjects as demography, energy, agriculture, and waste production. It issues an annual *State of the World* report, published as a large paperback book, as well as shorter reports (Worldwatch "Papers"). *State of the World* is considered by its writers to be a kind of "report card" on the health status of the world.[61] Fifty thousand copies of the 1987 report were printed in the United States, and some 200 colleges used it as a text in one or more courses.[62] One hundred thousand copies of the 1988 report have been printed. The book is also translated into most of the major languages of the world. Information of this kind provides individuals, organizations, and governments a useful tool for anticipating the future, even though it reports on the current state of affairs.

Finally, we should note that not all work on the future is done by groups or organizations. Intellectuals in many walks of life scrutinize current trends and predict future ones. Some intellectuals reside in universities, but others exist as free-lance writers or pursue intellectual interests as a sideline to another profession.[63] Many "intellectual workers" have strong institutional ties and responsibilities that limit their ability to inquire; the strength of unattached intellectuals is that they do not have such institutional ties and can thus more freely examine issues. They are often more willing to give an unvarnished analysis or a bold opinion. Their ability to pursue broad questions without regard for vested interests is of great value to humanity. Hence, those societies that nurture the development of intellectuals provide an important human resource. Rachel Carson is an example of an unattached intellectual whose work has been influential.[64] Alvin Toffler, a futurist, has had far more influence on public opinion than most professors. Still, academic futurists remain important contributors to thought about technology and the future, and they are very influential by virtue of their contact with society's budding intellectuals and future citizens.

When we add up the resources for thinking about the future, we are struck by two things. First, these organizations and individuals afford a relatively thin bulwark for protecting mankind, even though a single person can make a powerful difference, as was the case with Rachel Carson and Sherwood Rowland. Second, the sponsors of technologies hold enormous power and often have much narrower concerns than we might wish. Because a great deal of thinking about the future and directing of technology will be done by such technological sponsors, the selection of sponsors is important for shaping the future.

Giving Technology the Right Sponsors

The selection of a technology's sponsors is important. Who should be responsible for making the major decisions about a technology? Sponsorship is the most powerful device we have, more powerful than regulation (which usually can state only what it is illegal to do) and certainly more powerful than technology assessment. After all, who decides what will be done with the technology assessment?

Because sponsors engage in so many activities crucial to technology—invention, testing, promotion, propaganda—they must be carefully selected and carefully watched thereafter. Investment in a technology can generate whole communities and industries.[65] These organizations and networks, once set in motion, are very difficult to dislodge. Thus the creation of such institutions should be given more care. If we want to achieve a better fit between technology and human needs, we must do a better job of matching technology with its supportive institutions.

It is evident that our society does not think enough about this issue and that we have made some serious misassignments of sponsors; nuclear power is a case in point. Furthermore, in general the major concern in the United States is to give technology to organizations and managers who will develop it most rapidly. But this is not enough. Given the pervasive impacts of technology on our lives, we must consider more thoroughly whether we have the right institutions to handle the side effects of our technologies.

In the United States we could classify most institutions into three broad categories: government, business, and voluntary associations.[66] It is evident that for the most part we have given the management of technologies to business organizations. This choice has produced many benefits as well as quite a few drawbacks. We could learn some very useful lessons in technology management from more socialist countries, just as they are now trying to learn from us. Even though business management has notable successes (the telephone system), it also has some serious failures (nuclear power). In some cases (space exploration) we have given sponsorship largely to the government, again with a mixture of successes and failures. Finally, some technologies (for example, laetrile, polio vaccines, and birth-control pills), though privately manufactured, have been disseminated or supported largely by voluntary associations. This list of managers of technology could be expanded to include universities, private foundations, and "think tanks." This is the institutional material we have to work with.

We do not yet know enough about the way that sponsorship operates to prescribe which kinds of technologies ought to be given to which kinds of sponsors. In some cases the private sector is perfectly appropriate, although some after-the-fact adjustment and regulation may be required. Often the largest problem, as we have seen in the cases of nuclear power and trucking, is that insufficient or badly conceived regulation produces very serious hazards. But overregulation can also be dangerous, particularly when it occurs in an adversarial climate. Similarly, government can be a very dangerous sponsor when insufficient funds are given to a program, as happened with the space shuttle program. Voluntary associations are a sort of "court of last resort" for many technologies, and this can mean the support both of useful technologies (such as rocketry and birth control) and of those of more questionable value (such as laetrile). Maintaining the vigor of voluntary associations will mean that we will still have errors of both kinds.

If technologies are dependent on sponsorship, then we have two kinds of problems with which we must cope. On the one hand, a technology may not have adequate support: the technology may be a very good idea, but the commercial and governmental forces supporting it may not be sufficient to make it thrive. This situation apparently exists with birth-control technologies, which are no longer favored for research and development by large American pharmaceutical firms.[67] We are not very good at strengthening such technologies, and one of the problems is that government is often unduly influenced by industry opinions and vice versa. When industry and government turn thumbs down on a valuable technology, there seems to be little our society can do to change the status of that technology. Such a situation has arisen with solar power (passive as well as active), with "alternative" technologies generally, and with mass transportation.[68] We need to develop some better ways of coping with such a situation.

On the other hand, a technology may have, in terms of its real value, far too much support. As a result these "muscle-bound" technologies cannot be changed even when there is widespread acknowledgment of their ill effects. They "keep right on a 'hurtin,'" as the song goes. The biggest problem with such technologies is that along with "financial muscle" comes "persuasive muscle."[69] Thus the tobacco industry has waged a relentless campaign against the idea that cigarette smoking is bad for people.[70] Nuclear power, the agricultural chemical industry, and above all the automobile industry have enormous power to shape attitudes through their paid scientific talent. A. T. & T. had enormous power, which it used to shape telecommunications in the United States. We have seen the kinds of campaigns waged by the CFC industry. Karl Marx was quite right to see such overpromotion as inherent in capitalism, but we have discovered that socialist societies have similar problems, for their large technical establishments are also likely to have large lobbying and propaganda powers.

Thus there is what I call the "dinosaur" problem. Consider the following scenario. A man wakes up one morning to find a dinosaur in his front yard. He asks the dinosaur to leave, but the animal refuses. He decides to threaten the dinosaur and goes into the house to get a high-powered rifle. When he gets outside, the dinosaur refuses to budge. "Then I'll shoot you," the man warns, pointing the weapon. "No you won't," the creature replies, "you don't want a dead dinosaur in your front yard." Large technologies with powerful sponsors are very much like this: It's difficult to get them to budge, and killing them would leave an incredible mess. Regulation is a partial solution, but it is a long and difficult process, takes a lot of public support, and is often ineffective.

For each of these kinds of error the problems are fewer if we monitor the situation as it proceeds. Undernourished technologies can be strengthened (we'll find a way; we did it for atomic power), and juggernauts-in-the-making can be cut back, *if we watch them carefully*. My impression, though, is that we seldom do. The institutions that should be watching are either not very good at watching, or they don't have enough power, once they spot a problem, to do anything about it. And then the systems—superhighways, power networks, gas pipelines—are put in place, and what follows is what Ogburn calls "adaptive culture": heavy investments in support services, in training, and in public "education," and the appearance of other technologies that shape themselves to fit the changed conditions. With the advent of automobiles, supermarkets replaced the corner store, shopping centers replaced "downtown," and the trolley tracks got ripped out. Once this has happened, going back becomes too painful to contemplate, so no one does contemplate it. And so we keep driving automobiles, causing air pollution, traffic jams, and energy wastage. Allan Schnaiberg has correctly referred to the system as a kind of "treadmill."[71] How do we get off it?

The Need for an
Inquiring Attitude

We need an inquiring attitude toward our technology. Every use of a technology could be considered an experiment.[72] But not all action is an experiment, because experimenting demands a degree of watchfulness. In American society we try all sorts of things, but the monitoring, feedback, and appropriate reaction to the observations is often missing. We need to assess a technology before implementation, during implementation, and once it is in place; constant vigilance ought to accompany risky actions. In the laboratory, experimental trials are routinely watched, but in the outside world we often fail to take note of the technological trials in progress because we just don't take the time and effort to do it. And thus we have to suffer the consequences of a process out of control.

Let's look at the problem in a broader perspective. We are dealing here with an issue of consciousness, the ability to be aware of what is going on. But whereas the study of technology is ancient, the study of consciousness is a relatively new activity. Freud's study of the mechanisms of defense took place only at the end of the nineteenth century.[73] Irving Janis's study of group awareness is barely two decades old.[74] We are working on studying the consciousness of societies, but we have a long way to go before coming to any useful results.[75] Control of technology and the study of consciousness are closely linked. We cannot control technology effectively until we understand the limits of our ability to monitor it.

The flow of social information and the effective use of technology go hand in hand. Still, many people in American society suffer from information overload and the accompanying lack of time for contemplation. We seem to have little time to devote to such matters as the state of technological development. Although the issue of the "pace of life" is beyond the scope of this chapter, it is an important factor in our understanding of why technology is not better monitored.[76] People in a hurry are likely to pay less attention to side effects of technology than those with more leisure. Our ability to process information is limited, as are our attention span and our ability to remember even phone numbers![77] But a society's ability to maintain control of its technology is closely linked with the ability and interest of its citizens to monitor events.

Again and again we find the proof of this observation in our efforts to protect ourselves from the ravages of technology. Rachel Carson and Ralph Nader stand out as examples of individual citizens who decided to take personal action to expose technological problems. In both cases we find that both the official sponsors of the technology (the agricultural-chemical and automobile industries) and the responsible regulatory agencies were indifferent or hostile to citizens' concerns. Even though in each case these critics were guilty of certain exaggerations or misleading arguments, the key point is that attention to these issues had to come from outside the establishment, which, in spite of its formal responsibility for the technology, failed to pay sufficient attention to these particular issues.

Such personal courage and resolution is of course most admirable, but we also need to notice the more significant fact that only the effective flow of information allowed these critics to take a stand. Both Carson and Nader were "recruited" by an idea; we need to encourage the maintenance of a society in which such generative creativity is common, in which thoughts "enlist" people. The best guarantee of protection from technology's ills is a citizenry capable of obtaining and understanding information. This leads us to note again the importance of a good liberal education that includes scientific and technological literacy. People who don't know are not likely to criticize. We need to reverse the trend toward a "plug-in society" in which all that is understood about a technology is how to turn it on.

Another bulwark of the free flow of information is the ability to organize privately for the public good. The work of Carson and Nader was promoted by private organizations—by the environmental movement and by Public Citizen, Ralph Nader's organization. These organizations and the movements that surround them are certainly open to criticism,[78] but the value of their existence and their efforts is indisputable. If they did not exist, they would have to be invented.

Private organizations gather, process, and disseminate huge amounts of information about technology and this kind of private generation and dissemination of information is extremely important. Even the information citizens get from the media must often first be preprocessed by such organizations.[79] Until recently one of the difficulties of environmental protection in the Soviet Union was the lack of such organizations; now this is changing.[80]

Voluntary institutions are a prime generator of thought in an open society. Their importance is that they can be formed rapidly and can thus provide social support to critical and constructive thought about technology. In the United States this same impulse to organize is responsible for the rapid implementation of many technologies, which seem to attract entrepreneurs as soon as inventions appear.[81] Of course such free-wheeling has its limits, as does any system.[82]

The largest drawback to constant monitoring is the time it requires. Even though constant monitoring of technology may eliminate the worst mistakes, it requires a willingness to remain involved during long, sometimes difficult periods of development. Unlike most citizens, organized sponsors tend to have long spans of attention. They have a vested interest in the technologies and are likely to hang on longer when things get rough. Citizens, however, are more impatient, are less likely to keep at it. This attitude—this willingness to keep at it—is a tremendous hazard once the technology becomes dangerous. It is possible to believe too much in generative rationality; the belief that "science created the problem, so science can fix it" is dangerous and simplistic. The private sponsor who has invested heavily in the now-misbehaving technology may assume that some kind of technological solution must exist, but "betting on the come" does not always work. Often there is no technological solution.

This, then, is the basic problem with technology development as experimentation: Having the right degree of commitment is difficult; too weak a commitment will lead to premature withdrawal, but heavy commitment may result in holding on after it would be wise to quit. Having discrete stages of operation, each of which includes an evaluation, would help in achieving some balance of commitment. For instance, when we build a prototype power plant, we would evaluate the prototype. Then we would build a pilot plant, and evaluate that. Finally, we go to full-scale operation and evaluate that. The ability to stop the whole thing ought to exist at each level. But how rare are honest evaluations in the real world! And thus how rare it is to see a project that hindsight later declares stupid stopped at an intermediate stage![83]

Social Intelligence: How Societies Think About Their Problems

Suppose we wished to characterize the performance of social information-processing systems. One way to do this might be to classify such systems according to their ability to handle technological problems. We might identify three styles of coping: pathological, calculative, and generative.[84]

1. A *pathological system* is not able to handle technological hazards well even under normal conditions. A system like this actively works to circumvent regulations and laws enacted to protect society, and it places pressure on individuals and organizations that dissent. Such a system currently exists in many third-world countries, where local corruption and interference from foreign multinational corporations leads to depletion of rain forests, exhaustion of land, soil erosion, and other environmental problems.[85]

2. *Calculative systems* perform well under routine conditions. Laws are passed in response to recognized problems, and these laws are generally enforced. This is a great improvement over pathological systems, in which dishonest or illegal activities are often condoned. Under reasonably stable conditions, such a system is adequate. When complex technological interactions arise, however, or when the lags mentioned earlier become excessively long, this system is no longer adequate. The United States performs within a calculative range of competence; it attempts to avoid or contain certain environmental problems, although its actions are too slow on some occasions and too small on others.

3. A *generative system* is what we need to deal with many of our technological problems. Such a system would not only react successfully, but would do a much better job of anticipating future hazards. Emphasis would be on creating an adequate information base first, and then on creating an adequate repertoire of *scenarios* to provide for discussion and planning. Foresight and inquiry

would permeate the social decision-making process and would include private bodies as well as governmental ones. Such a system would include an effort to use planning in a way harmonious with other social goals, thereby avoiding the confrontation that characterizes so much American regulation.[86] No current national government has achieved a generative level of functioning, although many have adopted some generative strategies, such as the use of futuring activities, planning, brainstorming, the creation of special institutes for the future, and so forth.

Each of these systems in turn will have a characteristic response pattern to observations of technological hazards. These will range from a complete denial of the observations, through routine response, to more creative and thoughtful responses, which address not only the present problem, but underlying processes as well. The spectrum of responses to such observations is summarized in Table 17-1.

All societies, at one time or another, will use the complete spectrum of responses listed in Table 17-1. But each society will also have a characteristic response pattern. Pathological societies typically respond to observations of hazards with denial activities. Problems will be suppressed or encapsulated. Calculative societies use more repair activities. They will act, but only on problems identified immediately. Generative societies often use reform strategies, not only learning about the observations, but also improving the social intelligence process along the way.

In each society, folkways and institutions reinforce the standard patterns of activity. Norms are expressed in laws and legal institutions. Socialization inculcates modes of handling information. Journalistic and intellectual practices follow expectations of disclosure or suppression. As a society moves from pathological to generative functioning, freedom of thought and communication and demands for accountability increase. Special organizations for social intelligence are more common in generative societies than in the other two kinds.

It has been fascinating in the last few years to see the U.S.S.R. shifting from a pathological level toward a calculative level of social intelligence, with an underlying change in philosophy that is strongly generative. Overall, however, the changes in the world's societies toward a more generative model have been slow, much slower than the rapid erosion of topsoil, depletion of forests, and decline in environmental safety margins documented by the annual reports of the Worldwatch Institute. The apparatus we have for tracking changes in our world is very much less adequate than it should be. Human beings, unique among the animal species, can create institutions that have allowed us to have an unprecedented impact on our world through the development of science, medicine, and engineering. Institutions can also provide the ability to monitor and control the changes, but so far this ability has remained essentially latent. If we are to survive and prosper, we must create the appropriate monitoring and control institutions.

Human life is an experiment. For the experiment to succeed, humanity needs the ability to foresee and to act on this foresight. We realize now that we are living in a closed system. For the foreseeable future, we are stuck on a large sphere of rock, water, and air surrounded by the vastness of space. The surface of that sphere constitutes our whole world, and it is a world we are rapidly degrading. Only if we take

Table 17-1   Responses to Observations of Technological Hazards

A. DENIAL ACTIVITIES

Suppression: Observers are punished or dismissed, and observations are expunged from the record.

Encapsulation: Observers are ignored, and the validity of their observations is disputed or denied.

B. REPAIR ACTIVITIES

Public relations: Observations emerge publicly, but their significance is denied or disputed.

Local repairs: The problem is admitted and fixed, but its wider implications are denied.

C. REFORM ACTIVITIES

Dissemination: The problem is admitted to be a global one, and global action is taken on it.

Reorganization: Action on the problem leads to reconsideration and reform of the operational system.

---

into account, far more than we have done so far, the future impacts of our actions, can we continue to live good lives on this orb. From the physical point of view, the system in which we live is a closed loop. We must close the intellectual loop as well. We need the ability to see what we are doing and what impacts our activities will have. Only when our technology is accompanied by adequate thinking institutions is it likely to be a friend, rather than an addiction or a threat.

Conclusion

Coping with modern technology requires a set of intellectual institutions that are still in a process of development in our society. Allowing our society to operate the kinds of technologies it does is like putting a child at the controls of a nuclear reactor. This mismatch between technologies and institutions is dangerous. Developing ways to correct this situation rapidly should be a major concern for us. We need a much better examination of the kinds of sponsors we allow to run our technological systems; the ones who run them now are not doing very well.

We must realize that with certain sponsors come certain kinds of dynamics. Profit-oriented firms will behave in certain ways; government agencies will behave in other ways. Better regulation makes a difference, and monitoring by citizen groups improves government accountability. To take responsibility means that we accept the consequences for our actions. We cannot be surprised by the behavior of sponsors once we have studied them. Once we know how sponsors behave, we must then take this knowledge and use it to shape the way that technologies are supported by institutions.

We also need an expanded sense of responsibility. One kind of responsibility accepts its role in past actions; what we need, however, is not someone to take the blame, but rather someone who will look forward. Society needs to develop foresight; it needs to appreciate its role as the guarantor of the welfare of future generations. This role will not come easily to us. Our inquiries are only beginning. More profound study is required to examine in depth the subjects that we have only briefly considered here. Even at this point, however, it is clear that the answer to the question "Do we have the right kinds of institutions for dealing with high technology?" is obviously "no." It is time to redesign the institutions in our society so that they will be capable of coping with the kinds of challenges we have examined here. One of the reasons for this book is to stimulate thought about such a redesign. We had better get on with it soon.

Notes

1. For a useful survey of this entire issue, Isaac Asimov (Ed.), *Living In the Future* (New York: Beaufort Books, 1985) is first-rate.

2. Reginald Turnill, *The Observer's Spaceflight Dictionary* (London: Frederick Warne, 1978), p. 275.

3. William D. Marbach, Joyce Barnath, Mark Miller, and Andrew Nagorski "Anatomy of a Catastrophe," *Newsweek,* Sept. 1, 1986, pp. 26–28. The Soviets concealed an even more serious disaster in 1957 or 1958 that was uncovered by the Russian scientist Zhores Medvedev in his book *Nuclear Disaster in the Urals* (New York: Norton, 1979). Although there has been a recent admission of this explosion, the details have not been made public; see comments in Thomas Powers, "Chernobyl as a Paradigm of a Faustian Bargain," *Discover,* June 1986, pp. 33–35.

4. Heinz Gartmann, *Man Unlimited: Technology's Challenge to Human Endurance* (New York: Pantheon, 1957), pp. 7–40. A more recent review of such problems is Edith Weiner and Arnold Brown, "Human Factors: The Gap Between Humans and Machines," *The Futurist,* May-June 1989, pp. 9–11.

5. Alan Weisman, "Los Angeles Fights for Breath," *New York Times Magazine,* July 30, 1988, pp. 15–30, 33, 48–49.

6. See William L. Thomas (Ed.), *Man's Role in Changing the Face of the Earth* (Chicago: University of Chicago Press, 1970), 2 vols.; and Jean Dorst, *Before Nature Dies* (Baltimore: Penguin Books, 1971).

7. Lester R. Brown, William U. Chandler, Christopher Flavin, Jodi Jacobsen, Cynthia Pollack, Sandra Postel, Linda Starke, and Edward C. Wolf, *State of the World 1987* (New York:

Norton, 1987). This is an excellent summary of many aspects of this important subject by the Worldwatch Institute, which issues these reports annually.

8. Paul Ehrlich and Anne Ehrlich, *Extermination: The Causes and Consequences of the Disappearance of Species* (New York: Ballantine Books, 1981).

9. Norman Myers, *A Wealth of Wild Species* (Boulder, Colo.: Westview Press, 1983).

10. Gene Marine, *America the Raped* (New York: Avon, 1969). It should be borne in mind that this is a book about what has been done in the United States, a country with a strong environmental movement. Elsewhere, the destruction has often been worse, as in South America, for instance, where the removal of tropical rain forests threatens the destruction of hundreds of thousands of species and a change in global weather patterns. We should also note, however, that the destruction elsewhere in the world is in no small part due to the consumptive nature of our society—we represent a relatively small proportion of the world's population but consume a disproportionate amount of its resources.

11. Howard A. Wilcox, *Hot-House Earth* (New York: Praeger, 1975).

12. See Lester Brown, "Sustaining World Agriculture," in Brown, *State of the World 1987,* pp. 122–138. On vulnerability to disease, see Peter R. Day (Ed.), *The Genetic Basis of Epidemics in Agriculture,* Annals of the New York Academy of Sciences, Vol. 287 (1977). More broadly, see Kenneth A. Dahlberg (Ed.), *New Dimensions for Agriculture and Agricultural Research* (Totowa, N. J.: Rowman and Allenheld, 1986).

13. Jack Doyle, *Altered Harvest: Agriculture, Genetics, and the Fate of the World's Food Supply* (New York: Viking Press, 1985).

14. See Dana Cuff, "Designing Lives: Epistemological People in Architecture," in Ron Westrum (Ed.), *Organizations, Designs, and the Future: Conference Proceedings* (Ypsilanti: Eastern Michigan University, 1986), pp. 4–17.

15. Postoccupancy evaluations are occupants' systematic evaluations of the good and bad points of buildings. Firms such as Herman Miller Research have conducted such studies.

16. On planning generally, see Leonard W. Dobb, *The Plans of Men* (Hamden, Conn.: Archon, 1967). For problems with planning, see Peter Hall, *Great Planning Disasters* (Berkeley: University of California Press, 1982).

17. I began thinking along these lines after reading Friedrich Heer, *Intellectual History of Europe* (New York: Doubleday, 1968), 2 volumes, a remarkable work.

18. Lewis Coser, *Men of Ideas: A Sociologist's View* (New York: Free Press, 1970), pp. 145–170.

19. See, for instance, Crane Brinton, *The Anatomy of Revolution* (New York: Random House, 1952), pp. 155–214; and Coser, *Men of Ideas,* pp. 145–170.

20. It is also true, however, that few balanced assessments of the strengths of Communism have been made; virtually everything on the subject either extols Communism or attacks it.

21. Hilary Moss, "Time Delay in the Cause-Effect Chain, and Its Stunning Consequences," in his *The Technologic Trap* (Howick Aukland, New Zealand: Privately published, 1984), pp. 104–114.

22. Christopher Flavin, "Reassessing Nuclear Power," in Brown, *State of the World 1987,* pp. 57–80.

23. The problem of technological inertia is discussed in Langdon Winner, *Autonomous Technology: Technics-out-of-Control as a Theme in Political Thought* (Cambridge, Mass.: MIT Press, 1955), pp. 279–305.

24. David Collingridge and Peter James, "Technology, Organizations and Incrementalism: High Rise Building in the U.K.," *Technology Analysis and Strategic Management,* Vol. I, No. 1 (1989), pp. 79–97.

25. See, for instance, Chris Argyris and Donald Schon, *Organizational Learning: A Theory of Action Perspective* (Reading, Mass.: Addison-Wesley, 1978); and Chris Argyris, Donald Putnam, and Diana Mclain, *Action Science: Concepts, Methods, and Skills for Research and Intervention* (San Francisco: Jossey-Bass, 1985).

26. See C. West Churchman, *The Design of Inquiring Systems: Basic Concepts of Systems and Organization* (New York: Basic Books, 1971).

27. Lloyd S. Etheredge and James Short, "Thinking About Government Learning," *Journal of Management Studies,* Vol. 20, No. 1 (1983), pp. 41–58.

28. Winfred B. Hirschmann, "Profit from the Learning Curve," *Harvard Business Review,* Vol. 42 (Jan.-Feb. 1964), pp. 125–139.

29. In this respect such books as Theodore Lowi's *The End of Liberalism: The Second Republic of the United States,* 2nd ed. (New York: Norton, 1979) are particularly valuable. In showing that the impacts of many "liberal" laws were quite different from what their proposers expected, Lowi has called attention to the need for a more thorough consideration of unanticipated consequences. Clearly, the Prohibition period can be considered a massive case study of the unanticipated consequences of legislation.

30. At the time this book was written the nature of the problem was still being researched. Even if in the light of subsequent research CFCs turn out to be harmless, the example still serves our purposes here, for the dynamics discussed likely hold true for other kinds of environmental problems as well.

31. This debate is very well covered by a superb article: Paul Brodeur, "Annals of Chemistry: In the Face of Doubt," *New Yorker,* June 9, 1986, pp. 70–87; see also Cynthia Pollock Shea, "Protecting Life on Earth: Steps to Save the Ozone Layer," Worldwatch Paper 87, (Washington, D.C., Dec. 1988).

32. In 1975 Dupont took out double-page advertisements in newspapers nationwide arguing that "to act without the facts—whether it be to alarm consumers, or to enact restrictive legislation—is irresponsible." (Brodeur, "Annals of Chemistry," p. 73.) The generally optimistic pronouncements of industry scientists about the CFC danger should make us wary in the future about technology assessments carried out by private firms.

33. Robert Suro, "Grass-Roots Groups Show Power Battling Pollution Close to Home," *New York Times,* July 2, 1989, pp. 1, 12.

34. Jonathan Lasch, Katherine Gillman, and David Sheridan, *A Season of Spoils: The Story of the Reagan Administration's Attack on the Environment* (New York: Pantheon Books, 1984), pp. 3–81.

35. Brodeur, "Annals of Chemistry," p. 83.

36. Telephone interview with Arlen Krueger, sensor scientist at NASA/Goddard, Greenbelt, Maryland, in July 1986.

37. See Ron Westrum, "Social Intelligence About Hidden Events: Its Significance for Scientific Research and Social Policy," *Knowledge: Creation, Diffusion, and Utilization,* Vol. 3, No. 3 (Mar. 1982), pp. 381–400.

38. Arlen Krueger, who is responsible for managing the TOMS sensor, told me that the Nimbus 7 itself was developed by Goddard to monitor pollution at the request of Congress (telephone interview, July 1986).

39. Office of Technology Assessment, *Elementary and Secondary Education for Science and Engineering: A Technical Memorandum* (Washington, D.C.: U.S. Government Printing Office, Dec. 1988), pp. 138–143. By contrast, we do an excellent job at the graduate level.

40. Alexander Taylor III, Gisela Bolte, and Sara White, "The Growing Gap in Retraining," *Time,* Mar. 28, 1983, pp. 50–51.

41. A good example is the unwillingness of American firms to invest in expensive development of solar-energy technology. Even though government was willing to furnish the money for the research, American firms do not wish to take the risks associated with production; see Matthew L. Wald, "U.S. Companies Losing Interest in Solar Energy," *New York Times,* Mar. 7, 1989, pp. 1, 48.

42. John H. McArthur and Bruce R. Scott, *Industrial Planning in France* (Boston: Harvard Graduate School of Business Administration, 1969).

43. David Dickson, *The New Politics of Science* (New York: Pantheon Books, 1984), pp. 232–243. This assessment of OTA is not a universal one.

44. Telephone conversation with Daryl Chubin, senior analyst with the Office of Technology Assessment, Feb. 17, 1988.

45. David Dickson, "Europeans Embrace Technology Assessment," *Science,* Vol. 231 (Feb. 7, 1986), pp. 541–542. In February 1988, Daryl Chubin indicated that the flood of foreign visitors to OTA was still high.

46. Even among think tanks, OTA's staff is not among the largest. The Rand Corporation has about 1,100 people and Stanford Research Institute has almost 3,000, compared to OTA's 150.

47. H. Wentworth Eldredge, "The Mark III Survey of University-Level Futures Courses," Ciba Foundation, *Symposium on the Future as an Academic Discipline* (Amsterdam: Elsevier, 1975), pp. 5–15.

48. Interdisciplinary studies generally tend to have low prestige because they are not considered to be as "rigorous" as disciplinary pursuits. The less removed from practice and the more the emphasis on analysis as opposed to synthesis, the higher the prestige of the activity. This system of status is very damaging to practical activities, for it tends to isolate the elements that are necessary for successful action; see Joseph R. Royce, *The Encapsulated Man* (New York: Van Nostrand, 1964).

49. Allan Schnaiberg, *The Environment: From Surplus to Scarcity,* (Oxford: Oxford University Press, 1980), pp. 277–291.

50. Universities' greatest strength may lie in protecting their own existence. Of 66 institutions of any kind that existed in the year 1530 and still exist today, 62 are universities; see review by Charles A. Kiesler of James L. Bess, (Ed.), "College and University Organization: Insights from the Behavioral Sciences," *Administrative Science Quarterly,* Vol. 32, No. 1 (Mar. 1987), pp. 146–148.

51. Yehezkel Dror, comment in Ciba Foundation *Symposium on the Future,* p. 98.

52. Daniel Bell (Ed.), "Toward the Year 2000: Work in Progress," *Daedalus,* Summer 1967.

53. James Traub, "Futurology: The Rise of the Predicting Profession," *Saturday Review,* Dec. 1979, pp. 24–31.

54. Others have assessed the usefulness of the World Future Society very differently; see, for instance, Marshall W. Gregory, "Do Future Studies Really Help?" *Change,* Jan./Feb. 1988, pp. 50–53. Mike McPhillips was kind enough to bring this article to my attention.

55. Alexander King, "The Club of Rome: A Case Study of Institutional Innovation," *Interdisciplinary Science Reviews,* Vol. 4, No. 1 (Mar. 1979), pp. 54–64; I found this essay in-

spiring. See also Alexander King, "The Club of Rome: Reaffirmation of a Mission," *Interdisciplinary Science Reviews,* Vol. 11, No. 1 (Mar. 1986), pp. 13–18.

56. Donella H. Meadows, Dennis L. Meadows, Joergen Randers, and William W. Behrens III, *Limits to Growth: A Report for the Club of Rome on the Predicament of Mankind* (New York: Universe Books, 1972). Other reports to the Club of Rome include Mihajlo Mesarovic and Eduard Pestel, *Mankind at the Turning Point* (New York: Signet, 1974); and Guenter Friedrichs and Adam Schaff, *Microelectronics and Society* (New York: Mentor, 1983).

57. H. S. D. Cole, C. Freeman, M. Jahoda, and K. L. R. Pavitt (Eds.), *Models of Doom* (New York: Universe Books, 1973).

58. Sam Cole, "Global Models: A Review of Recent Developments," *Futures,* Aug. 1987, pp. 403–428.

59. Sara Pacher, "The World According to Lester Brown," *Utne Reader,* Sept./Oct. 1987, pp. 84–93 (reprinted from *Mother Earth News*).

60. Telephone interview with Steve Dujack, Worldwatch Institute, Jan. 16, 1989.

61. Lester Brown, Christopher Flavin, and Edward C. Wolf, "Earth's Vital Signs," *The Futurist,* July-Aug. 1988, pp. 13–20.

62. Clearly the academic community values the information collected and lucidly presented by Brown and his associates in *The State of the World 1987* and *The State of the World 1988* (New York: Norton, 1987, 1988).

63. Interestingly, of the 23 people whose work Alvin Toffler excerpted in his collection *The Futurists* (New York: Random House, 1972), the majority are academics; those who are not, including Toffler himself, tended to have current or previous organizational affiliations in such capacities as editors, party leaders, or legislative assistants.

64. Rachel Carson, who wrote *Silent Spring,* was a writer, not a professor. A much more rigorous work, *Pesticides in the Living Landscape* (Madison: University of Wisconsin Press, 1964), was already under way and was published two years later by Robert L. Rudd, a professor of zoology at the University of California, Davis. Would Rudd's work, or environmental problems generally, have received the attention it did, had Carson not written on the subject? I very much doubt it.

65. For one set of examples, see David Noble, *America by Design: Science, Technology and the Rise of Corporate Capitalism* (Oxford: Oxford University Press, 1979). Noble provides a vivid description of the shaping of technological choices and standards by corporations in the United States between 1880 and 1930. Readers need not share the author's interpretations to benefit from the great wealth of information he presents.

66. Ron Westrum and Khalil Samaha, *Complex Organizations: Growth, Struggle, and Change* (Englewood Cliffs, N.J.: Prentice-Hall, 1984), pp. 4–11.

67. See the remarkable article by Carl Djerassi, "The Bitter Pill," *Science,* Vol. 245, (July 25, 1989), pp. 356–361. Djerassi's research group developed the first birth-control pill in Mexico in 1951.

68. On solar power, see Matthew L. Wald, "U.S. Companies Losing Interest in Solar Energy," *New York Times,* Mar. 7, 1989, pp. 1, 48.

69. Joel Primack and Frank von Hippel, *Advice and Dissent: Scientists in the Political Arena* (New York: Basic Books, 1974), documents the perversion of truth by scientists acting as advisors to the government. The extent of the perversion of truth in the private sector can only be imagined by comparison.

70. Eliot Marshall, "Tobacco Science Wars," *Science,* Vol. 236 (Apr. 17, 1987), pp. 250–251. More generally, see Peter Taylor, *The Smoke Ring: Tobacco, Money, and Multinational Politics* (New York: Pantheon Books, 1984).

71. Allan Schnaiberg, *The Environment: From Surplus to Scarcity,* p. 220.

72. Compare this viewpoint with that in Donald Campbell, "Reforms as Experiments," in E. L. Struening and M. Guttentag (Eds.), *Handbook of Evaluation Research,* Vol. 1 (1975), pp. 71–100. See also the critique in William N. Dunn, "Reforms as Arguments," and Campbell's reply, "Experiments as Arguments," both in *Knowledge: Creation, Diffusion, and Utilization,* Vol. 3, No. 3 (Mar. 1982), pp. 293–338.

73. Sigmund Freud, "The Dissection of Personality," in his *Complete Introductory Lectures in Psychoanalysis* (New York: Norton, 1966), pp. 521–544.

74. Irving L. Janis, *Victims of Groupthink* (Boston: Houghton Mifflin, 1972).

75. See, for instance, Amatai Etzioni, *The Active Society* (New York: Free Press, 1968); Warren Breed, *The Self-Guiding Society* (New York: Free Press, 1971); Armand Mauss, *Social Problems as Social Movements* (Philadelphia: Lippincott, 1975); and Westrum, "Social Intelligence," pp. 381–400.

76. Consider, for instance, the pace of business life in the early republic versus current pressures to perform and achieve; see Arthur Cole, "The Tempo of Mercantile Life in Colonial America," *Business History Review,* Vol. 32 (Autumn, 1959), pp. 277–299. Compare with David E. Sanger, "Wall Street's Tomorrow Machine," *New York Times,* Oct. 19, 1986, Business Section, pp. 1, 30; and M. Mitchell Waldrop, "Computers Amplify Black Monday," *Science,* Vol. 238 (Oct. 30, 1987), pp. 602–604. See also the very interesting article by Jean Stoetzel, "La Pression Temporelle," *Sondages,* Vol. 15, No. 3 (1953), pp. 11–23.

77. George A. Miller, "The Magic Number Seven Plus or Minus Two: Some Limits on Our Capacity to Process Information," *Psychological Review,* Vol. 63, No. 1, (1956), pp. 81–97.

78. For a well-rounded assessment of the environmental movement, see David L. Sills, "The Environmental Movement and Its Critics," *Human Ecology,* Vol. 3, No. 1 (Jan. 1975), pp. 1–41. For a somewhat more negative view, see David Vogel, "The Politics of the Environment 1970–1987: A Big Agenda," *Wilson Quarterly,* Vol. XI, No. 4 (Autumn 1987), pp. 51–68.

79. Gaye Tuchman, *Making News: A Study in the Construction of Reality* (New York: Free Press, 1980), pp. 82–103.

80. For the former situation, see Thane Gustafson, "Technology Assessment, Soviet Style," *Science,* Vol. 208 (June 20, 1980), pp. 1343–1348. For the effect of *glasnost,* see the articles of Bill Keller in the *New York Times:* "In Soviet, Old Commissars and New Causes Compete," Sept. 27, 1987, Section I, pp. 1, 9; "Storm of Protest Rages over a Leningrad Dam," Sept. 27, 1987, Section I, p. 9; and "No Longer Merely Voices in the Russian Wilderness," Dec. 27, 1987, Section IV, p. 14.

81. See Peter Drucker, *Innovation and Entrepreneurship: Practice and Principles* (New York: Harper & Row, 1985); and Everett Rogers and Judith Larsen, *Silicon Valley Fever: The Growth of High-Technology Culture* (New York: Basic Books, 1984).

82. For instance, in a number of cases, Americans have developed a technology, only to find a lack of capital or the will to take it into production. The Japanese and others have often quite literally capitalized on these failures of nerve and have made world-class products from ideas initiated elsewhere. See Christopher Freeman, *Technology Policy and Economic Performance: Lessons from Japan* (New York: Pinter, 1987), pp. 31–54.

83. Cf. Collingridge and James, "Technology, Organizations, and Incrementalism," p. 96.

84. This classification is based on Ron Westrum, "Organizational and Inter-Organizational Thought," paper presented to the World Bank Conference on "Safety Control and Risk Management," Washington, D.C. Oct. 1988.

85. Ahmed Idris-Soven, Elizabeth Idris-Soven, and Mary K. Vaughn (Eds.), *The World as a Company Town* (The Hague: Mouton, 1978).

86. Kazuhiko Kawamura, "A Comparative Perspective of Risk Management in the United States and Japan," *IEEE Technology and Society Magazine,* Sept. 1987, pp. 3–11.

Index